Rev. J. G. Wood

Insect Architecture

Rev. J. G. Wood

Insect Architecture

ISBN/EAN: 9783743323582

Manufactured in Europe, USA, Canada, Australia, Japa

Cover: Foto ©berggeist007 / pixelio.de

Manufactured and distributed by brebook publishing software
(www.brebook.com)

Rev. J. G. Wood

Insect Architecture

INSECT ARCHITECTURE

BY

JAMES RENNIE

NEW EDITION, MUCH ENLARGED

BY THE REV. J. G. WOOD, M.A.

AUTHOR OF "HOMES WITHOUT HANDS," ETC.

.

WITH NEARLY TWO HUNDRED ILLUSTRATIONS.

LONDON:

BELL AND DALDY, YORK STREET,

COVENT GARDEN.

1869.

EDITOR'S PREFACE TO THE NEW EDITION.

THE proprietors of this interesting work having felt that much additional knowledge of the subject has been obtained since the book was written, have asked me to supply characteristic examples of Insect Architecture which were not to be found in its pages. I have accordingly added a considerable number of such examples, in most cases accompanied by figures drawn from the specimens described. I have not been at liberty to alter or expunge, and am not, therefore, responsible for any portion of the letterpress except those passages which are enclosed in brackets [ ]. Some of the specimens from which the figures have been drawn are in my own collection, but the greater part are to be found in the British Museum.

CONTENTS.

CHAPTER VI.

CHAPTER VII.

CHAPTER VIII.

CHAPTER IX.

CHAPTER X.

CHAPTER XI.

CHAPTER XII.

Contents.

CHAPTER XII.—continued.

Muff-Shaped Tents 250
Leaf-Mining Caterpillars 252
Bark-Mining Caterpillars 257
Grubs of Beetles 258
Wasp-Beetle 261
Stag-Beetle 262

CHAPTER XIII.

CHAPTER XIV.

CHAPTER XV.

CHAPTER XVI.

CHAPTER XVII.

CHAPTER XVIII.

CHAPTER XIX.

CHAPTER XX.

ILLUSTRATIONS.

INSECT ARCHITECTURE.

CHAPTER I.

INTRODUCTION.

IT can never be too strongly impressed upon a mind anxious for the acquisition of knowledge, that the commonest things by which we are surrounded are deserving of minute and careful attention. The most profound investigations of Philosophy are necessarily connected with the ordinary circumstances of our being, and of the world in which our every-day life is spent. With regard to our own existence, the pulsation of the heart, the act of respiration, the voluntary movement of our limbs, the condition of sleep, are among the most ordinary operations of our nature; and yet how long were the wisest of men struggling with dark and bewildering speculations before they could offer anything like a satisfactory solution of these phenomena, and how far are we still from an accurate and complete knowledge of them! The science of Meteorology, which attempts to explain to us the philosophy of matters constantly before our eyes, as dew, mist, and rain, is dependent for its illustrations upon a knowledge of the most complicated facts, such as the influence of heat and electricity upon the air; and this knowledge is at present so imperfect, that even these common occurrences of the weather, which men have been observing and reasoning upon for ages, are by no means satisfactorily explained, or reduced to the precision that every science should aspire to. Yet, however difficult it may be entirely

B

to comprehend the phenomena we daily witness, everything in nature is full of instruction. Thus the humblest flower of the field, although, to one whose curiosity has not been excited, and whose understanding has, therefore, remained uninformed, it may appear worthless and contemptible, is valuable to the botanist, not only with regard to its place in the arrangement of this portion of the Creator's works, but as it leads his mind forward to the consideration of those beautiful provisions for the support of vegetable life, which it is the part of the physiologist to study and to admire.

This train of reasoning is peculiarly applicable to the economy of insects. They constitute a very large and interesting part of the animal kingdom. They are everywhere about us. The spider weaves his curious web in our houses; the caterpillar constructs his silken cell in our gardens; the wasp that hovers over our food has a nest not far removed from us, which she has assisted to build with the nicest art; the beetle that crawls across our path is also an ingenious and laborious mechanic, and has some curious instincts to exhibit to those who will feel an interest in watching his movements; and the moth that eats into our clothes has something to plead for our pity, for he came, like us, naked into the world, and he has destroyed our garments, not in malice or wantonness, but that he may clothe himself with the same wool which we have stripped from the sheep. An observation of the habits of these little creatures is full of valuable lessons, which the abundance of the examples has no tendency to diminish. The more such observations are multiplied, the more are we led forward to the freshest and the most delightful parts of knowledge; the more do we learn to estimate rightly the extraordinary provisions and most abundant resources of a creative Providence; and the better do we appreciate our own relations with all the infinite varieties of nature, and our dependence, in common with the ephemeron that flutters its little hour in the summer sun, upon that Being in whose scheme of existence the humblest as well as the highest creature has its destined purposes. " If you speak of a stone," says St. Basil, one of the Fathers of the Church, " if you speak of a fly, a gnat, or a bee, your conversation will be a sort of demonstration of His power whose hand formed them, for the wisdom of the workman is commonly perceived in that which is of little size. He who has stretched out the heavens, and dug up the bottom of the sea, is also He who has pierced a passage through the sting of the bee for the ejection of its poison."

If it be granted that making discoveries is one of the most satisfactory of human pleasures, then we may without hesitation affirm, that the study of insects is one of the most delightful branches of natural history, for it affords peculiar facilities for its pursuit. These facilities are found in the almost inexhaustible variety which insects present to the curious observer. As a proof of the extraordinary number of insects within a limited field of observation, Mr. Stephens informs us, that in the short space of forty days, between the middle of June and the beginning of August, he found, in the vicinity of Ripley, specimens of above two thousand four hundred species of insects, exclusive of caterpillars and grubs, —a number amounting to nearly a fourth of the insects ascertained to be indigenous. He further tells us, that, among these specimens, although the ground had, in former seasons, been frequently explored, there were about one hundred species altogether new, and not before in any collection which he had inspected, including several new genera ; while many insects reputed scarce were in considerable plenty.* The localities of insects are, to a certain extent, constantly changing ; and thus the study of them has, in this circumstance, as well as in their manifold abundance, a source of perpetual variety. Insects, also, which are plentiful one year, frequently become scarce, or disappear altogether, the next—a fact strikingly illustrated by the uncommon abundance, in 1826 and 1827, of the seven-spot lady-bird (*Coccinella septempunctata*), in the vicinity of London, though during the two succeeding summers this insect was

* Stephens' Illustrations, vol. i., p. 72, note.

comparatively scarce, while the small two-spot lady-bird (*Coccinella bipunctata*) was plentiful.

There is, perhaps, no situation in which the lover of nature and the observer of animal life may not find opportunities for increasing his store of facts. It is told of a state prisoner, under a cruel and rigorous despotism, that when he was excluded from all commerce with mankind, and was shut out from books, he took an interest and found consolation in the visits of a spider; and there is no improbability in the story. The operations of that persecuted creature are among the most extraordinary exhibitions of mechanical ingenuity; and a daily watching of the workings of its instinct would beget admiration in a rightly-constituted mind. The poor prisoner had abundant leisure for the speculations in which the spider's web would enchain his understanding. We have all of us, at one period or other of our lives, been struck with some singular evidence of contrivance in the economy of insects, which we have seen with our own eyes. Want of leisure, and probably want of knowledge, have prevented us from following up the curiosity which for a moment was excited. And yet some such accident has made men naturalists, in the highest meaning of the term. Bonnet, evidently speaking of himself, says, " I knew a naturalist, who, when he was seventeen years of age, having heard of the operations of the ant-lion, began by doubting them. He had no rest till he had examined into them : and he verified them, he admired them, he discovered new facts, and soon became the disciple and the friend of the Pliny of France "* (Réaumur). It is not the happy fortune of many to be able to devote themselves exclusively to the study of nature, unquestionably the most fascinating of human employments ; but almost every one may acquire sufficient knowledge to be able to derive a high gratification from beholding the more common operations of animal life. His materials for contemplation are always before him. Some weeks ago we made an excursion to West Wood, near Shooter's Hill, expressly for the purpose of observing the

* Contemplation de la Nature, part ii. ch. 42.

insects we might meet with in the wood : but we had not
got far among the bushes, when heavy rain came on. We
immediately sought shelter among the boughs of some thick
underwood, composed of oak, birch, and aspen ; but we
could not meet with a single insect, not even a gnat or a fly,
sheltered under the leaves. Upon looking more narrowly,
however, into the bushes which protected us, we soon found
a variety of interesting objects of study. The oak abounded
in galls, several of them quite new to us ; while the leaves
of the birch and the aspen exhibited the curious serpentine
paths of the minute mining caterpillars. When we had
exhausted the narrow field of observation immediately around
us, we found that we could considerably extend it, by breaking
a few of the taller branches near us, and then examining
their leaves at leisure. In this manner two hours glided
quickly and pleasantly away, by which time the rain had
nearly ceased ; and though we had been disappointed in our
wish to ramble through the wood, we did not return without
adding a few interesting facts to our previous knowledge of
insect economy.*

It will appear, then, from the preceding observations,
that cabinets and collections, though undoubtedly of the
highest use, are by no means indispensable, as the observer
of nature may find inexhaustible subjects of study in every
garden and in every hedge. Nature has been profuse
enough in affording us materials for observation, when we
are prepared to look about us with that keenness of inquiry,
which curiosity, the first step in the pursuit of knowledge,
will unquestionably give. Nor shall we be disappointed
in the gratification which is thus within our reach. Were
it no more, indeed, than a source of agreeable amusement,
the study of insects comes strongly recommended to the
notice of the well-educated. The pleasures of childhood
are generally supposed to be more exquisite, and to contain
less alloy, than those of riper years ; and if so, it must be

* The original observations in this volume which are marked by the
initials J. R.. are by J. Rennie, A.M., A.L.S.. and those which are enclosed in
brackets are by the Rev. J G. Wood, M.A., F.L.S.

because then everything appears new and dressed in fresh beauties : while in manhood, and old age, whatever has frequently recurred begins to wear the tarnish of decay. The study of nature affords us a succession of "ever-new delights," such as charmed us in childhood, when everything had the attractions of novelty and beauty; and thus the mind of the naturalist may have its own fresh and vigorous thoughts, even while the infirmities of age weigh down the body.

It has been objected to the study of insects, as well as to that of Natural History in general, that it tends to withdraw the mind from subjects of higher moment; that it cramps and narrows the range of thought; and that it destroys, or at least weakens, the finer creations of the fancy. Now, we should allow this objection in its fullest extent, and even be disposed to carry it further than is usually done, if the collecting of specimens only, or, as the French expressly call them, chips (*échantillons*), be called a study. But the mere collector is not, and cannot be, justly considered as a naturalist; and, taking the term naturalist in its enlarged sense, we can adduce some distinguished instances in opposition to the objection. Rousseau, for example, was passionately fond of the Linnæan botany, even to the driest minutiæ of its technicalities; and yet it does not appear to have cramped his mind, or impoverished his imagination. If Rousseau, however, be objected to as an eccentric being, from whose pursuits no fair inference can be drawn, we give the illustrious example of Charles James Fox, and may add the names of our distinguished poets, Goldsmith, Thomson, Gray, and Darwin, who were all enthusiastic naturalists. We wish particularly to insist upon the example of Gray, because he was very partial to the study of insects. It may be new to many of our readers, who are familiar with the 'Elegy in a Country Churchyard,' to be told that its author was at the pains to turn the characteristics of the Linnæan orders of insects into Latin hexameters, the manuscript of which is still preserved in his interleaved copy of the 'Systema Naturæ.'

Further, to use the somewhat exaggerated words of Kirby and Spence, whose work on Entomology is one of the most instructive and pleasing books on the science, 'Aristotle among the Greeks, and Pliny the Elder among the Romans, may be denominated the fathers of Natural History, as well as the greatest philosophers of their day ; yet both these made insects a principal object of their attention : and in more recent times, if we look abroad, what names greater than those of Redi, Malpighi, Vallisnieri, Swammerdam, Leeuwenhock, Réaumur, Linnæus, De Geer, Bonnet, and the Hubers ? and at home, what philosophers have done more honour to their country and to human nature than Ray, Willughby, Lister, and Derham ? Yet all these made the study of insects one of their most favourite pursuits."*

And yet this study has been considered, by those who have superficially examined the subject, as belonging to a small order of minds ; and the satire of Pope has been indiscriminately applied to all collectors, while, in truth, it only touches those who mistake the means of knowledge for the end :—

> "O! would the sons of men once think their eyes
> And reason given them but to study Flies !
> See Nature, in some partial, narrow shape,
> And let the Author of the whole escape ;
> Learn but to trifle ; or, who most observe,
> To wonder at their Maker, not to serve."†

Thus exclaims the Goddess of Dulness, sweeping into her net all those who study nature in detail. But if the matter were rightly appreciated, it would be evident that no part of the works of the Creator can be without the deepest interest to an inquiring mind ; and that a portion of creation which exhibits such extraordinary manifestations of design as is shown by insects must have attractions for the very highest understanding.

An accurate knowledge of the properties of insects is of great importance to man, merely with relation to his

* Introduction to Entomology, vol. i.
† Dunciad, book iv.

own comfort and security. The injuries which they inflict upon us are extensive and complicated; and the remedies which we attempt, by the destruction of those creatures, both insects, birds, and quadrupeds, who keep the ravages in check, are generally aggravations of the evil, because they are directed by an ignorance of the economy of nature. The little knowledge which we have of the modes by which insects may be impeded in their destruction of much that is valuable to us, has probably proceeded from our contempt of their individual insignificance. The security of property has ceased to be endangered by quadrupeds of prey, and yet our gardens are ravaged by aphides and caterpillars. It is somewhat startling to affirm that the condition of the human race is seriously injured by these petty annoyances; but it is perfectly true that the art and industry of man have not yet been able to overcome the collective force, the individual perseverance, and the complicated machinery of destruction which insects employ. A small ant, according to a most careful and philosophical observer, opposes almost invincible obstacles to the progress of civilization in many parts of the equinoctial zone. These animals devour paper and parchment; they destroy every book and manuscript. Many provinces of Spanish America cannot, in consequence, show a written document of a hundred years' existence. " What development," he adds, " can the civilization of a people assume, if there be nothing to connect the present with the past—if the depositories of human knowledge must be constantly renewed—if the monuments of genius and wisdom cannot be transmitted to posterity ?" * Again, there are beetles which deposit their larvæ in trees in such formidable numbers that whole forests perish beyond the power of remedy. The pines of the Hartz have thus been destroyed to an enormous extent; and in North America, at one place in South Carolina, at least ninety trees in every hundred, upon a tract of two thousand acres, were swept away by a small black, winged bug. And yet, according to Wilson, the historian of American birds, the people of

* Humboldt, Voyage, lib. vii., ch. 20.

the United States were in the habit of destroying the red-headed woodpecker, the great enemy of these insects, because he occasionally spoilt an apple.* The same delightful writer and true naturalist, speaking of the labours of the ivory-billed woodpecker, says, " Would it be believed that the larvæ of an insect or fly, no larger than a grain of rice, should silently, and in one season, destroy some thousand acres of pine-trees, many of them from two to three feet in diameter, and a hundred and fifty feet high ? In some places the whole woods, as far as you can see around you, are dead, stripped of the bark, their wintry-looking arms and bare trunks bleaching in the sun, and tumbling in ruins before every blast."† The subterraneous larva of some species of beetle has often caused a complete failure of the seed-corn, as in the district of Halle in 1812.‡ The corn-weevil, which extracts the flour from grain, leaving the husk behind, will destroy the contents of the largest storehouses in a very short period. The wire-worm and the turnip-fly are dreaded by every farmer. The ravages of the locust are too well known not to be at once recollected as an example of the formidable collective power of the insect race. The white ants of tropical countries sweep away whole villages with as much certainty as a fire or an inundation; and ships even have been destroyed by these indefatigable republics. Our own docks and embankments have been threatened by such minute ravagers.

The enormous injuries which insects cause to man may thus be held as one reason for ceasing to consider the study of them as an insignificant pursuit; for a knowledge of their structure, their food, their enemies, and their general habits, may lead, as it often has led, to the means of guarding against their injuries. At the same time we derive from them both direct and indirect benefits. The honey of the bee, the dye of the cochineal, and the web of the silk-worm, the advantages of which are obvious, may well be balanced

---

* Amer. Ornith., i., p. 144.　　　　　　† Ibid., iii., p. 21.
‡ Blumenbach; see also Insect Transformations, p. 231.

against the destructive propensities of insects which are offensive to man. But a philosophical study of natural history will teach us that the direct benefits which insects confer upon us are even less important than their general uses in maintaining the economy of the world. The mischiefs which result to us from the rapid increase and the activity of insects are merely results of the very principle by which they confer upon us numberless indirect advantages. Forests are swept away by minute beetles; but the same agencies relieve us from that extreme abundance of vegetable matter which would render the earth uninhabitable were this excess not periodically destroyed. In hot countries the great business of removing corrupt animal matter, which the vulture and hyæna imperfectly perform, is effected with certainty and speed by the myriads of insects that spring from the eggs deposited in every carcase by some fly seeking therein the means of life for her progeny. Destruction and reproduction, the great laws of nature, are carried on very greatly through the instrumentality of insects; and the same principle regulates even the increase of particular species of insects themselves. When aphides are so abundant that we know not how to escape their ravages, flocks of lady-birds instantly cover our fields and gardens to destroy them. Such considerations as these are thrown out to show that the subject of insects has a great philosophical importance —and what portion of the works of nature has not? The habits of all God's creatures, whether they are noxious, or harmless, or beneficial, are worthy objects of our study. If they affect ourselves, in our health or our possessions, whether for good or for evil, an additional impulse is naturally given to our desire to attain a knowledge of their properties. Such studies form one of the most interesting occupations which can engage a rational and inquisitive mind; and, perhaps, none of the employments of human life are more dignified than the investigation and survey of the workings and the ways of nature in the minutest of her productions.

The exercise of that habit of observation which can alone make a naturalist—" an out-of-door naturalist," as Daines

Barrington calls himself—is well calculated to strengthen even the most practical and merely useful powers of the mind. One of the most valuable mental acquirements is the power of discriminating among things which differ in many minute points, but whose general similarity of appearance usually deceives the common observer into a belief of their identity. The study of insects, in this point of view, is most peculiarly adapted for youth. According to our experience, it is exceedingly difficult for persons arrived at manhood to acquire this power of discrimination; but, in early life, a little care on the part of the parent or teacher will render it comparatively easy. In this study the knowledge of things should go along with that of words. "If names perish," says Linnæus, "the knowledge of things perishes also:"* and, without names, how can any one communicate to another the knowledge he has acquired relative to any particular fact, either of physiology, habit, utility, or locality? On the other hand, mere catalogue learning is as much to be rejected as the loose generalizations of the despisers of classification and nomenclature. To name a plant, or an insect, or a bird, or a quadruped rightly, is one step towards an accurate knowledge of it; but it is not the knowledge itself. It is the means, and not the end, in natural history, as in every other science.

If the bias of opening curiosity be properly directed, there is not any branch of natural history so fascinating to youth as the study of insects. It is, indeed, a common practice in many families to teach children, from their earliest infancy, to treat the greater number of insects as if they were venomous and dangerous, and, of course, meriting to be destroyed, or at least avoided with horror. Associations are by this means linked with the very appearance of insects, which become gradually more inveterate with advancing years; provided, as most frequently happens, the same system be persisted in, of avoiding or destroying almost every insect which is unlucky enough to attract observation.

* Nomina si pereant, perit et cognitio rerum.

How much rational amusement and innocent pleasure is thus thoughtlessly lost; and how many disagreeable feelings are thus created, in the most absurd manner! " In order to show that the study or (if the word be disliked) the observation of insects is peculiarly fascinating to children, even in their early infancy, we may refer to what we have seen in the family of a friend, who is partial to this, as well as to all the departments of natural history. Our friend's children, a boy and girl, were taught, from the moment they could distinguish insects, to treat them as objects of interest and curiosity, and not to be afraid even of those which wore the most repulsive appearance. The little girl, for example, when just beginning to walk alone, encountered one day a large staphylinus (*Goërius olens?* STEPHENS; vulgo, *the devil's coach-horse*), which she fearlessly seized, and did not quit her hold, though the insect grasped one of her fingers in his formidable jaws. The mother, who was by, knew enough of the insect to be rather alarmed for the consequences, though she prudently concealed her feelings from the child. She did well; for the insect was not strong enough to break the skin, and the child took no notice of its attempts to bite her finger. A whole series of disagreeable associations with this formidable-looking family of insects was thus averted at the very moment when a different mode of acting on the part of the mother would have produced the contrary effect. For more than two years after this occurrence the little girl and her brother assisted in adding numerous specimens to their father's collection, without the parents ever having cause, from any accident, to repent of their employing themselves in this manner. The sequel of the little girl's history strikingly illustrates the position for which we contend. The child happened to be sent to a relative in the country, where she was not long in having carefully instilled into her mind all the usual antipathies against " everything that creepeth on the earth ;" and though she afterwards returned to her paternal home, no persuasion or remonstrance could ever again persuade her to touch a common beetle, much less a staphylinus, with its tail turned up in a threatening attitude,

and its formidable jaws ready extended for attack or defence.*
We do not wish that children should be encouraged to expose
themselves to danger in their encounters with insects. They
should be taught to avoid those few which are really noxious
—to admire all—to injure none.

The various beauty of insects—their glittering colours,
their graceful forms—supplies an inexhaustible source of
attraction. Even the most formidable insects, both in appearance and reality,—the dragon-fly, which is perfectly
harmless to man, and the wasp, whose sting every human
being almost instinctively shuns,—are splendid in their appearance, and are painted with all the brilliancy of natural
hues. It has been remarked that the plumage of tropical
birds is not superior in vivid colouring to what may be
observed in the greater number of butterflies and moths.†
" See," exclaims Linnæus, " the large, elegant painted wings
of the butterfly, four in number, covered with delicate feathery
scales! With these it sustains itself in the air a whole day,
rivalling the flight of birds and the brilliancy of the peacock.
Consider this insect through the wonderful progress of its
life,—how different is the first period of its being from the
second, and both from the parent insect! Its changes are
an inexplicable enigma to us: we see a green caterpillar,
furnished with sixteen feet, feeding upon the leaves of a
plant; this is changed into a chrysalis, smooth, of golden
lustre, hanging suspended to a fixed point, without feet, and
subsisting without food ; this insect again undergoes another
transformation, acquires wings, and six feet, and becomes a
gay butterfly, sporting in the air, and living by suction upon
the honey of plants. What has nature produced more worthy
of our admiration than such an animal, coming upon the stage
of the world, and playing its part there under so many different masks ?" The ancients were so struck with the transformations of the butterfly, and its revival from a seeming
temporary death, as to have considered it an emblem of the
soul, the Greek word *psyche* signifying both the soul and a

* J. R., in Mag. of Natural History, vol. i., p. 334.
† Miss Jermyn's Butterfly Collector, p. 11.

butterfly; and it is for this reason that we find the butterfly introduced into their allegorical sculptures as an emblem of immortality. Trifling, therefore, and perhaps contemptible, as to the unthinking may seem the study of a butterfly, yet when we consider the art and mechanism displayed in so minute a structure,—the fluids circulating in vessels so small as almost to escape the sight—the beauty of the wings and covering—and the manner in which each part is adapted for its peculiar functions,—we cannot but be struck with wonder and admiration, and allow, with Paley, that "the production of beauty was as much in the Creator's mind in painting a butterfly as in giving symmetry to the human form."

A collection of insects is to the true naturalist what a collection of medals is to the accurate student of history. The mere collector, who looks only to the shining wings of the one, or the green rust of the other, derives little knowledge from his pursuit. But the cabinet of the naturalist becomes rich in the most interesting subjects of contemplation, when he regards it in the genuine spirit of scientific inquiry. What, for instance, can be so delightful as to examine the wonderful variety of structure in this portion of the creation; and, above all, to trace the beautiful gradations by which one species runs into another? Their differences are so minute, that an unpractised eye would proclaim their identity; and yet, when the species are separated, and not very distantly, they become visible even to the common observer. It is in examinations such as these that the naturalist finds a delight of the highest order. While it is thus one of the legitimate objects of his study to attend to minute differences of structure, form, and colouring, he is not less interested in the investigation of habits and economy; and in this respect the insect world is inexhaustibly rich. We find herein examples of instinct to parallel those of all the larger animals, whether they are solitary or social; and innumerable others besides, altogether unlike those manifested in the superior departments of animated nature. These instincts have various direc-

tions, and are developed in a more or less striking manner to
our senses, according to the force of the motive by which
they are governed. Some of their instincts have for their
object the preservation of insects from external attack : some
have reference to procuring food, and involve many remarkable stratagems; some direct their social economy, and
regulate the condition under which they live together either
in monarchies or republics, their colonizations, and their
migrations; but the most powerful instinct which belongs
to insects has regard to the preservation of their species.
We find, accordingly, that as the necessity for this preservation is of the utmost importance in the economy of nature,
so for this especial object many insects, whose offspring,
whether in the egg or the larva state, are peculiarly exposed
to danger, are endued with an almost miraculous foresight,
and with an ingenuity, perseverance, and unconquerable
industry, for the purpose of avoiding those dangers, which
are not to be paralleled even by the most singular efforts of
human contrivance. The same ingenuity which is employed
for protecting either eggs, or caterpillars and grubs, or pupæ
and chrysalides, is also exercised by many insects for their
own preservation against the changes of temperature to
which they are exposed, or against their natural enemies.
Many species employ those contrivances during the period
of their hybernation, or winter sleep. For all these purposes
some dig holes in the earth, and form them into cells; others
build nests of extraneous substances, such as bits of wood
and leaves; others roll up leaves into cases, which they close
with the most curious art; others build a house of mud,
and line it with the cotton of trees, or the petals of the most
delicate flowers; others construct cells, of secretions from
their own bodies; others form cocoons, in which they
undergo their transformation ; and others dig subterraneous
galleries, which, in their complexity of arrangement, in
solidity, and in complete adaptation to their purposes, vie
with the cities of civilised man. The contrivances by which
insects effect these objects have been accurately observed
and minutely described, by patient and philosophical in-

quirers, who knew that such employments of the instinct
with which each species is endowed by its Creator offered
the most valuable and instructive lessons, and opened to
them a wide field of the most delightful study. The construction of their habitations is certainly among the most
remarkable peculiarities in the economy of insects ; and it is
of this subject that we propose to treat under the general
name, which is sufficiently applicable to our purpose, of
Insect Architecture.

In the descriptions which we shall give of Insect Architecture, we shall employ as few technical words as possible :
and such as we cannot well avoid, we shall explain in their
places ; but, since our subject chiefly relates to the reproduction of insects, it may be useful to many readers to introduce
here a brief description of the changes which they undergo.

It was of old believed that insects were produced spontaneously by putrefying substances ; and Virgil gives the
details of a process for *creating* a swarm of bees out of the
carcase of a bull ; but Redi, a celebrated Italian naturalist,

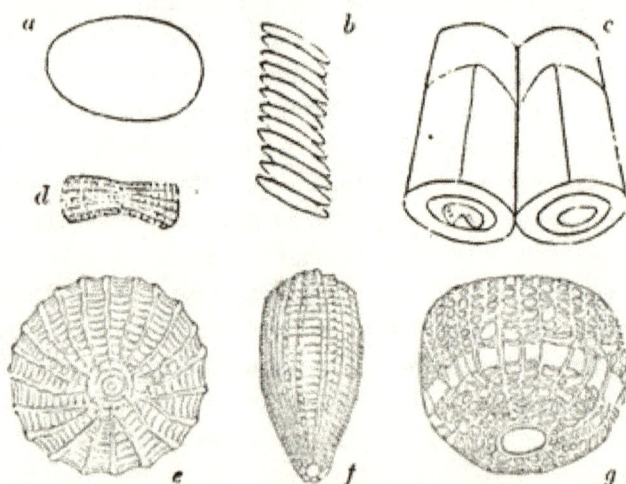

Magnified eggs, of *a, Geometra armillata,* b, of an unknown water insect; c. of the
lacquey moth ; d, of a caddis-fly (*Phryganea striata*) ; e, of red underwing moth (*Catocala
nupta*) ; f, of *Pontia Brassicæ* ; g, of the Clifden Nonpareil moth.

proved by rigid experiments that they are always, in such
cases, hatched from *eggs* previously laid. Most insects,
indeed, lay eggs, though some few are viviparous, and some

propagate both ways. The eggs of insects are very various in form, and seldom shaped like those of birds. We have here figured those of several species, as they appear under the microscope.

When an insect first issues from the egg, it is called by naturalists *larva*, and, popularly, a caterpillar, a grub, or a maggot. The distinction, in popular language, seems to be, that *caterpillars* are produced from the eggs of moths or butterflies; *grubs* from the eggs of beetles, bees, wasps, &c.: and *maggots* (which are without feet) from blow-flies, house-flies, cheese-flies, &c., though this is not very rigidly adhered to in common parlance. Maggots are also sometimes called *worms*, as in the instance of the meal-worm; but the common earth-worm is not a larva, nor is it by modern naturalists ranked among insects.

There are, however, certain larvæ, as those of the Cicada, the crickets, the water-boatman (*Notonecta*), the cock-

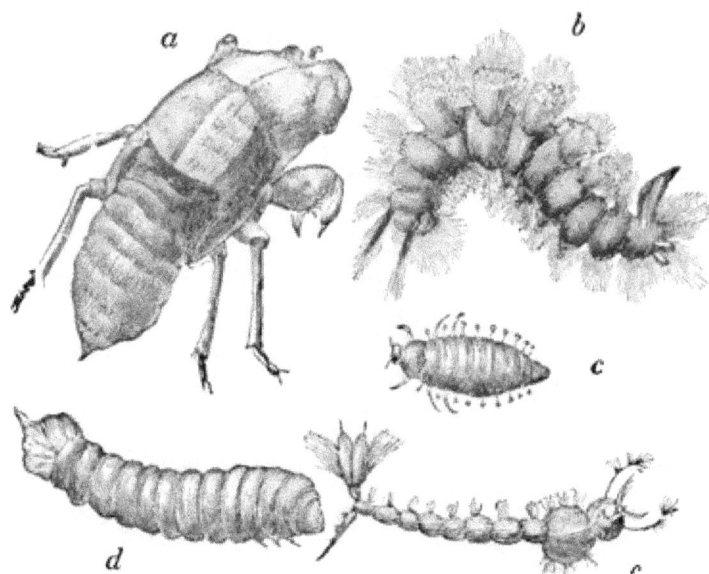

a, Ametabolous pupa of Cicada; *b*, caterpillar of tussock moth (*Laria fascelina*); *c*, larva of the poplar beetle (*Chrysomela populi*); *d*, larva of Sirex; *e*, larva of the common gnat.

roach, &c., which resemble the perfect insects in form, excepting that they are destitute of wings; but in the pupa

state these appear in a rudimentary condition, at least in
such species as have wings in the mature stage of existence.
The pupæ are active and eat. Insects, the larvæ and pupæ
of which are so similar to the adults, are termed *Ametabolous*
(*a*, without, μεταβολη, change); those the larvæ of which
undergo changes of a marked character, *Metabolous* (*Insecta
ametabóla* and *Insecta metabola*, Burmeister).

Larvæ are remarkably small at first, but grow rapidly.
The full-grown caterpillar of the goat-moth (*Cossus ligniperda*) is thus seventy-two thousand times heavier than when
it issues from the egg; and the maggot of the blow-fly is, in
twenty-four hours, one hundred and fifty-five times heavier
than at its birth. Some larvæ have feet, others are without;
none have wings. They cannot propagate. They feed
voraciously on coarse substances; and as they increase in
size, which they do very rapidly, they cast their skins three
or four times. In defending themselves from injury, and in
preparing for their change by the construction of secure
abodes, they manifest great ingenuity and mechanical skill.
The figures on the preceding page exemplify various forms
of insects in this stage of their existence.

When larvæ are full grown, they cast their skins for the
last time, undergo a complete change of form, excepting in

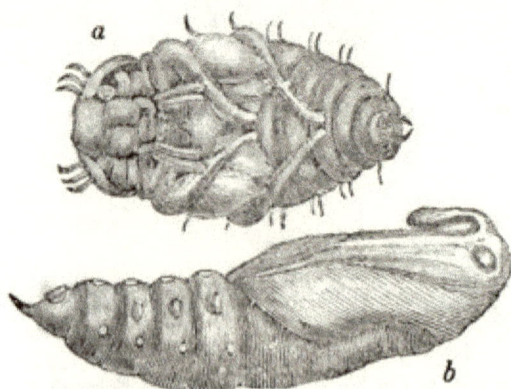

a, Pupa of a Water-Beetle (*Hydrophilus*); *b*, pupa of *Sphinx Ligustri*.

the case of ametabolous larvæ, cease to eat, and remain nearly
motionless. The inner skin of the larva now becomes con-

verted into a membranous or leathery covering, which wraps the insect closely up like a mummy: in this condition it is termed *Pupa*, from its resemblance to an infant in swaddling bands. Nympha, or nymph, is another term given to insects in this stage;* moreover from the pupæ of many of the butterflies appearing gilt as if with gold, the Greeks called them *Chrysalides*, and the Romans *Aureliæ*, and hence naturalists frequently call a pupa *chrysalis*, even when it is not gilt. We shall see, as we proceed, the curious contrivances resorted to for protecting insects in this helpless state. The following are examples of insects in the *imago*, or perfect state.

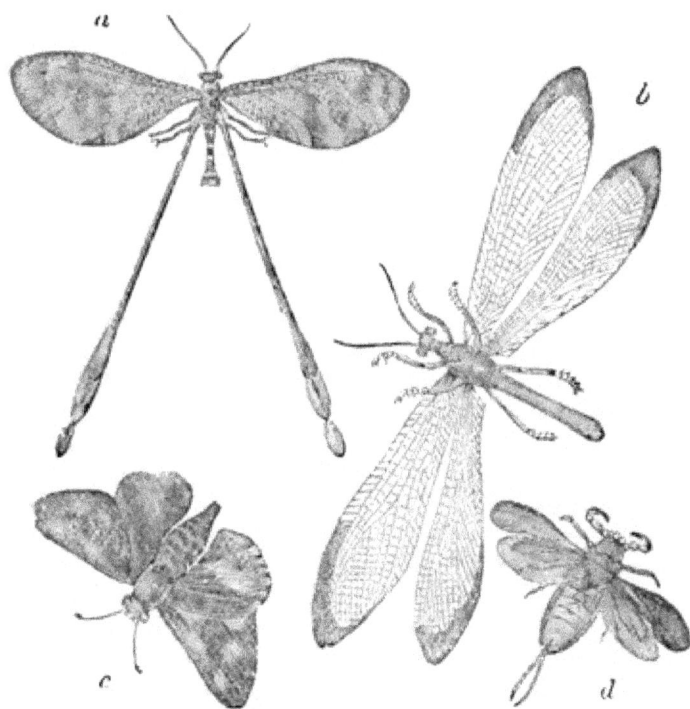

Insects in the Imago or perfect state.

a, Nemopteryx coa, Leach.—*b, Myrmeleon formicalynx*, Fabricius.—*c, Hesperia comma*, Fabricius.—*d, Nepa cinerea*, Linnæus.

After a certain time, the insect which has remained in its pupa-case, like a mass of jelly without shape, is gradually

* Generally to ametabolous pupæ.

preparing for its final change, when it takes the form of a perfect insect. This state was called by Linnæus *Imago,* because the insect, having thrown off its mask, becomes a perfect *image* of its species. Of some, this last portion of their existence is very short, others live through a year, and some exist for longer periods. They feed lightly, and never increase in size. The chief object of all is to perpetuate their species, after which the greater number quickly die. It is in this state that they exercise those remarkable instincts for the preservation of their race, which are exhibited in their preparations for the shelter of their eggs, and the nourishment of their larvæ.

CHAPTER II.

STRUCTURES FOR PROTECTING EGGS.—MASON-WASPS ; MASON-
BEES ; MINING-BEES.

THE provisions which are made by the different species of
insects for protecting their eggs, appear in many cases
to be admirably proportioned to the kind of danger and
destruction to which they may be exposed. The eggs themselves, indeed, are not so liable to depredation and injury as
the young brood hatched from them ; for, like the seeds of
plants, they are capable of withstanding greater degrees both
of heat and cold than the insects which produce them.
According to the experiments of Spallanzani, the eggs of
frogs that had been exposed to various degrees of artificial
heat were scarcely altered in their productive powers by a
temperature of 111° of Fahrenheit, but they became corrupted
after 133°. He tried the same experiment upon tadpoles and
frogs, and found they all died at 111°. Silkworms died at a
temperature of 108°, while their eggs did not entirely cease
to be fertile till 144°. The larvæ of flesh-flies perished,
while the eggs of the same species continued fertile, at about
the same comparative degrees of heat as in the preceding
instances. Intense cold has a still less effect upon eggs than
extreme heat. Spallanzani exposed the eggs of silkworms to
an artificial cold 23° below zero, and yet, in the subsequent
spring, they all produced caterpillars. Insects almost invariably die at the temperature of 14°, that is, at 18° below
the freezing point.* The care of insects for the protection
of their eggs is not entirely directed to their preservation in
the most favourable temperature for being hatched, but to
secure them against the numerous enemies which would
attempt their destruction ; and, above all, to protect the grubs,

* See Spallanzani's Tracts, by Dalyell, vol. i.

when they are first developed, from those injuries to which they are peculiarly exposed. Their prospective contrivances for accomplishing these objects are in the highest degree curious.

Most persons have more or less acquaintance with the hives of the social species of bees and wasps; but little is generally known of the nests constructed by the solitary species, though in many respects these are not inferior to the others in displays of ingenuity and skill. We admire the social bees, labouring together for one common end, in the same way that we look with delight upon the great division of labour in a well-ordered manufactory. As in a cotton-mill some attend to the carding of the raw material, some to its formation into single threads, some to the gathering these threads upon spindles, others to the union of many threads into one,—all labouring with invariable precision because they attend to a single object;—so do we view with delight and wonder the successive steps by which the hive-bees bring their beautiful work to its completion,—striving, by individual efforts, to accomplish their general task, never impeding each other by useless assistance, each taking a particular department, and each knowing its own duties. We may, however, not the less admire the solitary wasp or bee, who begins and finishes every part of its destined work; just as we admire the ingenious mechanic who perfects something useful or ornamental entirely by the labour of his own hands,—whether he be the patient Chinese carver, who cuts the most elaborately-decorated boxes out of a solid piece of ivory, or the turner of Europe, who produces every variety of elegant form by the skilful application of the simplest means.

Our island abounds with many varieties of solitary wasps and bees; and their nests may therefore be easily discovered by those who, in the proper seasons, are desirous of observing the peculiarities of their architecture.

MASON-WASPS.

In September, 1828, a common species of solitary mason-wasp (*Odynerus,* LATR.) was observed by us (J. R.) on the east wall of a house at Lee, in Kent, very busy in excavating a hole in one of the bricks, about five feet from the ground. Whether there might not have been an accidental hole in the brick before the wasp commenced her labours, is unknown, as she had made considerable progress in the work when first observed: but the brick was one of the hardest of the yellow sort made in this neighbourhood. The most remarkable circumstance in the process of hewing into the brick was the care of the insect in removing to a distance the fragments which from time to time she succeeded in detaching. It did not appear to suit her design to wear down the brick, particle by particle, as the furniture beetle (*Anobium*

Odynerus.—Natural size.

pertinax) does in making its pin-hole galleries in old wood. Our wasp-architect, on the contrary, by means of her strong *tranchant*-toothed jaws, severed a piece usually about the bigness of a mustard-seed. It might have been supposed that these fragments would have been tossed out of the hole as the work proceeded, without further concern; as the mole tosses above ground the earth which has been cleared out of its subterranean gallery. The wasp was of a different opinion; for it was possible that a heap of brick chips, at the bottom of the wall, might lead to the discovery of her nest by some of her enemies, particularly by one or other of the numerous tribe of what are called *ichneumon*-flies. This name is given to them, from the similarity of their habit of destroying eggs, to that of the little animal which proves so formidable an enemy to the multiplication of the crocodile of Egypt. They may be also denominated *cuckoo*-flies, because, like that bird, they thrust their egg into the nest of another species. These flies are continually prowling about and prying into every corner, to find, by stealth, a nidus for their eggs. It might have been some such consideration as this which induced the wasp to carry off the fragments as they were successively detached. That concealment was the motive, indeed, was proved : for one of the fragments which fell out of the hole by accident, she immediately sought for at the bottom of the wall, and carried off like the rest. It

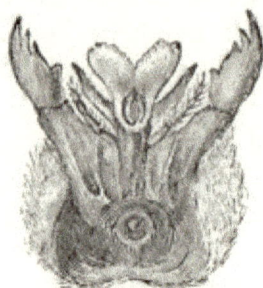

Mandibles—Jaws of Mason-Wasp.—Greatly magnified.

was no easy matter to get out one of the fragments, as may readily be conceived when the size of the insect is compared with that of the entrance of which this (●) is the exact size, as taken from the impression of a bit of dough upon the hole when finished. It was only by seizing the fragment with her jaws, and retreating backwards, that the matter could be accomplished; though, after the interior of the excavation was barely large enough to admit of her turning round, she more than once attempted to make her exit head-foremost, but always unsuccessfully. The weight of the fragments removed did not appear to impede her flight, and she generally returned to her task in about two or three minutes.

Within two days the excavation was completed; but it required two other days to line it with a coating of clay, to deposit the eggs, two in number, and, no doubt, to imprison a few live spiders or caterpillars for the young when hatched —a process which was first observed by Ray and Willughby,[*] but which has since been frequently ascertained. In the present instance, this peculiarity was not seen ; but the little architect was detected in closing up the entrance, which was formed of a layer of clay more than double the thickness of

[*] Ray, Hist. Insect., 254.

the interior lining. In November following, we hewed away the brick around this nest, and found the whole excavation was rather less than an inch in depth.

Notwithstanding all the precautions of the careful parent to conceal her nest it was found out by one of the cuckoo flies (*Tachina larvarum ?*)—probably a common species very similar to the house-fly, but rather larger, which deposited

Cuckoo-Fly (*Tachina larvarum ?*). – Natural size.

an egg there; and the grub hatched from it, after devouring one of the wasp-grubs, formed itself a cocoon (*a*), as did the

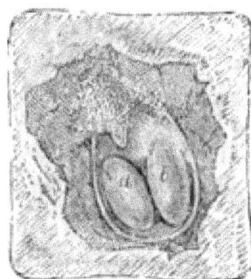

Mason-Wasp's Nest and Cocoons.— About one-third the natural size.

other undevoured grub of the wasp (*b*). Both awaited the return of summer to change into winged insects, burst their cerements, and proceed as their parents did.

Another mason-wasp (*Odynerus murarius*, LATR.), differ-

Mason-Wasp (*Odynerus murarius*).—Natural size.

ing little in appearance from the former, may often be seen frequenting sandy banks exposed to the sun, and constructing its singular burrows. The sort of sand-bank which it selects is hard and compact; and though this may be more difficult to penetrate, the walls are not liable to fall down upon the little miner. In such a bank, the mason-wasp bores a tubular gallery two or three inches deep. The sand upon which Réaumur found some of these wasps at work was almost as hard as stone, and yielded with difficulty to his nail; but the wasps dug into it with ease, having recourse, as he ascertained, to the ingenious device of moistening it by letting fall two or three drops of fluid from their mouth, which rendered the mass ductile, and the separation of the grains easy to the double pickaxe of the little pioneers.

When this wasp has detached a few grains of the moistened sand, it kneads them together into a pellet about the size of

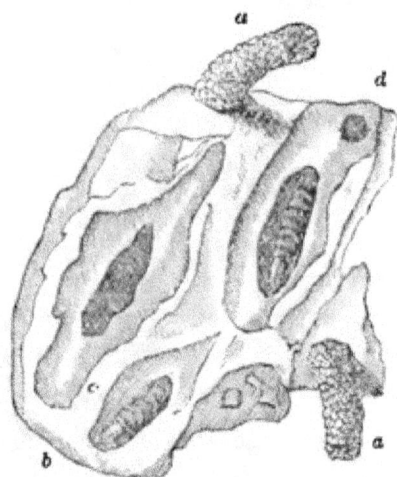

Nests, &c., of Mason-Wasps.—About half the natural size.

a, The tower of the nest; b, the entrance after the tower is removed; c, the cell; d, the cell, with a roll of caterpillars prepared for the larva.

one of the seeds of a gooseberry. With the first pellet which it detaches, it lays the foundation of a round tower, as an outwork, immediately over the mouth of its nest. Every pellet which it afterwards carries off from the interior is added to the wall of this outer round tower, which advances in height as the hole in the sand increases in depth. Every two or three minutes, however, during these operations, it

takes a short excursion, for the purpose, probably, of re-
plenishing its store of fluid wherewith to moisten the sand.
Yet so little time is lost, that Réaumur has seen a mason-
wasp dig in an hour a hole the length of its body, and at the
same time build as much of its round tower. For the greater
part of its height this round tower is perpendicular; but
towards the summit it bends into a curve, corresponding to
the bend of the insect's body, which in all cases of insect
architecture, is the model followed. The pellets which form
the walls of the tower are not very nicely joined, and numerous vacuities are left between them, giving it the appearance
of filigree-work. That it should be thus slightly built is
not surprising, for it is intended as a temporary structure for
protecting the insect while it is excavating its hole, and as a
pile of materials, well arranged and ready at hand, for the
completion of the interior building,—in the same way that
workmen make a regular pile of bricks near the spot where
they are going to build. This seems, in fact, to be the main
design of the tower, which is taken down as expeditiously as
it had been reared. Réaumur thinks that, by piling in the
sand which has previously been dug out, the wasp intends to
guard her progeny for a time from being exposed to the too
violent heat of the sun; and he has even sometimes seen that
there were not sufficient materials in the tower, in which case
the wasp had recourse to the rubbish she had thrown out
after the tower was completed. By raising a tower of the
materials which she excavates, the wasp produces the same
shelter from external heat as a human creature would who
chose to inhabit a deep cellar of a high house. She further
protects her progeny from the ichneumon-fly, as the
engineer constructs an outwork to render more difficult the
approach of an enemy to the citadel. Réaumur has seen
this indefatigable enemy of the wasp peep into the mouth of
the tower, and then retreat, apparently frightened at the
depth of the cell which he was anxious to invade.

The mason-wasp does not furnish the cell she has thus
constructed with pollen and honey, like the solitary bees,
but with living caterpillars, and these always of the same

species—being of a green colour, and without feet. She fixes the caterpillars together in a spiral column : they cannot alter their position, although they remain alive. They are an easy prey to their smaller enemy ; and when the grub has eaten them all up, it spins a case, and is transformed into a pupa, which afterwards becomes a wasp. The number of caterpillars which is thus found in the lower cavity of the mason-wasp's nest is ordinarily from ten to twelve. The mother is careful to lay in the exact quantity of provision which is necessary to the growth of the grub before he quits his retreat. He works through his store till his increase in this state is perfected, and he is on the point of undergoing a change into another state, in which he requires no food. The careful purveyor, cruel indeed in her choice of a supply, but not the less directed by an unerring instinct, selects such caterpillars as she is conscious have completed their growth, and will remain thus imprisoned without increase or corruption till their destroyer has gradually satisfied the necessities of his being. " All that the worm of the wasp," says Réaumur, " has to do in his nest, from his birth to his transformation, is to eat." There is another species of wasp which does not at once enclose in its nest all the sustenance which its larva will require before transformation, but which from time to time imprisons a living caterpillar, and when that is consumed, opens the nest and introduces another.

[The upper figure in the accompanying illustration exhibits two of the curious towers built by this interesting insect and drawn of their natural size.

The insect is one of the most plentiful in England, and can be found on sunny days, flitting about sand-banks and making its curious habitations. The length is nearly half an inch, and the colour is black, variegated with five yellow bands upon the abdomen.

The lower figure represents the habitations of one of the British solitary wasps, *Pompilus punctum*, and is given in order to show a curious resemblance in the structure. The specimen from which the sketch was taken was found under the eaves of a roof which protected a bee-hive. The cells were thirteen in number, very carefully constructed of earth, and several of them were closed. Although these cells were not fossorial in their nature, several other species of the same genus are as accomplished burrowers as any insect. *Pompilus plumbeus*, for example, another black species, burrows into

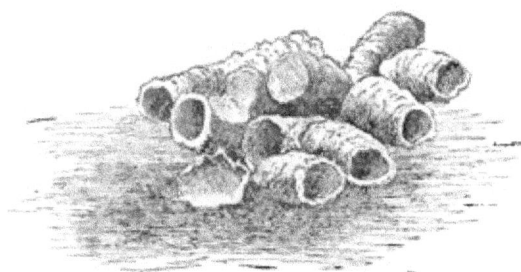

sand, and is very plentiful on our more southern shores. It may usually be found hovering about sand-banks, and flitting about with such agility that it is by no means an easy insect to catch. The male is peculiarly apt to evade the stroke of the entomologist's net.

Then there is *Pompilus rufipes*, which is a black insect, but distinguished by the conspicuous red colour of the hind legs. This is very fond of our coasts, and may be found wherever the soil is suitable for its excavations. Many species of this genus carry off spiders for the purpose of provisioning their nests. Several species, which live far inland, prefer light and dry earth to sand, and make therein their burrows, preferring our little white spider as the provision for their young. Although the same insect may be often observed to carry the same kind of prey to its home, it does not at all follow that no change is ever made.

But the most remarkable example of this fact may be found in a very common swift-winged insect, black in colour, with a reddish patch on the end of the abdomen. Its name is *Trachytes pompiliformis*, and it generally stocks its nest with small caterpillars. Mr. F. Smith, however, has taken it when in the act of carrying off a small species of grass-hopper—certainly the very last insect that would be thought of as likely to be immured by a captor which must be scarcely larger than itself.

This insect is to be found in most warm and sandy situations, and may be looked for at the end of summer and beginning of autumn. It may be easily known by its red spot on the abdomen, and the large, transverse head; it is wider than the thorax.

One species of mining-bee, not often found in England, chooses some very singular insects wherewith to feed its young. Its name is *Philanthus triangulum*, and it is a very fierce, waspish-looking creature, with a large wide head, wider even than the thorax, sharp and powerful jaws, and with broad wings. The head and thighs are black, with a few spots of a yellowish white, and the abdomen is yellow, with a black spot in the middle of each segment. Its length rather exceeds half an inch. The actions of this insect do not belie its looks, for it is a fierce and active creature, seizing upon various bees and dragging them into its tunnel.

Mr. F. Smith discovered the metropolis of this usually scarce insect at Sandown Bay, in the Isle of Wight, and has given an interesting description of its habits. He states that although it is so ferocious towards other insects, it appears to be perfectly harmless as far as man is concerned, allowing itself to be handled without even attempting to use its sting. Indeed, he was quite unable to provoke the insect to do so. Various bees were captured by the Philanthus, and the favour-ites seemed to belong to the genus Andrena, itself a burrowing bee, and the common hive-bee. The Philanthus seemed perfectly indifferent whether they attacked the comparatively small and feeble Andrena, or the formidable hive-bee, taking them as they came, and caring nothing for the sting. The Philanthus that burrowed on the top of the cliff, seemed to prefer the hive-bee, because the red clover attracted greater numbers of that insect. Those that made their burrows at the top of the cliff, took the Andrenæ. Of course, the Philanthus is obliged to catch more of the Andrenæ than the hive bees. Only one species of this genus is known in England ; it is to be found in July and August.

There is a very large genus of rather small humming-bees, many of which are popularly mistaken for wasps, on account of their sharply pointed and yellow banded abdomen ; they belong to the genus Crabro, and are extremely variable in the material into which they burrow, and the insects with which they feed their young. Some species burrow into dry bramble sticks. If the reader should wish to obtain specimens of them, as well as other burrowers, he will find bramble, rose, and jessamine sticks most prolific in them. The best plan is to collect a quantity of these sticks and put them into glass tubes, with the ends stopped with wire gauze ; there is then an absolute certainty of identifying the insect with its habitation. The spring is the best time for collecting. Sometimes these creatures are afflicted with parasites, which also are detained in the tube, so as to yield valuable information to the captor.

Some species burrow in sand-banks and feed their young with gnats, others burrow into dead timber, and stock their tunnels with flies of various kinds. One very useful species, *Crabro lævis*, burrows in sand-banks, and provisions its nest with the noxious turnip-fleas (*Halticæ*), great numbers of which are needed to stock a single burrow. It is rather a social insect, many burrows being often found near each other. The turnip-flea has so hard a shell, that the young Crabro seems hardly capable of eating it. Mr. Smith, however, has remarked that another burrowing-bee stocks its nest with certain weevils that are almost too hard to be pierced with a pin, and that the shell is probably softened by the damp ground. The greater number of this group, however, are burrowers into the ground, and stock their nests with flies of some kind.

Another species of this large genus, *Crabro luteipalpis*, is fond of making its burrows in the mortar of old walls, preferring those spots where nails have been drawn, making the process of burrowing easier for the insect. It is not uncommon in the outskirts of London. All gardeners, especially those who cultivate roses, ought to encourage this very little insect, and welcome its presence, for it provisions its nest with the aphides, or green blight, which infect the rose-trees, and which have destroyed so many promising plants.

The female, which is the larger of the two sexes, measures only three lines in length. The colour is shining black, and the head is rather squared.

Among other burrowing species of this genus we may mention *Crabro varius*, a rather long and slender insect, black in colour, with yellow spottings about the thorax. It prefers very hard fine sand, such as is found in partially excavated sandbanks, and provisions its burrows with gnats. It is tolerably plentiful.

Our last example of the earth-burrowers belonging to this genus is *Crabro Wesmoeli*, which chooses similar localities, being mostly found in sand-banks. It carries off flies of different kinds for the food of its young. There is a very common insect, closely allied to the last mentioned genus, whose horns are worthy of notice. This is called *Typoxylon figulus*. It is a small creature, with a large head and slender abdomen. Its colour is black, and on the edges of the segments of the abdomen there is a little silvery shining down.

It generally burrows into light earth, though it sometimes drives its tunnel into decayed wood. In either case, it provisions its nest with spiders, flying into the hedges, pulling the unfortunate spiders out of their webs, and carrying them into the burrow. One burrow contains a series of cells, which are separated from each other by partitions of sand, the particles of which are firmly cemented together by some glutinous substance secreted by the insect. Some species of this genus burrow into the pith of the bramble and other shrubs.

One of the most determined of our British burrowers is the insect which is known by the name of *Ammophila sabulosa.* It has a large, squared head, wider than the thorax, a very long and slender body, and short though powerful wings. The colour is black, with a slight rust-red tint on the base of the abdomen.

When the female has dug her burrow, she sets off in search of a caterpillar of proportionate size, and having conveyed it into her dwelling, she affixes an egg to the imprisoned larva, and goes off in search of another, carefully stopping up the entrance with stones. In fine weather she will fill one burrow with caterpillars in a few hours, and then begin another nursery for the future young. This species appears always to make use of caterpillars, but another allied species prefers spiders. Mr. F. Smith mentions that he has found in a high sand-bank as many as twenty females apparently hibernating together till suitable weather enabled them to pursue their usual economy.

There really seems to be scarcely any genus of insect that is not seized upon by one or other of these burrowers, and packed away in a half living state to form food for their young. There is one of these solitary burrowing wasps called the *Astata boops,* deriving its specific name from its large round eyes, which in the male completely unite at the back of the head. The abdomen is shaped something like a boy's peg-top, or a symmetrical turnip, the peg of the top, or the point of the turnip, corresponding to the top of the abdomen. Its length is about half an inch, and its colour is black, with a rust-red patch on the end of the abdomen.

There is a remarkably pretty, and very variable, sand-wasp, which is plentiful in most parts of the country. The colour is black, and the abdomen is banded by four yellow bars. Its feet are also yellow. Mr. Smith has written a very interesting account of the proceedings of this insect.

" Having frequently observed the habits of the type of this genus, *Mellinus arvensis,* and reared it from the larva state, a few observations are here recorded. When the parent insect has formed a burrow of the required length, and enlarged the extremity into a chamber of proper dimensions, she issues forth in search of the proper nutriment for her young; this consists of various dipterous insects: species of various genera are equally adapted to her purpose—*Muscidæ, Syrphidæ,* &c., are captured.

" It is amusing to see four or five females lie in wait upon a patch of cow-dung until some luckless fly settles on it. When this happens, a cunning and gradual approach is made; a sudden attempt would not succeed. The fly is the insect of quickest flight, therefore a degree of intrigue is necessary. This is managed by running past the victim slowly, and apparently in an unconcerned manner, until the poor fly is caught unawares, and carried off by the Mellinus to its burrow. The first fly being deposited, an egg is laid. The necessary number of flies are soon secured, and her task is completed. Sometimes she is interrupted by rainy weather, and it is some days ere she can store up the quantity required.

" A larva found feeding became full-fed in ten days. Six flies were devoured, the heads, harder parts of the throat, portions of the abdomen, and the legs, being left untouched. The larva spins a tough, thin, brown silken cocoon, passes the winter and spring in the larva state, changes to the nymph on the approach of summer, and appears about the beginning of autumn in the perfect state."

There is a genus of hymenopterous insects known by the name of *Scolia,* which are remarkable for their fossorial powers. The species represented in the engraving is called *Scolia Xantiana,* and is a native of California.

When the female Scolia is about to fulfil the great object for which she came into the world, she looks about for a suitable spot, where the ground is not too hard, and digs a perpendicular burrow of some depth, enlarging it at the bottom, and digging horizontally, so that the general shape of the burrow somewhat resembles that of a boot. When the burrow is completed, the insect flies off in search of food for its young, and presently returns, bearing with her a grub, which she clasps tightly under her chest, so that her wings may be at liberty. She then takes the grub to the bottom of the tunnel, deposits an egg upon it, and if the grub be a small one, goes off to fetch another. When a sufficiency of food has been obtained, she covers up the grub and egg and

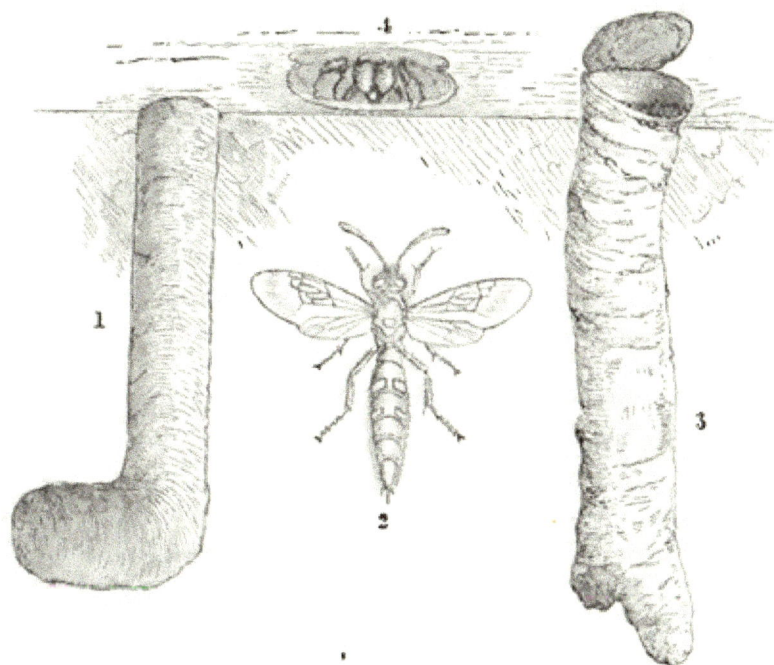

leaves the latter to its fate. In due time it is hatched, and begins straightway to feed upon its unfortunate fellow-prisoner. When all the food is gone, it is old enough to assume the perfect form, and when it finally becomes a perfect insect, it makes its way into the open air, and straightway looks out for a mate.

An European species of this genus, which is called *Scolia flavifrons*, is remarkable for the four large, round spots on the upper surface of the abdomen. This species always feeds its young on the grub of a beetle, one of the lamellicorn group, and in this case the grub is so large that one is sufficient.

In the illustration, the left hand figure shows a section of the burrow of *Scolia Xantiana*, and exhibits the enlarged portion of the tunnel in which are placed the young Scolia and the unfortunate grub which has to serve it for food. The insect itself is seen in the centre.

For figures 3 and 4 the reader is referred to the heading " Spiders."

There is another British insect which feeds its young with flies, and which catches them in a manner somewhat similar to that which has recently been narrated when treating of the Mellinus. The insect in question is called *Oxybelus unuglumis*, and is a very pretty species. Its length is seldom much more than a quarter of an inch, and its colour is black, with some silvery hair about the face, and with some spots and bands of white, more or less yellowish, upon the pointed abdomen. The male is usually smaller than the female, but compensates for this want of size by his more brilliant colouring.

Mr. F. Smith has described to me the method employed by this insect in catching flies. In the air it would not have a chance of success, and so it proceeds after a fashion very much like that which is adopted by the hunting-spider. Choosing some spot where flies are likely to settle, such as a bare, sunny bank, the Oxybelus alights upon it and begins to run about without any apparent motive. At first the flies are rather alarmed, but after a while they become accustomed to the rapid movements of their foe, and allow it to come nearer and nearer the cause of its perambulations. As soon as it has succeeded in drawing within a few inches of a fly, the Oxybelus leaps upon it, just like the hunting spider on its prey. and flies off before the victim knows that an attack is even meditated.

The burrow of this species is made in hard white sand.

Several species of the genus *Cerceris* are noted, not only as burrowers, but for the exceeding variety of the food which they store in their dwellings. The most common species, *Cerceris arenaria*, makes its tunnel in hard, sandy spots, and is usually to be found about the middle of July and August.

The length of this insect rather exceeds half an inch, and its colour is black, profusely spotted and barred with yellow. It is rather slenderly made, and gives little external indications of the great strength which it possesses.

This insect prefers to stock its nest with weevils of different kinds—a most singular choice, when the hardness of the exterior is taken into consideration. The well-known nut-weevil (*Balaninus nucum*), with its hard, round body, and long mouth, is frequently taken by this species of Cerceris, and Mr. Smith further mentions that he has captured it in the act of taking the weevil called *Otiorhynchus sulcatus* to its nest.

This beetle is among the most noxious of our garden foes, and the more so because its ravages are unseen. In its larval state it infests the roots of many of our succulent plants and flowers, and has a habit of eating away the plant just at the junction of the root and stem. Even flowers in pots are apt to be infested by this insect, and often die without the cause of their death being discovered. It is about half an inch in length, white, and is destitute of feet, their office being performed by bundles of stiff hairs, which are dispersed round the body.

In its perfect state it is about the third of an inch in length, the colour is black, covered with a coating of very fine and short grey hairs, and along its back are a number of short longitudinal grooves. From this latter circumstance it derives its name of " *sulcatus*," or grooved.

The exterior of this beetle is extremely hard, even exceptionally so among the hard-bodied weevils. It is extremely difficult to get a pin through the body, and the entomologist is often obliged to bore a hole with a stout needle before the pin can be inserted. Yet, the Cerceris uses this insect as the food of its young, and stores them away in its burrow. That the young should eat them seems as impossible as if a lobster or a box-tortoise had been inserted in their place. It is, however, thought by most practical entomologists that the shell of the weevil is softened by lying in the damp ground, and that as the young is not hatched for several days after the burrow is sealed up, the hard wing cases have time to soften.

Another species of the same genus, *Cerceris interrupta,* has the curious habit of making its burrow in the hardest ground which it can penetrate, and is generally to be found in well used footpaths. This species also uses weevils for the food of its young, but prefers those small weevils which are classed under the genus *Apion,* and which are readily known by their pear-shaped bodies and rather elongated heads. There are about seventy species of *Apion,* so that the Cerceris has plenty of choice.]

MASON-BEES.

It would not be easy to find a more simple, and, at the same time, ingenious specimen of insect architecture than the nests of those species of solitary bees which have been justly called mason-bees (*Megachile,* LATREILLE). Réaumur, who was struck by the analogies between the proceedings of insects and human arts, first gave to bees, wasps, and caterpillars those names which indicate the character of their labours; and which, though they may be considered a little fanciful, are at least calculated to arrest the attention. The nests of mason-bees are constructed of various materials; some with sand, some with earth mixed with chalk, and some with a mixture of earthy substances and wood.

On the north-east wall of Greenwich Park, facing the road, and about four feet from the ground, we discovered (J. R.),

Mason-Bee (*Anthophora retusa*).—Natural size.

December 10th, 1828, the nest of a mason-bee, formed in the perpendicular line of cement between two bricks. Externally there was an irregular cake of dry mud, precisely as if a handful of wet road-stuff had been taken from a cart-rut and thrown against the wall; though, upon closer inspection, the cake contained more small stones than usually occur in the mud of the adjacent cart-ruts. We should in fact have passed it by without notice had there not been a circular hole on one side of it, indicating the perforation of some insect.

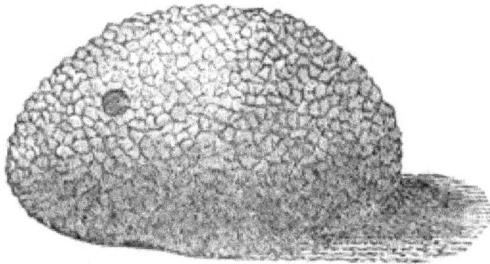

Exterior Wall of Mason-Bee's Nest.

This hole was found to be the orifice of a cell about an inch deep, exactly of the form and size of a lady's thimble, finely polished, and of the colour of plaster-of-paris, but stained in various places with yellow.

This cell was empty; but, upon removing the cake of mud, we discovered another cell, separated from the former by a partition about a quarter of an inch thick, and in it a living bee, from which the preceding figure was drawn, and which, as we supposed, had just changed from the pupa to the winged state, in consequence of the uncommon mildness of

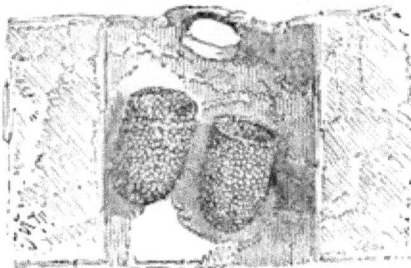

Cells of a Mason-Bee (*Anthophora retusa*).—One-third the natural size.

the weather. The one which had occupied the adjacent cell had no doubt already dug its way out of its prison, and would probably fall a victim to the first frost.

Our nest contained only two cells—perhaps from there not being room between the bricks for more.

[There are only four British species of this genus. One species, *A. acervorum*, seems perfectly indifferent whether it burrows into banks or into the mortar of old walls. If possible, the former locality seems to be the most favoured.

This species is notable for the many parasites who infect the habitation and destroy the inmates. Perhaps the very worst and most destructive of these parasites is the common earwig, which wreaks wholesale desolation in the nest. It creeps into the burrow, and if it finds a store of pollen laid up for the young, it will eat the pollen. But if the young grub be hatched it will eat the grub. If the inmate be in the pupal state, or even if it be ready to emerge in its perfect condition, the earwig will eat it.

There are two bees which are parasitic upon this unfortunate insect, both belonging the genus *Melecta*.

But the most destructive of these parasites appears to be an insect which belongs to the great family of *Chalcididæ*. These insects are of the hymenopterous order, are of very minute dimensions, and of the most brilliant colours. Indeed, if they were an inch or two in length, instead of the eighth or twelfth of an inch, they would not suffer in comparison with the most gorgeous inhabitants of tropical countries.

Their forms are most eccentric, some species having the abdomen small and round and set on a long foot-stalk, while others have that portion of the body placed so closely against the thorax, that the short footstalk is scarcely visible. Others have certain joints of the legs so large that a single joint equals the entire abdomen. Some have the ovipositor projecting boldly from the body, while others have it tucked up underneath, and others again have it quite short. But there is one point which distinguishes them all, namely, the almost veinless character of the wings.

Some of the *Chalcididæ* are parasitic upon insects in their earliest stages, actually depositing their eggs in those of moths and butterflies. Others are entirely parasitic upon parasites, laying their eggs in the aphidii, which are parasites of the aphis. Some of them haunt the galls, and contrive to make their young parasitic upon the immature cynipidæ which lie within the gall. The common small tortoise-shell butterfly is terribly infested with these little creatures, and we have bred hundreds of the gem-like *Chalcididæ* from the larvæ and pupæ of that butterfly.

One of the *Chalcididæ*, belonging to the genus *Melittobia*, is a parasite upon the *Anthophora*; and the curious part of the proceeding is, that it finds there another parasite, which becomes developed in the home of the bee: the *Melittobia* feeds indiscriminately upon the bee and parasite.

Although the *Melittobia* does not make such wholesale destruction as is wrought by the earwig when it gets into a nest, it does more damage to the bee, on account of its great numbers. Some three or four females will lay a great quantity of eggs within a nest, and from those eggs a hundred of the young will be developed. When the larvæ are fully grown, they quit their hold of their prey, and fall to the bottom of the cell, where they lie until they have assumed the perfect form. They then burst forth, together with those of the bee that may have escaped their attacks.]

An interesting account is given by Réaumur of another mason-bee (*Megachile muraria*), not a native of Britain, selecting earthy sand, grain by grain; her glueing a mass of these together with saliva, and building with them her cells from the foundation. But the cells of the Greenwich Park nest were apparently composed of the mortar of the brick wall; though the external covering seems to have been constructed as Réaumur describes his nest, with the occasional addition of small stones.

About the middle of May, 1829, we discovered the mine from which all the various species of mason-bees in the vicinity seemed to derive materials for their nests. (J. R.) It was a bank of brown clay, facing the east, and close by the margin of the river Ravensbourn, at Lee, in Kent. The frequent resort of the bees to this spot attracted the attention of some workmen, who, deceived by their resemblance to wasps, pointed it out as a wasps' nest; though they were not a little surprised to see so numerous a colony at this early season. As the bees had dug a hole in the bank, where they were incessantly entering and reappearing, we were of opinion that they were a peculiar sort of the social earth-bees (*Bombi*). On approaching the spot, however, we remarked that the bees were not alarmed, and manifested none of the irritation usual in such cases, the consequence of jealous affection for their young. This led us to observe their operations more minutely; and we soon discovered that on issuing from the hole each bee carried out in its mandibles a piece of clay. Still supposing that they were social earth-bees, we concluded that they were busy excavating a hollow for their nest, and carrying off the refuse to prevent discovery. The mouth of the hole was overhung, and partly concealed, by a large pebble. This we removed, and widened the entrance of the hole, intending to dig down and ascertain the state of the operations; but we soon found that it was of small depth. The bees, being scared away, began scooping out clay from another hole about a yard distant from the first. Upon our withdrawing a few feet from the first hole, they returned thither in preference, and continued assiduously digging and removing the clay. It became obvious, therefore, from their thus changing place, that they were not constructing a nest, but merely quarrying for clay as a building material. By catching one of the bees (*Osmia bicornis*) when it was loaded with its burden, we ascertained that the clay was not only carefully kneaded, but was also more moist than the mass from which it had been taken. The bee, therefore, in preparing the pellet, which was nearly as large as a garden-pea, had moistened it with its saliva, or some similar fluid, to render it, we may suppose, more tenacious, and better fitted for building. The reason of their digging a hole, instead of taking clay indiscriminately from the bank, appeared to be for the purpose of economizing their saliva, as the weather was dry, and the clay at the surface was parched and hard. It must have been this circumstance which induced them to prefer digging a hole, as it were, in concert, though each of them had to build a separate nest.

The distance to which they carried the clay was probably considerable, as there was no wall near, in the direction they all flew towards, upon which they could build; and in the same direction also, it is worthy of remark, they could have procured much nearer the very same sort of clay. Whatever might be the cause of their preference, we could not but admire their extraordinary industry. It did not require more than half a minute to knead one of the pellets of clay; and, from their frequent returns, probably not more than five minutes to carry it to the nest, and apply it where wanted. From the dryness of the weather, indeed, it was indispensable for them to work rapidly, otherwise the clay could not have been made to hold together. The extent of the whole labour of forming a single nest may be imagined, if we estimate that it must take several hundred pellets of clay for its completion. If a bee work fourteen or fifteen hours a-day, therefore, carrying ten or twelve pellets to its nest every hour, it will be able to finish the structure in about two or three days; allowing some hours of extra time for the more nice workmanship of the cells in which the eggs are to be deposited, and the young grubs reared.

That the construction of such a nest is not a merely agreeable exercise to the mason-bee has been sufficiently proved by M. Du Hamel. He has observed a bee (*Megachile muraria*) less careful to perform the necessary labour for the protection of her offspring than those we have described, but not less desirous of obtaining this protection, attempt to usurp the nest which another had formed. A fierce battle was invariably the consequence of this attempt; for the true mistress would never give place to the intruder. The motive for the injustice and the resistance was an indisposition to further labour. The trial of strength was probably, sometimes, of as little use in establishing the right as it is amongst mankind; and the proper owner, exhausted by her efforts, had doubtless often to surrender to the dishonest usurper.

The account which Réaumur has given of the operations of this class of bees differs considerably from that which we have here detailed; from the species being different, or from his bees not having been able to procure moist clay. On the contrary, sand was the chief material used by the mason-bees (*Megachile muraria*); which they had the patience to select from the walks of a garden, and knead into a paste or mortar, adapted to their building. They had consequently to expend a much greater quantity of saliva than our bees (*Osmia bicornis*), which worked with moist clay. Réaumur, indeed, ascertained that every individual grain of sand is moistened previous to its being joined to the pellet, in order to make it adhere more effectually. The tenacity of the mass is, besides, rendered stronger, he tells us, by adding a proportion of earth or garden-mould. In this manner, a ball of mortar is formed, about the size of a small shot, and carried off to the nest. When the structure of this is examined, it has all the appearance externally of being composed of earth and small stones or gravel. The ancients, who were by no means accurate naturalists, having observed bees carrying pellets of earth and small stones, supposed that they employed these to add to their weight, in order to steady their flight when impeded by the wind.

The nests thus constructed appear to have been more durable edifices than those which have fallen under our observation;—for Réaumur says they were harder than many sorts of stone, and could scarcely be penetrated with a knife. Ours, on the contrary, do not seem harder than a piece of sun-baked clay, and by no means so hard as brick. One circumstance appeared inexplicable to Réaumur and his friend Du Hamel, who studied the operations of these insects in concert. After taking a portion of sand from one part of the garden-walk, the bees usually took another portion from a spot almost twenty and sometimes a hundred paces off, though the sand, so far as could be judged by close examination, was precisely the same in the two places. We should be disposed to refer this more to the restless character of the insect than to any difference in the sand. We have observed a wasp paring the outside of a plank, for materials to form its nest; and though the plank was as uniform in the qualities of its surface, nay, probably more so than the sand could be, the wasp fidgeted about, nibbling a fibre from one, and a fibre from another portion, till enough was procured for one load. In the same way, the whole tribe of wasps and bees flit restlessly from flower to flower, not unfrequently revisiting the same blossom, again and again, within a few seconds. It appears to us, indeed, to be far from improbable, that this very restlessness and irritability may be one of the springs of their unceasing industry.

By observing, with some care, the bees which we found digging the clay, we discovered one of them (*Osmia bicornis*) at work upon a nest, about a gunshot from the bank. The place it had chosen was the inner wall of a coal-house, facing the south-west, the brick-work of which was but roughly finished. In an upright interstice of half an inch in width, between two of the bricks, we found the little architect assiduously building its walls. The bricklayer's mortar had either partly fallen out, or been removed by the bee, who had commenced building at the lower end, and did not build downwards, as the social wasps construct their cells.

The very different behaviour of the insect here, and at the quarry, struck us as not a little remarkable. When digging and preparing the clay, our approach, however near, produced no alarm; the work went on as if we had been at a distance; and though we were standing close to the hole, this did not scare away any of the bees upon their arrival to procure a fresh load. But if we stood near the nest, or even in the way by which the bee flew to it, she turned back or made a wide circuit immediately, as if afraid to betray the site of her domicile. We even observed her turning back, when we were so distant that it could not reasonably be supposed she was jealous of us; but probably she had detected some prowling insect depredator, tracking her flight with designs upon her provision for her future progeny. We imagined we could perceive not a little art

in her jealous caution, for she would alight on the tiles as
if to rest herself; and even when she had entered the coal-
house, she did not go directly to her nest, but again rested
on a shelf, and at other times pretended to examine several.
crevices in the wall, at some distance from the nest. But
when there was nothing to alarm her, she flew directly to
the spot, and began eagerly to add to the building.

It is in instances such as these, which exhibit the adaptation of instinct to circumstances, that our reason finds the
greatest difficulty in explaining the governing principle of
the minds of the inferior animals. The mason-bee makes
her nest by an invariable rule; the model is in her mind,
as it has been in the mind of her race from their first creation : they have learnt nothing by experience. But the
mode in which they accomplish this task varies according
to the situations in which they are placed. They appear to
have a glimmering of reason, employed as an accessory and
instrument of their instinct.

The structure, when finished, consisted of a wall of clay
supported by two contiguous bricks, enclosing six chambers,

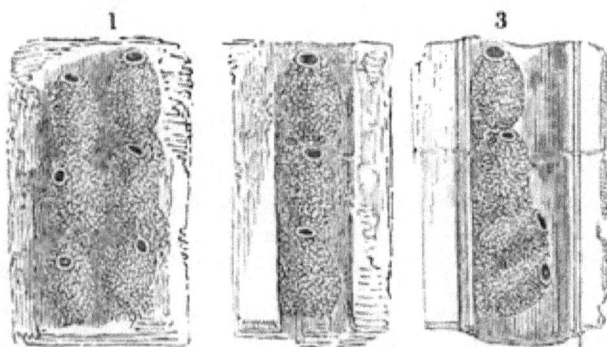

Cells of Mason-Bees, built, in the first and second figures, by *Osmia bicornis* between
bricks, and in the third, by *Megachile muraria* in the fluting of an old pilaster.—About
half the natural size.

within each of which a mass of pollen, rather larger than a
cherry-stone, was deposited, together with an egg, from
which in due time a grub was hatched. Contrary to what
has been recorded by preceding naturalists with respect to
other mason-bees, we found the cells in this instance quite

parallel and perpendicular; but it may also be remarked, that the bee itself was a species altogether different from the one which we have described above as the *Anthophora retusa*, and agreed with the figure of the one we caught quarrying the clay—(*Osmia bicornis*).

[In Mr. F. Smith's elaborate catalogue of the British hymenoptera there is a most interesting account of the habits of this insect, which is the most abundant species of the genus, and is spread not only over the whole of England, but over the continent, being found as far south as Italy and as far north as Lapland.

" In a hilly country, or at the sea-side, it chooses the sunny side of cliffs or sandy banks in which to form its burrows, but in cultivated districts, particularly if the soil be clayey, it selects a decayed tree, preferring the stump of an old willow. It lays up a store of pollen and honey for the larvæ, which when full grown, spins a tough dark brown cocoon, in which they remain in the larval state until the autumn, when the majority change to pupæ, and soon arrive at their perfect condition. Many, however, pass the winter in the larva state. In attempting to account for so remarkable a circumstance, all must be conjecture, but it is not of un-frequent occurrence. This species frequently makes its burrows in the mortar of old walls.

Another species (*Osmia bicolor*) sometimes makes its cells in very peculiar situations. When obliged to have recourse to its natural powers, it uses its limbs right well, attacks the hard sandy banks, and works at them with the greatest perseverance. But it will not work one stroke where it can avoid the necessity, and in many cases, it contrives to avoid work with much ingenuity.

Lying hidden under hedges, bushes, grass, and herbage, are sure to be shells of various snails, such as the common garden-snail, and the banded-snail, whose diversified shell is the delight of children. These shells the bee thinks are as good as ready-made burrows, and she uses them accordingly.

She goes to the end of the shell, carrying her materials with her, and then builds a cell, and fills it with pollen and honey. Another cell is then made, and yet another, until the shell is nearly filled. As the shell widens, the *Osmia* places two cells side by side, and when the insect has worked within a short distance of the mouth, she places the cells horizontally, so as to fill up the space. There are several specimens of these curious habitations in the British Museum.

When the whole series of cells is completed, the bee closes up the entrance with little morsels of earth, bits of stick and little stones, all strongly glued together with some very adhesive substance.

Another species (*Osmia parietina*) has much simpler habits, and is much easier satisfied with a dwelling. This insect merely looks out for a flattish stone lying on the ground, and crawls under it to see if there is any hollow. If so, it attaches the cocoons to the stone and leaves them. On one stone, seen in the British Museum, no less than two hundred and thirty cocoons were placed, although the stone is only ten inches in length by six in width.

This insect is almost wholly confined to the north of England.]

There was one circumstance attending the proceedings of this mason-bee which struck us not a little, though we could not explain it to our own satisfaction. Every time she left her nest for the purpose of procuring a fresh supply of materials, she paid a regular visit to the blossoms of a lilac-tree which grew near. Had these blossoms afforded a supply of pollen, with which she could have replenished her cells, we could have easily understood her design; but the pollen of the lilac is not suitable for this purpose, and that she had never used it was proved by all the pollen in the cells being yellow, whereas that of the lilac is of the same pale purple colour as the flowers. Besides, she did not return immediately from the lilac-tree to the building, but always went for a load of clay. There seemed to us, therefore, to be only two ways to explain the circumstance :—she must either have applied to the lilac-blossoms to obtain a refreshment of honey, or to procure glutinous materials to mix with the clay.

When employed upon the building itself, the bee exhibited the restless disposition peculiar to most hymenopterous * insects; for she did not go on with one particular portion of her wall, but ran about from place to place every time she came to work. At first, when we saw her running from the bottom to the top of her building, we naturally imagined that she went up for some of the bricklayer's mortar to mix with her own materials; but upon minutely examining the walls afterwards, no lime could be discovered in their structure similar to that which was apparent in the nest found in the wall of Greenwich Park.

Réaumur mentions another sort of mason-bee, which selects a small cavity in a stone, in which she forms her nest of garden-mould moistened with gluten, and afterwards closes the whole with the same material.

[In the accompanying illustration is shown a series of cells which are constructed by an insect which is closely related to

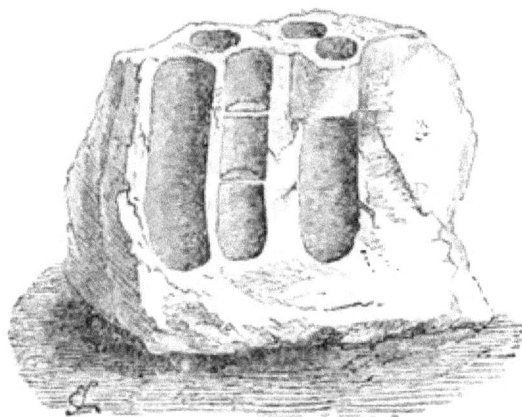

Cells of *Chalicodoma.*

the rose-cutter bee of our own country, to which it bears a close resemblance.

It is a native of South Africa, and its name is *Chalicodoma cælocerus.* The insect is about half an inch in length, and the colour of the head and body is black, that of the abdomen being brick red.

* The fifth order of Linnæus; insects with four transparent veined wings.

The nest is made of mud, which is collected by the patient insect and stuck against walls, trunks of trees, and similar localities. In this lump of mud the insect excavates a small number of burrows, each of which contains several cells. If the reader will refer to the central burrow, he will see that it is divided into three cells. The specimen from which this drawing is taken may be seen in the British Museum.

There is another South African insect which makes its mud nest, and fastens it against trees and walls. This is called *Synagris calida*, and its colour is almost dingy black, the only

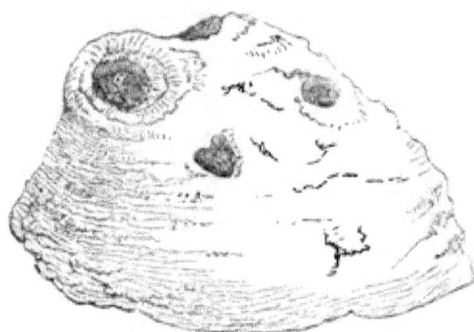

exception being the red tip to the abdomen. The holes seen in the engraving are the apertures through which the young brood has escaped into the world. The nest is represented of half its natural size.]

Mining-Bees.

A very small sort of bees (*Andrenæ*), many of them not larger than a house-fly, dig in the ground tubular galleries little wider than the diameter of their own bodies. Samouelle says, that all of them seem to prefer a southern aspect; but we have found them in banks facing the east, and even the north. Immediately above the spot where we have described the mason-bees quarrying the clay, we observed several holes, about the diameter of the stalk of a tobacco-pipe, into which those little bees were seen passing. The clay here was very hard : and on passing a straw into the hole as a director, and digging down for six or eight inches, a very smooth circular gallery was found, terminating in a thimble-shaped horizontal chamber, almost at right angles to the entrance and nearly twice as wide. In this chamber there was a ball of bright yellow pollen, as round as a garden pea, and rather larger, upon which a small white grub was feeding; and to which the mother bee had been adding, as she had just entered a minute before with her thighs loaded with pollen. That it

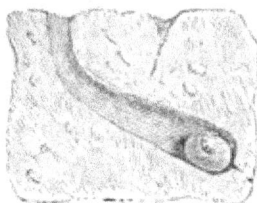

Cell of Mining-Bee (*Andrena*).—About half the natural size.

was not the male, the load of pollen determined; for the male has no apparatus for collecting or transporting it. The whole labour of digging the nest and providing food for the young is performed by the female. The females of the solitary bees have no assistance in their tasks. The males are idle; and the females are unprovided with labourers, such as the queens of the hive command.

Réaumur mentions that the bees of this sort, whose operations he had observed, piled up at the entrance of their galleries the earth which they had scooped out from the interior; and when the grub was hatched, and properly provided with food, the earth was again employed to close up the passage, in order to prevent the intrusion of ants, ichneumon-flies, or other depredators. In those which we have observed, this was not the case; but every species differs from another in some little peculiarity, though they agree in the general principles of their operations.

[The genus Andrena is an exceedingly large one, nearly seventy species being acknowledged in England alone. They choose various situations for their nest; a very favourite situation is a hard-trodden pathway; into this the bees burrow for some six or seven inches, and often drive their tunnels to a depth of ten inches. Digging up these habitations is not a very easy task, because the tunnel does not run straight, but turns aside when a stone or any similar obstacle comes in the way, and in getting out the stone the burrow is mostly broken. The only method of digging out the nest successfully is either by pushing a small twig up the hole, and using it as a guide, or by filling the entire hole with cotton wool, so as to prevent the earth from falling in.

The commonest species is *Andrena albicans*. Its length is rather less than half an inch, and its colour is black, with a thick coating of rich red hair on the upper part of the thorax. This species is plentiful on the continent, and is found as far south as Italy. But it is equally capable of enduring great cold, as it has been captured in the Arctic regions. Some-times the bee will not trouble itself to make a number of separate burrows, but will drive short supplementary tunnels from the side of the first burrow, so that they all open into one common entrance.

The Andrenæ are remarkable for the parasites with which they are infested, the most curious of which is that tiny strepsipterous insect called the Stylops.

One of the Andrenæ, called *Colletes Dariesana*, is remark-able for the character of its burrow. Like many of the insects which have already been described, it seems in-different whether it burrows in sandbanks or into the mortar of walls, provided that in the latter case the mortar is soft and friable.

The insect burrows a hole which is very deep in proportion to its size, the little bee being only the third of an inch in length, and the burrows some eight or ten inches in depth. When the mother *Colletes* has finished her tunnel, she lines the end of it with a thin kind of membrane, which has been well compared by Mr. F. Smith to goldbeater's skin. This lining is intended to enable the bee to store honey in the cell, as, if there were no such protection, the honey would soak in the ground and be lost.

Having stored up enough food for a single offspring, she shuts it off by a partition of the same membranous substance as the lining. Her next care is to make a thimble-like cup at the end, so as to have a double lining where the honey is to come, and then she puts a fresh supply in the new cell. This cell is then closed, and the bee proceeds with her work until she has made from six to eight cells in a single burrow. This insect suffers terribly from the depredations of the earwig, which completely empties the burrow both of food and of inhabitants. The colour of the insect is black, with a little reddish down on the upper part of the thorax, and some white on the legs. The abdomen is shining black, but each segment has a very narrow band of reddish down on its edge.

In 1850, Mr. F. Smith, to whose works such constant reference has been made, undertook the study of a genus of mining-bees belonging to this family. The species which he chiefly watched is *Halictus morio*, and his observations are peculiarly valuable, as showing the wonderful manner in which the economy of the race is managed. It is known that in these and many other insects, the pregnant females pass the winter in a state of hibernation, and begin to work in the following spring, and that therefore some arrangement must be needful that a supply of such queens should be kept up.

Mr. Smith found the case to stand thus. Early in April, the females appeared abundantly, and could be seen until June, but not a single male was to be found. During June and July, almost all the *Halicti* had disappeared, the reason being, that the queens had made their burrows, laid their eggs, stocked their cells, and then died, the duties of their life having been fulfilled. In the middle of August, the males began to appear, and in September the females of the first brood came out. They immediately set to work at their burrows, and laid their eggs. The ground, thoroughly warmed by the summer sun, soon hastened the young through their changes, and in an incredibly short time the insects of the second brood made their appearance. The females of this brood meet their mates, and then hide themselves until the following spring.

As in the case of Andrenæ, several tunnels are often made with one common entrance. The insect is very small, scarcely exceeding the sixth of an inch in length. The head and thorax are a dark green, the abdomen is white, and the legs are covered with silvery hairs. It is a plentiful insect, and is found haunting the holes of old walls.

Passing to another family of British mining-bees, we come to one species that is remarkable not only for its form, but for its economy. This is the *Eucera longicornis*, the only known species that inhabits England. In form it is chiefly remarkable from the fact that the antennæ of the male are as long as the entire body. The pupa of this insect is enclosed in a thin membrane, and when the male insect is about to emerge from its pupal shell, it has recourse to a rather curious expedient. At the base of the first joint of the front feet there is a bold notch. When the insect wishes to remove the thin membranous pellicle which envelopes the antennæ, it lays these organs in the notch, draws them through, and thus easily strips off the pellicle. The antennæ are most beautifully formed, the surface of each joint being marked with an elaborate pattern like network, so that they form beautiful objects for the microscope.

The soil preferred by the *Eucera* is of a clayey nature. When it has completed the burrow, it presses the soil at the extremity with all its might, and smooths it so carefully that the burrow becomes capable of holding honey without needing any lining. The insect is generally found about the end of May or beginning of June, and in some places is found in great numbers. The ground colour of the insect is black, but the body is covered with a coating of short dun hairs. The length rather exceeds half an inch.]

CHAPTER III.

CARPENTER-BEES.

AMONG the solitary bees are several British species, which come under that class called carpenter-bees by M. Réaumur, from the circumstance of their working in wood, as the mason-bees work in stone. We have frequently witnessed the operations of these ingenious little workers, who are particularly partial to posts, palings, and the wood-work of houses which has become soft by beginning to decay. Wood actually decayed, or affected by dry-rot, they seem to reject as unfit for their purposes; but they make no objections to any hole previously drilled, provided it be not too large; and, like the mason-bees, they not unfrequently take possession of an old nest, a few repairs being all that in this case is necessary.

When a new nest is to be constructed, the bee proceeds to chisel sufficient space for it out of the wood with her jaws. We say *her*, because the task in this instance, as in most others of solitary bees and wasps, devolves solely upon the female, the male taking no concern in the affair, and probably being altogether ignorant that such a work is going forward. It is, at least, certain that the male is never seen giving his assistance, and he seldom, if ever, approaches the neighbourhood. The female carpenter-bee has a task to perform no less arduous than the mason-bee; for though the wood may be tolerably soft, she can only cut out a very small portion at a time. The successive portions which she gnaws off may be readily ascertained by an observer, as she carries them away from the place. In giving the history of a mason-wasp (*Odynerus*), at page 22, we remarked the care with which she carried to a distance little fragments of brick, which she detached in the progress of excavation. We have recently watched a precisely similar procedure in the instance of a carpenter-bee forming a cell in a wooden post. (J. R.) The only difference was, that the bee did not fly so far away with her fragments of wood as the wasp did; but she varied the direction of her flight every time: and we could observe that, after dropping, the chip of wood which she had carried off, she did not return in a direct line to her nest, but made a circuit of some extent before wheeling round to go back.

On observing the proceedings of this carpenter-bee next day, we found her coming in with balls of pollen on her thighs; and on tracing her from the nest into the adjacent garden, we saw her visiting every flower which was likely to yield her a supply of pollen for her future progeny. This was not all; we subsequently saw her taking the direction of the clay quarry frequented by the mason-bees, as we have mentioned in page 41, where we recognised her loading herself with a pellet of clay, and carrying it into her cell in the wooden post. We observed her alternating this labour for several days, at one time carrying clay, and at another pollen; till at length she completed her task, and closed the entrance with a barricade of clay, to prevent the intrusion of any insectivorous depredator, who might make prey of her young; or of some prying parasite, who might introduce its own eggs into the nest she had taken so much trouble to construct.

Some days after it was finished, we cut into the post, and exposed this nest to view. It consisted of six cells of a somewhat square shape, the wood forming the lateral walls; and each was separated from the one adjacent by a partition of clay, of the thickness of a playing card. The wood was not lined with any extraneous substance, but was worked as smooth as if it had been chiseled by a joiner. There were five cells, arranged in a very singular manner—two being almost horizontal, two perpendicular, and one oblique.

The depth to which the wood was excavated in this instance was considerably less than what we have observed in other species which dig perpendicular galleries several inches deep in posts and garden-seats; and they are inferior in ingenuity to the carpentry of a bee described by Réaumur (*Xylocopa violacea*), which has not been ascertained to be a native of Britain, though a single indigenous species of the

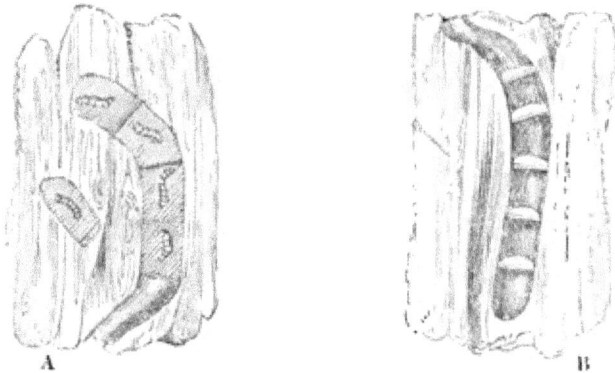

Cells of Carpenter-Bees, excavated in an old post.

In fig. A the cells contain the young grubs; in fig. B the cells are empty. Both figures are shown in section, and about half their natural size.

genus has been doubtingly mentioned, and is figured by Kirby and Spence, in their valuable 'Monographia.' If it ever be found here, its large size and beautiful violet-coloured wings will render mistakes impossible.

The violet carpenter-bee usually selects an upright piece of wood, into which she bores obliquely for about an inch; and then, changing the direction, works perpendicularly, and parallel to the sides of the wood, from twelve or fifteen inches, and half an inch in breadth. Sometimes the bee is contented with one or two of these excavations; at other times, when the wood is adapted to it, she scoops out three or four—a task which sometimes requires several weeks of incessant labour.

The tunnel in the wood, however, is only one part of the work; for the little architect has afterwards to divide the whole into cells, somewhat less than an inch in depth. It is necessary, for the proper growth of her progeny, that each should be separated from the other, and be provided with adequate food. She knows, most exactly, the quantity of food which each grub will require during its growth; and she therefore does not hesitate to cut it off from any additional supply. In constructing her cells, she does not employ clay, like the bee which we have mentioned above, but the sawdust, if we may call it so, which she has collected in gnawing out the gallery. It would not, therefore, have

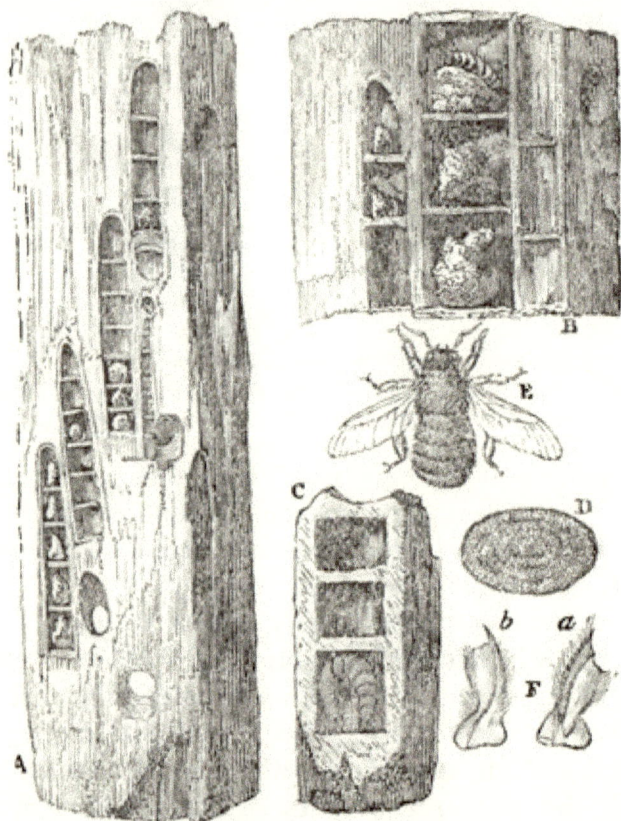

A represents a part of an espalier prop, tunnelled in several places by the violet Carpenter-Bee: the stick is split, and shows the nests and passages by which they are approached. B, a portion of the prop, half the natural size. C, a piece of thin stick, pierced by the Carpenter-Bee, and split, to show the nests. D, perspective view of one of the partitions. E, Carpenter-Bee (*Xylocopa violacea*). F, Teeth of the Carpenter-Bee, greatly magnified; *a*, the upper side; *b*, the lower side.

suited her design to scatter this about, as our carpenter-bee did. The violet-bee, on the contrary, collects her gnawings into a little store-heap for future use, at a short distance from her nest. She proceeds thus:—At the bottom of her excavation she deposits an egg, and over it fills a space nearly an inch high with the pollen of flowers, made into a paste with honey. She then covers this over with a ceiling composed of cemented sawdust, which also serves for the floor of the next chamber above it. For this purpose she cements round a wall a ring of wood-chips taken from her storeheap; and within this ring forms another, gradually contracting the diameter till she has constructed a circular plate, about the thickness of a crown-piece, and of considerable hardness. This plate of course exhibits concentric circles, somewhat similar to the annual circles in the cross section of a tree. In the same manner she proceeds till she has completed ten or twelve cells; and then she closes the main entrance with a barrier of similar materials.

Let us compare the progress of this little joiner with a human artisan—one who has been long practised in his trade, and has the most perfect and complicated tools for his assistance. The bee has learnt nothing by practice; she makes her nest but once in her life, but it is then as complete and finished as if she had made a thousand. She has no pattern before her—but the Architect of all things has impressed a plan upon her mind, which she can realize without scale or compasses. Her two sharp teeth are the only tools with which she is provided for her laborious work; and yet she bores a tunnel, twelve times the length of her own body, with greater ease than the workman who bores into the earth for water, with his apparatus of augurs adapted to every soil. Her tunnel is clean and regular; she leaves no chips at the bottom, for she is provident of her materials. Further, she has an exquisite piece of joinery to perform when her ruder labour is accomplished. The patient bee works her rings from the circumference to the centre, and she produces a shelf, united with such care with her natural glue, that a number of fragments are as solid as one piece.

The violet carpenter-bee, as may be expected, occupies several weeks in these complicated labours; and during that period she is gradually depositing her eggs, each of which is successively to become a grub, a pupa, and a perfect bee. It is obvious, therefore, as she does not lay all her eggs in the same place—as each is separated from the other by a laborious process—that the egg which is first laid will be the earliest hatched; and that the first perfect insect, being older than its fellows in the same tunnel, will strive to make its escape sooner, and so on of the rest. The careful mother provides for this contingency. She makes a lateral opening at the bottom of the cells; for the teeth of the young bees would not be strong enough to pierce the outer wood, though they can remove the cemented rings of sawdust in the interior. Réaumur observed these holes, in several cases; and he further noticed another external opening opposite to the middle cell, which he supposed was formed, in the first instance, to shorten the distance for the removal of the fragments of wood in the lower half of the building.

That bees of similar habits, if not the same species as the violet-bee, are indigenous to this country, is proved by Grew, who mentions, in his 'Rarities of Gresham College,' having found a series of such cells in the middle of the pith of an elder branch, in which they were placed lengthwise, one after another, with a thin boundary between each. As he does not, however, tell us that he was acquainted with the insect which constructed these, it might as probably be allied to the *Ceratina albilabris*, of which Spinola has given so interesting an account in the 'Annales du Muséum d'Histoire Naturelle' (x. 236). This noble and learned naturalist tells us, that one evening he perceived a female ceratina alight on the branch of a bramble, partly withered, and of which the extremity had been broken; and, after resting a moment, suddenly disappear. On detaching the branch, he found that it was perforated, and that the insect was in the very act of excavating a nidus for her eggs. He forthwith gathered a bundle of branches, both of the bramble and the wild-rose, similarly perforated, and took them home to examine them at leisure. Upon inspection, he found that the nests were furnished like those of the same tribe, with balls of pollen kneaded with honey, as a provision for the grubs.

The female ceratina selects a branch of the bramble or

wild-rose which has been accidentally broken, and digs into the pith only, leaving the wood and bark untouched. Her mandibles, indeed, are not adapted for gnawing wood; and, accordingly, he found instances in which she could not finish her nest in branches of the wild-rose, where the pith was not of sufficient diameter.

The insect usually makes her perforation a foot in depth, and divides this into eight, nine, or even twelve cells, each about five lines long, and separated by partitions formed by the gnawings of the pith, cemented by honey, or some similar glutinous fluid, much in the same manner with the *Xylocopa violacea*, which we have already described.

[This species is probably *Ceratina cærulea*, as the second species, *C. albilabris*, seems to have little claim to be considered as a British insect. It is plentiful in spots where it resides, but is very local. It can best be found by collecting all the specimens of bramble branches that have holes bored into the pith.

Mr. F. Smith says of this tiny bee, "Some years ago I observed a small bee most industriously employed in excavating a dead bramble stick. My attention was directed to the circumstance from observing some of the fallen pieces of pith on the ground immediately beneath. Occasionally fresh quantities of dust were pushed out. At length, the little creature came out of the stick as if to rest, and after sunning itself for a few minutes, it re-entered, and again commenced its labours. Later in the day, after stopping up the entrance, I cut off the branch and found in it a male and female ceratina."

The ceratina is only the sixth of an inch in length, and is deep shining blue in colour.

There are many other species of British bees which frequent the stems of bramble and other trees. One of them is known as *Prosopis signata*. The cells made by the bees of this genus are lined with a membrane, and are stocked with liquid honey. Some species will not take the trouble of boring a tunnel for themselves, but will make use of hollow stones, or similar localities, and place in them the silk-covered cocoons.

There are species of that versatile genus *Osmia* (*O. leucomelana*), in the habit of burrowing into dead bramble branches. The mother insect bores a hole some six inches in length, throwing the pieces of pith away, and then, depositing at the bottom an egg and a supply of food, she forms a cell by fixing across the burrow a stopper made of masticated leaves.

The stopper retains its place firmly, because the bee does not eat away the whole of the pith, but alternately widens and contracts the diameter of the burrow, each contracted portion being the termination of a cell. The perfect insect appears in the early summer of the following year.]

CARPENTER-WASPS.

As there are mason-wasps similar in economy to mason-bees, so are there solitary carpenter-wasps which dig galleries in timber, and partition them out into several cells by means

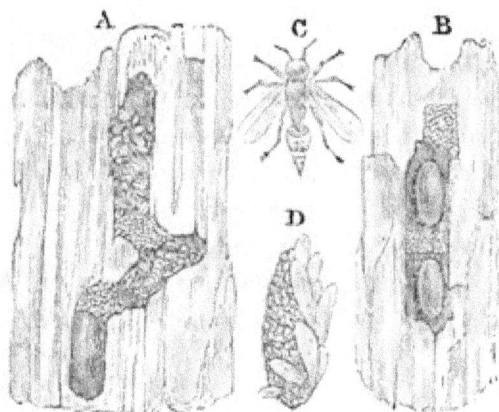

A B represent sections of old wooden posts, with the cells of the Carpenter-Wasp. In fig. A the young grubs are shown feeding on the insects placed there for their support by the parent wasp. The cells in fig. B contain cocoons. C, Carpenter-Wasp, natural size. D, cocoon of a Carpenter-Wasp, composed of sawdust and wings of insects.

of the gnawings of the wood which they have detached. This sort of wasp is of the genus *Eumenes*. The wood selected is generally such as is soft, or in a state of decay; and the hole which is dug in it is much less neat and regular than that of the carpenter-bees, while the division of the chambers is nothing more than the rubbish produced during the excavation.

The provision which is made for the grub consists of flies or gnats piled into the chamber, but without the nice order remarkable in the spiral columns of green caterpillars provided by the mason-wasp (*Odynerus murarius*). The most remarkable circumstance is, that in some of the species, when the grub is about to go into the pupa state, it spins a case (a cocoon), into which it interweaves the wings of the flies whose bodies it has previously devoured. In other species, the gnawings of the wood are employed in a similar manner.

[Some of the solitary wasps are also carpenters, and the genus *Crabro* has several species which are classed under this head. There is, for example, *Crabro claripes*, a little black insect with red and black abdomen, that burrows into dead bramble sticks, boring out the pith, and forming a series of cells in the narrow tube thus made. Sometimes this insect bores into decaying wood, but its general home is the bramble-stick. The same habits are common to several other British species of this genus, and the reader will find that old, decaying willow trees are chiefly visited by these pretty little insects. Their store of food, which they lay up for their young, mostly consists of dipterous insects, and various species of gnats are used for this purpose.

Another of the carpenter-wasps (*Pemphredon lugubris*) is really a useful insect. It makes its burrows in posts, rails, and similar localities, and provides its future young with a large stock of aphides. It has been seen to settle on a rose-bush, scrape off the branches a number of aphides, form them into a ball, and carry them off between its head and front legs.

The colour of this insect is dull black, from which circumstance it derives its name of *lugubris*. The head is large, and squared, and the abdomen is attached to the thorax by a large footstalk. Its length is about half an inch. It is a very common insect, and is believed to be the only British representation of its genus.

Several species do not take the trouble to form a burrow for themselves, but content themselves with building in holes ready made for them. Straws are favourite resorts of such insects, and in thatched buildings the straws of the roof are often filled with their cells.

One of these insects is a very little species, barely a quarter of an inch in length. Its colour is black, with some silver white hair on the face, and the legs are paler than the body. The abdomen has a long footstalk. Its scientific name is *Psen pallipes.* Like the insect which has just been described, it provisions its young with aphides.]

UPHOLSTERER-BEES.

In another part of this volume we shall see how certain caterpillars construct abodes for themselves, by cutting off portions of the leaves or bark of plants, and uniting them by means of silk into a uniform and compact texture; but this scarcely appears so wonderful as the prospective labours of some species of bees for the lodgment of their progeny. We allude to the solitary bees, known by the name of the leaf-cutting bees, but which may be denominated more generally *upholsterer-bees,* as there are some of them which use other materials beside leaves.

One species of our little upholsterers has been called the poppy-bee (*Osmia papaveris,* LATR.), from its selecting the scarlet petals of the poppy as tapestry for its cells. Kirby and Spence express their doubts whether it is indigenous to this country: we are almost certain that we have seen the nests in Scotland. (J. R.) At Largs, in Ayrshire, a beautiful sea-bathing village on the Firth of Clyde, in July, 1814, we found in a footpath a great number of the cylindrical perforations of the poppy-bee. [In his catalogue of British Hymenoptera, Mr. F. Smith makes the following remarks with regard to this insect. "The poppy-bee, *Anthocopa papaveris,* is closely allied to this genus (*Osmia*), and may indeed be placed before it as a connecting link with the *Osmia.* This interesting insect (*l'abeille Tapissiere*), of

Réaumur, has been supposed to inhabit this country, specimens having been placed in the collection at the British
Museum. But it was with much regret that I discovered,
when engaged upon the catalogue of British bees for the
Museum, and had occasion to examine each individual
specimen with care, that in the first place there was no
satisfactory evidence of the locality, and that in the next
place, all the males associated with the series were those of
Osmia adunca, of Panzear." For these reasons, this species
has been excluded from the list of British bees.] Réaumur
remarked that the cells of this bee which he found at Bercy,
were situated in a northern exposure, contrary to what he
had remarked in the mason-bee, which prefers the south.
The cells at Largs, however, were on an elevated bank,
facing the south, near Sir Thomas Brisbane's observatory.
With respect to exposure, indeed, no certain rule seems
applicable; for the nests of mason-bees which we found on
the wall of Greenwich Park faced the north-east, and we have
often found carpenter-bees make choice of a similar situation.
In one instance, we found carpenter-bees working indifferently
on the north-east and south-west side of the same post.

As we did not perceive any heaps of earth near the holes
at Largs, we concluded that it must either have been carried
off piecemeal when they were dug, or that they were old
holes re-occupied (a circumstance common with bees), and
that the rubbish had been trodden down by passengers.
Réaumur, who so minutely describes the subsequent operations of the bee, says nothing respecting its excavations.
One of these holes is about three inches deep, gradually
widening as it descends, till it assumes the form of a small
Florence flask. The interior of this is rendered smooth,
uniform, and polished, in order to adapt it to the tapestry
with which it is intended to be hung, and which is the next
step in the process.

The material used for tapestry by the insect upholsterer
is supplied by the flower-leaves of the scarlet field-poppy,
from which she successively cuts off small pieces of an oval
shape, seizes them between her legs, and conveys them to

F

the nest. She begins her work at the bottom, which she
overlays with three or four leaves in thickness, and the sides
have never less than two. When she finds that the piece
she has brought is too large to fit the place intended, she
cuts off what is superfluous, and carries away the shreds.
By cutting the fresh petal of a poppy with a pair of scissors,
we may perceive the difficulty of keeping the piece free from
wrinkles and shrivelling ; but the bee knows how to spread
the pieces which she uses as smooth as glass.

When she has in this manner hung the little chamber
all around with this splendid scarlet tapestry, of which she
is not sparing, but extends it even beyond the entrance,
she then fills it with the pollen of flowers mixed with honey,
to the height of about half an inch. In this magazine of
provisions for her future progeny she lays an egg, and over
it folds down the tapestry of poppy-petals from above. The
upper part is then filled in with earth ; but Latreille says
he has observed more than one cell constructed in a single
excavation. This may account for Réaumur's describing
them as sometimes seven inches deep ; a circumstance which
Latreille, however, thinks very surprising.

It will, perhaps, be impossible ever to ascertain, beyond
a doubt, whether the tapestry-bee is led to select the brilliant petals of the poppy from their colour, or from any
other quality they may possess, of softness or of warmth,
for instance. Réaumur thinks that the largeness, united
with the flexibility of the poppy-leaves, determines her
choice. Yet it is not improbable that her eye may be
gratified by the appearance of her nest ; that she may
possess a feeling of the beautiful in colour, and may look
with complacency upon the delicate hangings of the apartment which she destines for her offspring. Why should
not an insect be supposed to have a glimmering of the value
of ornament ? How can we pronounce, from our limited
notion of the mode in which the inferior animals think and
act, that their gratifications are wholly bounded by the
positive utility of the objects which surround them ? Why
does a dog howl at the sound of a bugle, but because it

offends his organ of hearing?—and why, therefore, may not a bee feel gladness in the brilliant hues of her scarlet drapery, because they are grateful to her organs of sight? All these little creatures work, probably, with more neatness and finish than is absolutely essential for comfort; and this circumstance alone would imply that they have something of taste to exhibit, which produces to them a pleasurable emotion.

The tapestry-bee is, however, content with ornamenting the interior only of the nest which she forms for her progeny. She does not misplace her embellishments with the error of some human artists. She desires security as well as elegance; and, therefore, she leaves no external traces of her operations. Hers is not a mansion rich with columns and friezes without, but cold and unfurnished within, like the desolate palaces of Venice. She covers her tapestry quite round with the common earth; and leaves her eggs enclosed in their poppy-case with a certainty that the outward show of her labours will attract no plunderer.

The poppy-bee may be known by its being rather more than a third of an inch long, of a black colour, studded on the head and back with reddish-grey hairs; the belly being grey and silky, and the rings margined with grey above, the second and third having an impressed transversal line.

A species of solitary bee (*Anthidium manicatum*, FABRICIUS), by no means uncommon with us, forms a nest of a peculiarly interesting structure. Kirby and Spence say, that it does not excavate holes, but makes choice of the cavities of old trees, key-holes, and similar localities; yet •it is highly probable, we think, that it may sometimes scoop out a suitable cavity when it cannot find one; for its mandibles seem equally capable of this, with those of any of the carpenter or mason-bees.

Be this as it may, the bee in question having selected a place suitably sheltered from the weather, and from

the intrusion of depredators, proceeds to form her nest,
the exterior walls of which she forms of the wool of
pubescent plants, such as rose-campion (*Lychnis coronaria*),
the quince (*Pyrus cydonia*), cats-ears (*Stachys lunata*), &c.
"It is very pleasant," says Mr. White, of Selborne, "to see
with what address this insect strips off the down, running
from the top to the bottom of the branch, and shaving it
bare with all the dexterity of a hoop-shaver. When it has
got a vast bundle, almost as large as itself, it flies away,
holding it secure between its chin and its fore-legs."* The
material is rolled up like a ribbon; and we possess a specimen in which one of these rolls still adheres to a rose-
campion stem, the bee having been scared away before
obtaining her load.

The manner in which the cells of the nest are made
seems not to be very clearly understood. M. Latreille
says, that, after constructing her nest of the down of
quince-leaves, she deposits her eggs, together with a store
of paste, formed of the pollen of flowers, for nourishing the
grubs. Kirby and Spence, on the other hand, tell us, that
"the parent bee, *after* having constructed her cells, laid an
egg in each, and filled them with a store of suitable food,
plasters them with a covering of vermiform masses, apparently composed of honey and pollen; and having done
this, aware, long before Count Rumford's experiments, what
materials conduct heat most slowly," she collects the down
from woolly plants, and "sticks it upon the plaster that
covers her cells, and thus closely envelops them with a
warm coating of down, impervious to every change of
temperature." "From later observations," however, they
are "inclined to think that these cells may possibly, as in
the case of the humble-bee, be in fact formed by the larva
previously to becoming a pupa, after having eaten the
provision of pollen and honey with which the parent bee
had surrounded it. The vermicular shape, however, of the
masses with which the cases are surrounded, does not seem

* Naturalist's Calendar, p. 100.

easily reconcilable with this supposition, unless they are considered as the excrement of the larva."*

Whether or not this second explanation is the true one, we have not the means of ascertaining; but we are almost certain the first is incorrect, as it is contrary to the regular procedure of insects to begin with the interior part of any structure, and work outwards. We should imagine, then, that the down is first spread out into the form required, and afterwards plastered on the inside to keep it in form, when probably the grub spins the vermicular cells previous to its metamorphosis.

It might prove interesting to investigate this more minutely; and as the bee is by no means scarce in the neighbourhood of London, it might not be difficult for a careful observer to witness all the details of this singular architecture. Yet we have repeatedly endeavoured, but without success, to watch the bees, when loaded with down, to their nests. The bee may be readily known from its congeners, by its being about the size of the hive-bee, but more broad and flattened, blackish-brown above, with a row of six yellow or white spots along each side of the rings, very like the rose-leaf cutter, and having the belly covered with yellowish-brown hair, and the legs fringed with long hairs of a rather lighter colour.

[This bee does not bore a tunnel for herself, but occupies that of some other insect. The nests of this insect are generally to be obtained from old willows, because these trees are so largely bored by the goat-moth caterpillar, and afford ample space for the larva. The woolly substance obtained from the plant is pressed against the sides of the burrow, so as to form a lining. She then makes a series of cells of a similar material, and the young larva, when it is about to change into the pupa state, envelops itself in a silken covering of a brown colour.

It is a curious fact, that the male of this insect is consi-

* Introduction to Entomology, vol. i. p. 435, 5th edit.

dorably larger than the female, thus reversing the usual order of things among insects. Only one species of this bee is known in England.]

A common bee belonging to the family of upholsterers is called the rose-leaf cutter (*Megachile centuncularis*, LATR.). The singularly ingenious habits of this bee have long attracted the attention of naturalists; but the most interesting description is given by Réaumur. So extraordinary does the construction of their nests appear, that a French gardener having dug up some, and believing them to be the work of a magician, who had placed them in his garden with evil intent, sent them to Paris to his master, for advice as to what should be done by way of exorcism. On applying to the Abbé Nollet, the owner of the garden was soon persuaded that the nests in question were the work of insects; and M. Réaumur, to whom they were subsequently sent, found them to be the nests of one of the upholsterer-bees, and probably of the rose-leaf cutter, though the nests in question were made of the leaves of the mountain ash (*Pyrus aucuparia*).

The rose-leaf cutter makes a cylindrical hole in a beaten pathway, for the sake of more consolidated earth (or in the cavities of walls or decayed wood), from six to ten inches deep, and does not throw the earth dug out from it into a heap, like the Andrenæ.* In this she constructs several cells about an inch in length, shaped like a thimble, and made of cuttings of leaves (not petals), neatly folded together, the bottom of one thimble-shaped cell being inserted into the mouth of the one below it, and so on in succession.

It is interesting to observe the manner in which this bee procures the materials for forming the tapestry of her cells. The leaf of the rose-tree seems to be that which she prefers, though she sometimes takes other sorts of leaves, particularly those with serrated margins, such as the birch, the perennial mercury (*Mercurialis perennis*), mountain-ash,

* See p. 50.

&c. She places herself upon the outer edge of the leaf which she has selected, so that its margin may pass between her legs. Turning her head towards the point, she commences near the footstalk, and with her mandibles cuts out a circular piece with as much expedition as we could do with a pair of scissors, and with more accuracy and neatness than could easily be done by us. As she proceeds, she keeps the cut portion between her legs, so as not to

Rose-leaf cutter Bees, and Nest lined with rose-leaves.

impede her progress; and using her body for a *trammel*, as a carpenter would say, she cuts in a regular curved line. As she supports herself during the operation upon the portion of the leaf which she is detaching, it must be obvious, when it is nearly cut off, that the weight of her body might tear it away, so as to injure the accuracy of its curvilineal shape. To prevent any accident of this kind, as soon as she suspects that her weight might tear it, she poises herself on her wings, till she has completed the incision. It has been said, by naturalists, that this manœuvre of poising herself on the wing, is to prevent her falling to the ground, when the piece gives way; but as no winged insect requires to take any such precaution, our explanation is probably the true one.

With the piece which she has thus cut out, held in a bent position perpendicularly to her body, she flies off to her nest, and fits it into the interior with the utmost neatness and ingenuity; and, without employing any paste or glue, she trusts, as Réaumur ascertained, to the spring the leaf takes in drying, to retain it in its position. It requires from nine to ten pieces of leaf to form one cell, as they are not always of precisely the same thickness. The interior surface of each cell consists of three pieces of leaf, of equal size, narrow at one end, but gradually widening at the other, where the width equals half the length. One side of each of the pieces is the serrated margin of the leaf from which it was cut, and this margin is always placed outermost, and the cut margin innermost. Like most insects, she begins with the exterior, commencing with a layer of tapestry, which is composed of three or four oval pieces, larger in dimensions than the rest, adding a second and a third layer proportionately smaller. In forming these, she is careful not to place a joining opposite to a joining, but with all the skill of a consummate artificer, lays the middle of each piece of leaf over the margins of the others, so as by this means both to cover and strengthen the junctions. By repeating this process, she sometimes forms a fourth or a fifth layer of leaves, taking care to bend the leaves at the narrow extremity or closed end of the cell, so as to bring them into a convex shape.

When she has in this manner completed a cell, her next business is to replenish it with a store of honey and pollen, which, being chiefly collected from thistles, forms a beautiful rose-coloured conserve. In this she deposits a single egg, and then covers in the opening with three pieces of leaf, so exactly circular, that a pair of compasses could not define

their margin with more accuracy. In this manner the
industrious and ingenious upholsterer proceeds till the whole
gallery is filled, the convex extremity of the one fitting into
the open end of the next, and serving both as a basis and as
the means of strengthening it. If, by any accident, the
labour of these insects is interrupted or the edifice deranged,
they exhibit astonishing perseverance in setting it again to
rights. Insects, indeed, are not easily forced to abandon
any work which they may have begun.

The monkish legends tell us that St. Francis Xavier,
walking one day in a garden, and seeing an insect, of the
Mantis genus, moving along in its solemn way, holding up
its two fore legs, as in the act of devotion, desired it to sing
the praises of God. The legend adds that the saint immediately heard the insect carol a fine canticle with a loud
emphasis. We want no miraculous voice to record the
wonders of the Almighty hand, when we regard the insect
world. The little rose-leaf cutter, pursuing her work with
the nicest mathematical art—using no artificial instruments
to form her ovals and her circles—knowing that the elastic
property of the leaves will retain them in their position—
making her nest of equal strength throughout, by the most
rational adjustment of each distinct part—demands from us
something more than mere wonder; for such an exercise of
instinctive ingenuity at once directs our admiration to the
great Contriver, who has so admirably proportioned her
knowledge to her necessities.

CHAPTER IV.

CARDER-BEES ; HUMBLE-BEES ; SOCIAL-WASPS.

THE bees and wasps, whose ingenious architecture we have already examined, are solitary in their labours. Those we are about to describe live in society. The perfection of the social state among this class of insects is certainly that of the hive-bees. They are the inhabitants of a large city, where the arts are carried to a higher excellence than in small districts enjoying little communication of intelligence. But the bees of the villages, if we may follow up the parallel, are not without their interest. Such are those which are called carder-bees and humble-bees.

CARDER-BEES.

The nests of the bees which Réaumur denominates carders (*Bombus muscorum*, LATR.) are by no means uncommon, and are well worth the study of the naturalist. During the hay harvest, they are frequently met with by mowers in the open fields and meadows; but they may sometimes be discovered in hedge-banks, the borders of copses, or among moss-grown stones. The description of the mode of building adopted by this bee has been copied by most of our writers on insects from Réaumur; though he is not a little severe on those who write without having ever had a single nest in their possession. We have been able to avoid such a reproach ; for we have now before us a very complete nest of carder-bees, which differs from those described by Réaumur, in being made not of moss, but withered grass. With this exception, we find that his account agrees accurately with our own observations. (J. R.)

The carder-bees select for their nest a shallow excavation about half a foot in diameter; but when they cannot find one to suit their purpose, they undertake the Herculean task of digging one themselves. They cover this hollow with a dome of moss—sometimes, as we have ascertained, of withered grass. They make use, indeed, of whatever materials may be within their reach; for they do not attempt to bring anything from a distance, not even when they are deprived of the greater portion by an experimental naturalist. Their only method of transporting materials to the building

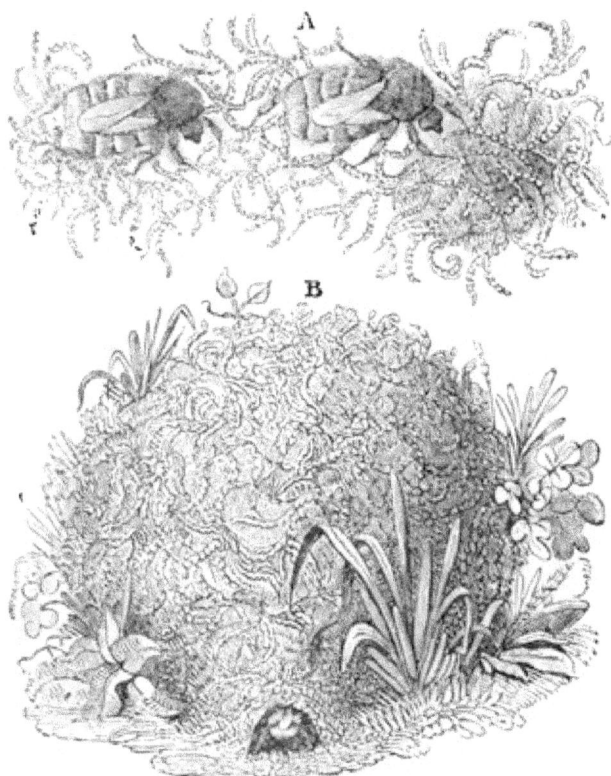

Fig. A represents two Carder-Bees heckling moss for their Nests;
B, exterior view of the Nest of the Carder-Bee.

is by pushing them along the ground—the bee, for that purpose, working backwards, with its head turned from the nest. If there is only one bee engaged in this labour, as usually happens in the early spring, when a nest is founded by a solitary female who has outlived the winter, she transports her little bundles of moss or grass by successive backward pushes, till she gets them home.

In the latter part of the season, when the hive is populous and can afford more hands, there is an ingenious division of this labour. A file of bees, to the number sometimes of half a dozen, is established, from the nest to the moss or grass which they intend to use, the heads of all the file of bees being turned from the nest and towards the material. The last bee of the file lays hold of some of the moss with her mandibles, disentangles it from the rest, and having *carded* it with her fore legs into a sort of felt or small bundle, she pushes it under her body to the next bee, who passes it in the same manner to the next, and so on till it is brought to the border of the nest,—in the same way as we sometimes see sugar-loaves conveyed from a cart to a warehouse, by a file of porters throwing them from one to another.

The elevation of the dome, which is all built from the interior, is from four to six inches above the level of the field. Beside the moss or grass, they frequently employ coarse wax to form the ceiling of the vault, for the purpose of keeping out rain, and preventing high winds from destroying it. Before this finishing is given to the nest, we have remarked, that on a fine sunshiny day the upper portion of the dome was opened to the extent of more than an inch, in order, we suppose, to forward the hatching of the eggs in the interior; but on the approach of night this was carefully covered in again. It was remarkable that the opening which we have just mentioned was never used by the bees for either their entrance or their exit from the nest, though they were all at work there, and, of course, would have found it the readiest and easiest passage; but they invariably made their exit and their entrance through the covert-way or gallery which opens at the bottom of the nest, and, in some nests, is about a foot long and half an inch wide. This is, no doubt, intended for concealment from field-mice, polecats, wasps, and other depredators.

On removing a portion of the dome and bringing the interior of the structure into view, we find little of the architectural regularity so conspicuous in the combs of a common bee-hive : instead of this symmetry, there are only a few egg-shaped, dark-coloured cells, placed somewhat irregularly, but approaching more to the horizontal than to the vertical position, and connected together with small amorphous* columns of brown wax. Sometimes there are two or three of these oval cells placed one above another, without anything to unite them.

These cells are not, however, the workmanship of the old bees, but of their young grubs, who spin them when they are about to change into nymphs. But, from these cases, when they are spun, the enclosed insects have no means of escaping, and they depend for their liberation on the old bees gnawing off the covering, as is done also by ants in the same circumstances. The instinct with which they know the precise time when it is proper to do this is truly wonderful. It is no less so, that these cocoons are by no means useless when thus untenanted, for they subsequently serve for honey-pots, and are indeed the only store-cells in the nest. For this purpose the edge of the cell is repaired and strengthened with a ring of wax.

The true breeding-cells are contained in several amorphous masses of brown-coloured wax, varying in dimensions,

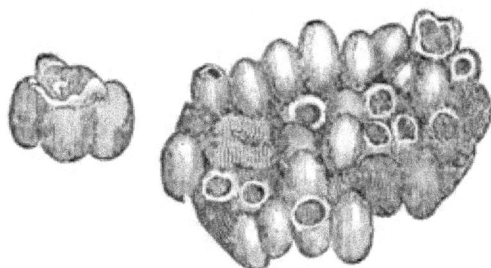

Breeding-Cells.

but of a somewhat flat and globular shape. On opening any of these, a number of eggs or grubs are found, on whose

* Shapeless.

account the mother bee has collected the masses of wax, which also contain a supply of pollen moistened with honey, for their subsistence.

The number of eggs or grubs found in one spheroid of wax varies from three to thirty, and the bees in a whole nest seldom exceed sixty. There are three sizes of bees, of which the females are the largest; but neither these nor the males are, as in the case of the hive-bee, exempt from labour, the females, indeed, always found the nests, since they alone survive the winter, all the rest perishing with cold. In each nest, also, are several females, that live in harmony together.

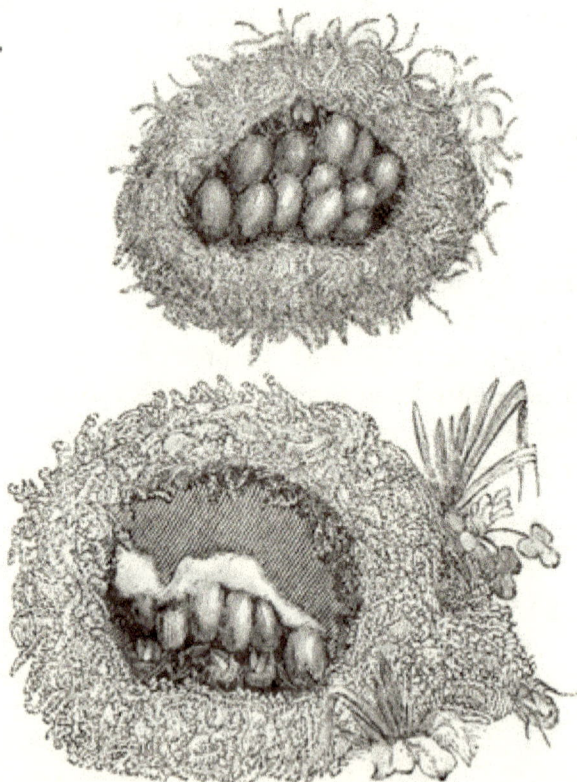

Interior views of Carder-Bee's Nest.

The carder-bees may be easily distinguished from their congeners (of the same genus), by being not unlike the colour of the withered moss with which they build their nests, having the fore part of their back a dull orange, and hinder part ringed with different shades of greyish yellow. They are not so large as the common humble-bee (*Bombus terrestris*, Latr.), but rather shorter and thicker in the body than the common hive-bee (*Apis mellifica*).

Lapidary-Bees.

A bee still more common, perhaps, than the carder is the orange-tailed bee, or lapidary (*Bombus lapidaria*), readily known by its general black colour and reddish orange tail. It builds its nest sometimes in stony ground, but prefers a heap of stones such as are gathered off grass fields or are piled up near quarries. Unlike the carder, the lapidary carries to its nest bits of moss, which are very neatly arranged into a regular oval. These insects associate in their labours; and they make honey with great industry. The individuals of a nest are more numerous than the carders, and likewise more pertinaciously vindictive. About two years ago we discovered a nest of these bees at Compton-Bassett, in Wiltshire, in the centre of a heap of limestone rubbish; but owing to the brisk defensive warfare of their legionaries, we could not obtain a view of the interior. It was not even safe to approach within many yards of the place; and we do not exaggerate when we say that several of them pursued us most pertinaciously about a quarter of a mile. (J. R.)

Humble-Bees.

The common humble-bee (*Bombus terrestris*) is precisely similar in its economy to the two preceding species, with this difference, that it forms its nest underground like the common wasp, in an excavated chamber, to which a winding passage leads, of from one to two feet, and of a diameter sufficient to allow of two bees passing. The cells have no covering beside the vault of the excavation and patches of coarse wax similar to that of the carder-bee.

[The accompanying illustration represents a group of cells made by this species. As may be seen by reference to the

engraving, they are not placed with any regularity, but seem
to be tossed about at random.

Some of the cells contain larvæ, in others, those closely
sealed, lie the pupæ in different stages of development, and

some of the cells are filled with a very fragrant and sweet
honey, which, however, is injurious to many persons, giving
them severe and persistent headaches, even though taken in
small quantities.]

Social-Wasps.

The nest of the common wasp (*Vespa vulgaris*) attracts
more or less the attention of everybody; but its interior
architecture is not so well known as it deserves to be, for
its singular ingenuity, in which it rivals even that of the
hive-bee (*Apis mellifica*). In their general economy the
social or republican wasps closely resemble the humble-
bee (*Bombus*), every colony being founded by a single

female who has survived the winter, to the rigours of which all her summer associates of males and working wasps uniformly fall victims. Nay, out of three hundred females which may be found in one vespiary, or wasp's nest, towards the close of autumn, scarcely ten or a dozen survive till the ensuing spring, at which season they awake from their hybernal lethargy, and begin with ardour the labours of colonization.

It may be interesting to follow one of these mother wasps through her several operations, in which she merits more the praise of industry than the queen of a bee-hive, who does nothing, and never moves without a numerous train of obedient retainers, always ready to execute her commands and to do her homage. The mother wasp, on the contrary, is at first alone, and is obliged to perform every species of drudgery herself.

Her first care, after being roused to activity by the returning warmth of the season, is to discover a place suitable for her intended colony; and, accordingly, in the spring, wasps may be seen prying into every hole of a hedge-bank, particularly where field-mice have burrowed. Some authors report that she is partial to the forsaken galleries of the mole; but this does not accord with our observations, as we have never met with a single vespiary in any situation likely to have been frequented by moles. But though we cannot assert the fact, we think it highly probable that the deserted nest of the field-mouse, which is not uncommon in hedge-banks, may be sometimes appropriated by a mother wasp as an excavation convenient for her purpose. Yet, if she does make choice of the burrow of a field-mouse, it requires to be afterwards considerably enlarged in the interior chamber, and the entrance gallery very much narrowed.

The desire of the wasp to save herself the labour of excavation, by forming her nest where other animals have burrowed, is not without a parallel in the actions of quadrupeds, and even of birds. In the splendid continuation of Wilson's American Ornithology, by Charles L. Bonaparte

G

(whose scientific pursuits have thrown around that name a beneficent lustre, pleasingly contrasted with his uncle's glory), there is an interesting example of this instinctive adoption of the labours of others. " In the trans-Mississippian territories of the United States, the burrowing-owl resides exclusively in the villages of the marmot, or prairie-dog, whose excavations are so commodious as to render it unnecessary that the owl should dig for himself, as he is said to do where no burrowing animals exist.* The villages of the prairie-dog are very numerous and variable in their extent,—sometimes covering only a few acres, and at others spreading over the surface of the country for miles together. They are composed of slightly-elevated mounds, having the form of a truncated cone, about two feet in width at the base, and seldom rising as high as eighteen inches from the surface of the soil. The entrance is placed either at the top or on the side, and the whole mound is beaten down externally, especially at the summit, resembling a much-used footpath. From the entrance, the passage into the mound descends vertically for one or two feet, and is thence continued obliquely downwards until it terminates in an apartment, within which the industrious prairie-dog constructs, on the approach of cold weather, a comfortable cell for his winter's sleep. The cell, which is composed of fine dry grass, is globular in form, with an opening at top, capable of admitting the finger ; and the whole is so firmly compacted, that it might without injury be rolled over the floor."†

In case of need the wasp is abundantly furnished by nature with instruments for excavating a burrow out of the solid ground, as she no doubt most commonly does—digging the earth with her strong mandibles, and carrying it off or pushing it out as she proceeds. The entrance-gallery is about an inch or less in diameter, and usually runs in a winding or zigzag direction, from one to two feet

* The Owl observed by Vieillot in St. Domingo digs itself a burrow two feet in depth, at the bottom of which it deposits its eggs upon a bed of moss.
† American Ornithology, by Charles Lucien Bonaparte, vol. i. p. 69.

in depth. In the chamber to which this gallery leads, and which, when completed, is from one to two feet in diameter, the mother wasp lays the foundations of her city, beginning with the walls.

The building materials employed by wasps were long a matter of conjecture to scientific inquirers; for the bluish-grey papery substance of the whole structure has no resemblance to any sort of wax employed by bees for a similar purpose. Now that the discovery has been made, we can with difficulty bring ourselves to believe that a naturalist so acute and indefatigable as M. Réaumur, should have, for twenty years, as he tells us, endeavoured, without success, to find out the secret. At length, however, his perseverance was rewarded. He remarked a female wasp alight on the sash of his window, and begin to gnaw the wood with her mandibles; and it struck him at once that she was procuring materials for building. He saw her detach from the wood a bundle of fibres about a tenth of an inch in length, and finer than a hair; and as she did not swallow these, but gathered them into a mass with her feet, he could not doubt that his first idea was correct. In a short time she shifted to another part of the window-frame, carrying with her the fibres she had collected, and to which she continued to add, when he caught her, in order to examine the nature of her bundle; and he found that it was not yet moistened nor rolled into a ball, as is always done before employing it in building. In every other respect it had precisely the same colour and fibrous texture as the walls of a vespiary. It struck him as remarkable that it bore no resemblance to wood gnawed by other insects, such as the goat-moth caterpillar, which is granular like sawdust. This would not have suited the design of the wasp, who was well aware that fibres of some length form a stronger texture. He even discovered, that before detaching the fibres, she bruised them (*les charpissoit*) into a sort of lint (*charpie*) with her mandibles. All this the careful naturalist imitated by bruising and paring the same wood of the window-sash with his penknife, till he succeeded in making a little bundle of fibres scarcely to be distinguished from that collected by the wasp.

We have ourselves frequently seen wasps employed in procuring their materials in this manner, and have always observed that they shift from one part to another more than once in preparing a single load—a circumstance which we ascribe entirely to the restless temper peculiar to the whole order of hymenopterous insects. Réaumur found that the wood which they preferred was such as had been long exposed to the weather, and is old and dry. White of Selborne, and Kirby and Spence, on the contrary, maintain that wasps obtain their paper from sound timber, hornets only from that which is decayed.* Our own observations, however, confirm the statement of Réaumur with respect to wasps, as, in every instance which has fallen under our notice, the wood selected was very much weathered; and in one case, an old oak post in a garden at Lee, in Kent, half destroyed by dry-rot, was seemingly the resort of all the wasps in the vicinity. In another case, the deal bond in a brick wall, which had been built thirty years, is at this moment (June, 1829) literally striped with the gnawings of wasps, which we have watched at the work for hours together. (J. R.)

[Different species of wasps use different materials for their nest. *Vespa vulgaris* always uses decayed wood, while *V. germanica* and other species use sound wood. Owing to the colour, the distinction between the nests of these insects is evident at a glance.

The bundles of ligneous fibres thus detached are moistened before being used, with a glutinous liquid, which causes them to adhere together, and are then kneaded into a sort of paste, or *papier maché.*

The method employed by the wasp in making its nest has been so admirably described by Mr. S. Stone, that we cannot do better than copy his description, which appeared in " Beeton's Annual " of 1865.

* Réaumur, vol. vi. bottom of page 182 ; Hist. of Selb. ii. 228 ; and Introd. to Entomol. i. 504, 5th edition.

" Having found a place suitable—the deserted burrows of the field-mice being perhaps more generally selected than any other by the underground species, the chamber formed by that animal for its nest being exactly the kind of place required by the insect—it proceeds to attach its web to the centre of the roof of the chamber. This consists, in the first instance, of a pedicle, or footstalk, about half an inch in length, at the extremity of which a single cell is formed, which is presently surrounded by others.

" Simultaneously with the formation of these cells, an umbrella-shaped covering is prepared above them. More cells are added, an egg being deposited in each of them as soon as formed, while constant additions are made to the covering until it has assumed a globular form, with only an aperture sufficiently large for the insect to pass in and out. Before the completion of the first covering, a second, just large enough to enclose it, is begun, and while this is in progress a third is commenced, and then a fourth, and so on. When young wasps have been produced in sufficient numbers to carry on the work without the assistance of the parent, an event which usually takes place in about six weeks from the commencement of the nest, she does not again leave home, but occupies herself solely in the task of depositing eggs as fast as cells can be formed by the workers for their reception.

" There are two methods by which the nests are enlarged by the workers after the queen has given up the task of building ; some species choosing one, some adopting the other. One consists in forming a series of regular sheets or layers, which are made to overlap each other like the slates or tiles on the roof of a building, in the same way as is pursued by the queen of every species so long as she continues to be the architect. When a few of these sheets have been completed, that is, when they have been made to assume a spherical form, with only a small aperture for ingress and egress, each internal sheet is cut away, nearly but not quite, as fast as additional ones are formed externally, the shell or covering therefore slightly increasing in thickness as the nest increases in size. Thus architects among the human race are careful to proportion the thickness, and consequently the strength of the walls to the magnitude of the building designed to be erected.

" The other method consists in forming hollow pieces, or raising, as it were, blisters all over the plain surface which the queen has left ; and upon these other blisters, and so on continually ; cutting away, as in the former case, the under skin on the formation of the outer one. The latter method is adopted by the workers of *V. crabro, V. vulgaris,* and *V. germanica;* the former by *V. Norvegica, V. sylvestris, V. rufa,* and probably by *V. arborea.* Cutting away the inner portions of the coverings is a necessary process in order to make room for the increased size of the comb or combs. The material cut away is not thrown by as useless, but is worked up afresh ; indeed this is effected in, and by, the very act of removing it : it is then either used in enlarging the combs or it is brought out and employed in making additions to the outside.

" As the nest increases in size, it is obvious that the cavity in which it is placed must be proportionably enlarged ; accordingly, each wasp, as it emerges from the aperture, may be observed to bring out with it a small lump of earth which it has scraped from the walls of the chamber, care being taken to keep a clear space of about a quarter of an inch between the covering of the nest, and the walls of the chamber. About the same space also occurs between the combs, which are placed horizontally, with the mouth of the cells downwards ; supporting columns or pillars being constructed at regular intervals so as to keep them at a proper distance apart, thus allowing the insects room to pass between them for the purpose of feeding the grubs. Supporting columns or pillars are also placed between the roof of the chamber and the crown of the nest, connecting the one with the other ; and these supports are constantly strengthened as the increasing weight of the nest renders such a precaution necessary.

" The material of which the wasps' nests are composed is a sort of paper manufactured chiefly from wood by the insects themselves; one species using sound wood for the purpose, another that which has become decayed. This they scrape by means of their jaws from posts, rails, gates, hurdles, &c., in which act it becomes mixed with some peculiar fluid with which they are provided; it then possesses nearly the same properties as the pulp from which paper is made, but is of firmer consistence. This is gathered in a small lump under the chest, to which it adheres, and in that way is carried to the nest.

" The operators having, after the exhibition of a considerable amount of fickleness in the choice, fixed upon a suitable place for commencing, or recommencing operations—for these remarks have reference to a nest already somewhat advanced in the building—place themselves along the edge of a yet unfinished piece, then walking slowly backward, spread the material as they go, along this edge, where it forms a thick streak; they then go forward to the point at which they began to spread the composition, again marching slowly backward, press this streak between their jaws, which acts as a pair of pincers, thus thinning it out throughout its whole length. They then go forward a second time, pressing it still thinner, and then a third, and so on, until they have rendered it sufficiently thin. Before this is accomplished, the operators have generally to go five or six times over their work. They do not return to the same spot with their next burden, but seek a fresh one, and thus allow the work they recently executed to become dry and firm, previous to making further additions to it. Possibly the material first ' used up ' was from wood of a dark colour; the next may be from light-coloured wood, and the next from that of an intermediate colour; and this it is which gives so much beauty to the coverings of the nests of these insects.

" *Vespa crabro* and *V. vulgaris* are the only species which use decayed wood or touchwood in the fabrication of their nests; the other species employ sound wood, varied occasionally by sound vegetable fibre obtained from plants of different kinds.

" From the upper combs in a nest, workers are produced;

from the lower ones, queens or females; and from the intermediate ones, males. Workers become developed early in the season, males not till an advanced period; and young females or queens not until towards the close of the season.

"The nests of *V. crabro, V. vulgaris,* and *V. germanica,* when of full size, measure not unfrequently twelve inches in diameter, the communities working on, in a favourable season, until the month of November; while the labours of the other species close, and the communities break up towards the end of August; their nests scarcely attaining to half the size of those above mentioned."

The accompanying illustration exhibits the nest of the common wasp in an early stage. The first cover has been

Nest of Wasp in an early stage.

completed, and a second is in course of progress. We have now before us a beautiful series of wasps' nests, in their various stages, prepared by Mr. Stone, in order to show the progressive enlargement of the edifice.

First, there is the single cell attached to a small part that had penetrated the roof of the burrow. Next comes a more advanced stage, in which three cells are made, and the roof is just begun, being not quite half an inch in diameter. Then come five cells, and a tolerably large roof; and then twelve cells, with a complete roof.

The next stage is that which is represented in the illustration, where the group of cells is seen suspended from its slender footstalk, and a second covering is in progress. By degrees the nest enlarges until the second layer or tier of

cells is begun, while the first tier is occupied in the centre
by the pupæ, sheltered by their little silk doors, and on the
circumference by the larvæ, whose cells are still open in
order to allow themselves to be fed by the nurse-wasps.

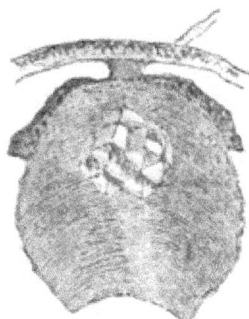

Section of the same Nest, showing the first tier of Cells.

In these nests, the difference between the homes of *Vespa
vulgaris* and those of *V. germanica* is very strongly marked,
the former being yellowish brown, and the latter grey.
One nest of *V. germanica*, is remarkable for being thickly
studded with the long, white eggs of some insect, probably a
parasite, which has gained admittance to the burrow, in spite
of the care of its guardians. It may be here mentioned, that
V. germanica is by far the most common species of wasp in
England.

The illustration at p. 90 represents a completed nest of
V. germanica. The rough, thick covering is seen outside, and
within are the tiers of cells, each layer being supported by
pillars from the layer immediately above. These pillars are
always formed at the angle where these cells touch each
other, so as to obtain as strong a foundation as possible. Only a very small space is left between the combs,
just enough room, in fact, for the nurse-wasps to pass as
they feed the young. The reader will remember that
the young wasps all hang with their heads downwards,
being held in their places by a sort of clasper at the end of
the tail.]

When the foundress-wasp has completed a certain number
of cells, and deposited eggs in them, she soon intermits her

building operations, in order to procure food for the young
grubs, which now require all her care. In a few weeks these
become perfect wasps, and lend their assistance in the
extension of the edifice; enlarging the original coping of the
foundress by side walls, and forming another platform of
cells, suspended to the first by columns, as that had been
suspended to the ceiling.

In this manner several platforms of combs are constructed,
the outer walls being extended at the same time; and, by

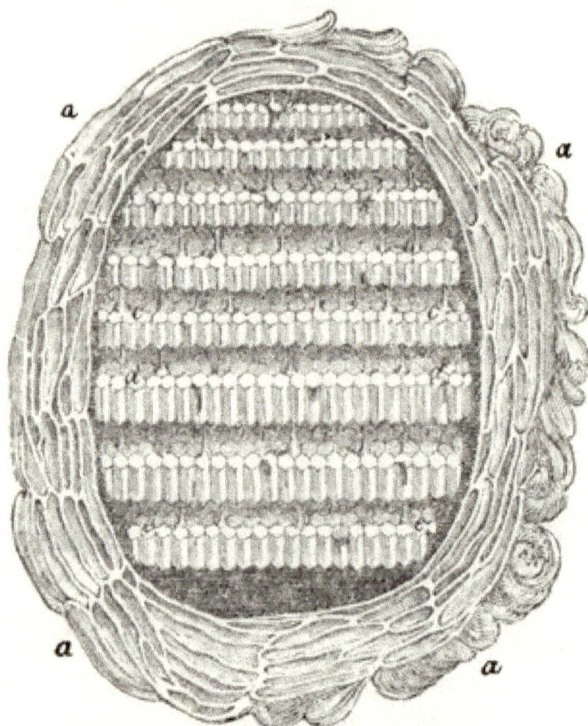

Section of the Social-Wasp's Nest.
a a, the external wall; *b, c c.* five small terraces of cells for the neuter wasps,
d d, e e, three rows of larger cells for the males and females.

the end of the summer, there is generally from twelve to
fifteen platforms of cells. Each contains about 1060 cells—
forty-nine being contained in an inch and a half square, and,
of course, making the enormous number of about 16,000
cells in one colony. Réaumur, upon these data, calculates

that one vespiary may produce every year more than 30.000 wasps, reckoning only 10.000 cells, and each serving successively for the cradle of three generations. But, although the whole structure is built at the expense of so much labour and ingenuity, it has scarcely been finished before the winter sets in, when it becomes nearly useless, and serves only for the abode of a few benumbed females, who abandon it on the

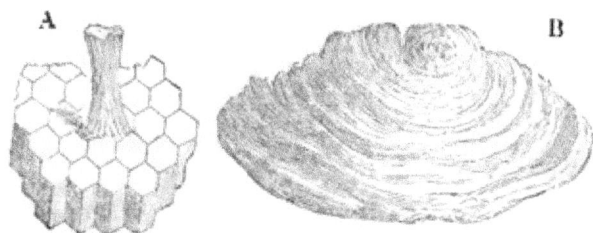

A, represents one of the rods from which the terraces are suspended. B, a portion of the external crust.

approach of spring, and never return ; for wasps do not, like mason-bees, ever make use of the same nest for more than one season.

Both Réaumur and the younger Huber studied the proceedings of the common wasp in the manner which has been so successful in observing bees—by means of glazed hives, and other contrivances. In this, these naturalists were greatly aided by the extreme affection of wasps for their young ; for though their nest is carried off, or even cut in various directions, and exposed to the light, they never desert it, nor relax their attention to their progeny. When a wasp's nest is removed from its natural situation, and covered with a glass hive, the first operation of the inhabitants is to repair the injuries it has suffered. They carry off with surprising activity all the earth or other matters which have fallen by accident into the nest ; and when they have got it thoroughly cleared of everything extraneous, they begin to secure it from further derangement, by fixing it to the glass with papyraceous columns, similar to those which we have already described. The breaches which the nest may have suffered are then repaired, and the thickness of the walls is

augmented, with the design, perhaps, of more effectually
excluding the light.

The nest of the hornet is nearly the same in structure
with that of the wasp; but the materials are considerably
coarser, and the columns to which the platforms of cells
are suspended are larger and stronger, the middle one being
twice as thick as any of the others. The hornet, also, does
not build underground, but in the cavities of trees, or in the

Hornet's Nest in its first stage.

thatch or under the eaves of barns. Réaumur once found
upon a wall a hornet's nest which had not been long begun,
and had it transferred to the outside of his study-window;
but in consequence, as he imagined, of the absence of the
foundress-hornet at the time it was removed, he could not get
the other five hornets, of which the colony consisted, either
to add to the building or repair the damages which it had
sustained.

M. Réaumur differs from our English naturalists, White,

and Kirby and Spence, with respect to the materials employed by the hornet for building. The latter say that it employs decayed wood; the former, that it uses the bark of the ash-tree, but takes less pains to split it into fine fibres than wasps do; not, however, because it is destitute of skill; for in constructing the suspensory columns of the platforms, a paste is prepared little inferior to that made by wasps. We cannot, from our own observations, decide which of the above statements is correct, as we have only once seen a hornet procuring materials, at Compton-Bassett, in Wiltshire; and in that case it gnawed the inner bark of an elm which had been felled for several months, and was, consequently, dry and tough. Such materials as this would account for the common yellowish-brown colour of a hornet's nest. (J. R.)

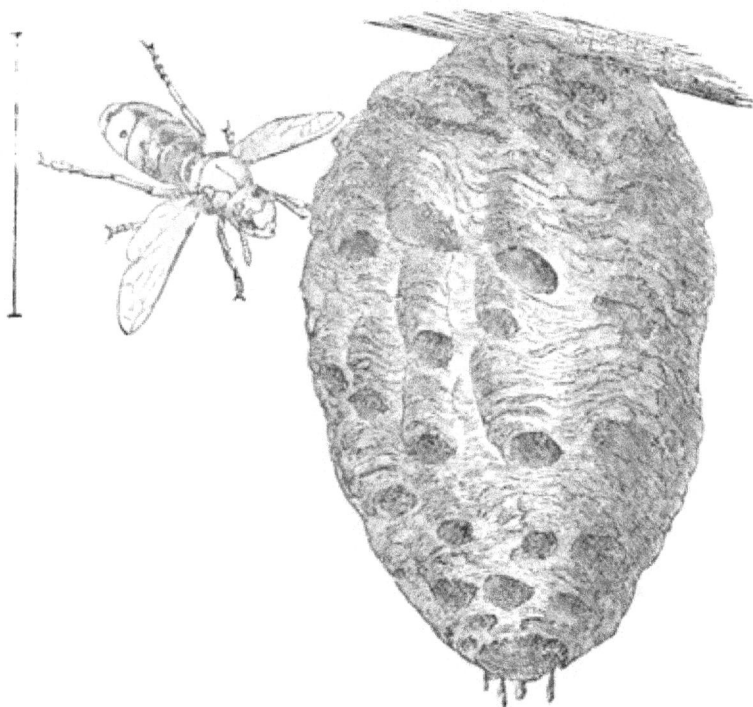

[The accompanying figure represents a completed hornet's nest as it appears when suspended from a beam. Hornets often choose for their home the space between the roof and the ceiling of summer-houses, and the nests that are made in such localities are mostly large and handsome. The reader should notice the blisters by means of which the insect enlarges its habitation.]

When hornets make choice of a tree for their domicile, they select one which is in a state of decay, and already partly hollowed; but they possess the means, in their sharp and strong mandibles, of extending the excavation to suit their purposes; and Réaumur frequently witnessed their operations in mining into a decayed tree, and carrying off what they had gnawed. He observed, also, that in such cases they did not make use of the large hole of the tree for an entrance, but went to the trouble of digging a gallery, sufficient for the passage of the largest hornet in the nest, through the living and undecayed portion of the tree. As this is perforated in a winding direction, it is no doubt intended for the purpose of protecting the nest from the intrusion of depredators, who could more easily effect an entrance if there were not such a tortuous way to pass through.

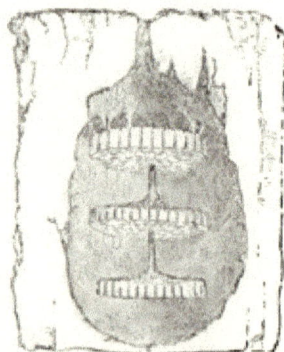

Hornet's Nest in a hollow tree.

[Here is an illustration of a hornet's nest as it appears in the hollow of a tree. Industrious as is this insect, it never takes needless trouble, and alters its nest according to circumstances. As has already been seen, the combs are defended by a complete cover when the nest is placed in an open situation. But when it is built in the hollow of a tree there is no cover at all, the insect evidently knowing that the wooden wall with which the cells are surrounded, affords a sufficient protection. In cases where a cover is made, the hornets do not form only a single entrance, as is the case with the wasp, but have a large number of small entrances in different parts of the wall. Some of these entrances can be seen in the illustration on page 93.

Hornets are in one sense more industrious than wasps. When night falls, the wasps betake themselves to their home, and sleep throughout the night. But, if the moon be up, the hornet is sure to work throughout the entire night, and will often do so, even when no moon is visible.]

One of the most remarkable of our native social wasps is the tree-wasp (*Vespa Britannica*), which is not uncommon in the northern, but is seldom to be met with in the southern parts of the island. Instead of burrowing in the ground like the common wasp (*Vespa vulgaris*), or in the hollows of trees like the hornet (*Vespa crabro*), it boldly swings its nest from the extremity of a branch, where it exhibits some resemblance, in size and colour, to a Welsh wig hung out to dry. We have seen more than one of these nests on the same tree, at Catrine, in Ayrshire, and at Wemyss Bay, in Renfrewshire. The tree which the Britannic wasp prefers is the silver fir, whose broad flat branch serves as a protection to the suspended nest both from the sun and the rain. We have also known a wasp's nest of this kind in a gooseberry-bush, at Red-house Castle, East Lothian. The materials and structure are nearly the same as those employed by the common wasp, and which we have already described. (J. R.)

[We have before us a beautiful example of a nest made by this species of wasp. There are no less than three consecutive coverings quite entire, while another is about three-fourths completed, and a fifth is just begun. The illustration exhibits a very perfect specimen.]

A singular nest of a species of wasp is figured by Réaumur, but is apparently rare in this country, as Kirby and Spence

mention only a single nest of similar construction, found in
a garden at East-Dale. This nest is of a flattened globular
figure, and composed of a great number of envelopes, so as
to assume a considerable resemblance to a half-expanded
Provence rose. The British specimen mentioned by Kirby and

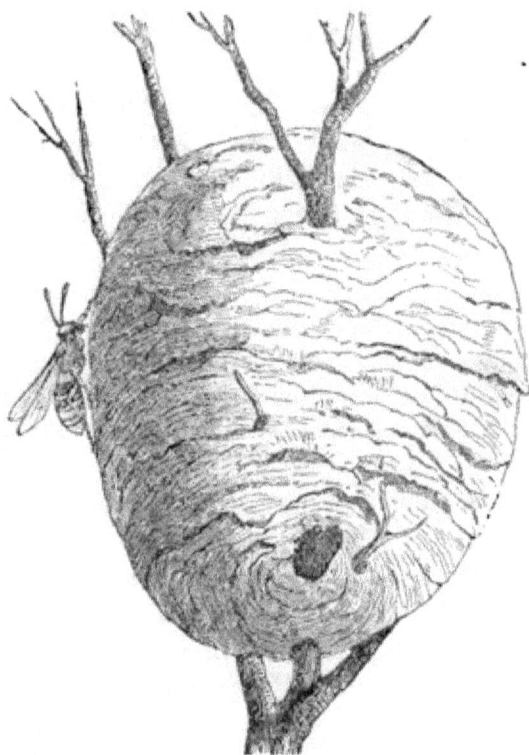

Spence had only one platform of cells; Réaumur had two;
but there was a large vacant space, which would probably
have been filled with cells, had the nest not been taken away
as a specimen. The whole nest was not much larger than a
rose, and was composed of paper exactly similar to that
employed by the common ground-wasp.*

* In the Mag. of Nat. Hist. 1839, p. 458, Mr. Shuckard gives an account
of the nest of a wasp, which he regards as *Vespa Britannica*,—remarkable for
the material of which it was constructed, and for the locality in which it was
found. This nest, which was exhibited at a meeting of the Entomological
Society, was found near Croydon, built in a sparrow's nest, and attached to the

[This is probably the nest of *V. rufa*. We possess several specimens of the nest, one of which corresponds tolerably closely with the edifice described in the work.]

There is another species of social-wasp (*Epipone nidulans*, LATR.) meriting attention from the singular construction of

Wasp's Nest.

its nest. It forms one or more terraces of cells, similar to those of the common wasp, but without the protection of an outer wall, and quite exposed to the weather. Swammerdam

lining feathers. "The smallness of the nest," says Mr. Shuckard, "and also of the tier of cells, as well as the peculiar material of which it appeared composed, led to a discussion, the tendency of which seemed to support the opinion that it was most probably the nest of a *Polistes*, a social-wasp not yet found in this country, but if not of *Polistes*, certainly not yet determined or known." The nest was ovate, about an inch and a half long, with a tier of cells internally, originating from a common pedicle. It appeared to be constructed "of the agglutinated particles of a soft white wood, probably willow, very imperfectly triturated;" whence it had externally a rough granulated appearance. It was sprinkled with black specks, arising perhaps from the intermixture of more decayed portions of the wood; and was of a very fragile texture. "The nature of the material, and its unfinished execution, as well as the situation in which it was found, appear to me to be its own peculiarities, and I must necessarily consider it merely an accidental variation in material and locality from the usual nests of the *Vespa Britannica* of Leach."

H

found a nest of this description attached to the stem of a nettle. Réaumur says that they are sometimes attached to the branch of a thorn or other shrub, or to stalks of grass;—peculiarities which prove that there are several species of these wasps.

The most remarkable circumstance in the architecture of this species of vespiary is, that it is not horizontal, like those formerly described, but nearly vertical. The reason appears to be, that if it had been horizontal, the cells must have been frequently filled with rain ; whereas, in the position in which it is placed, the rain runs off without lodging. It is, besides, invariably placed so as to face the north or the

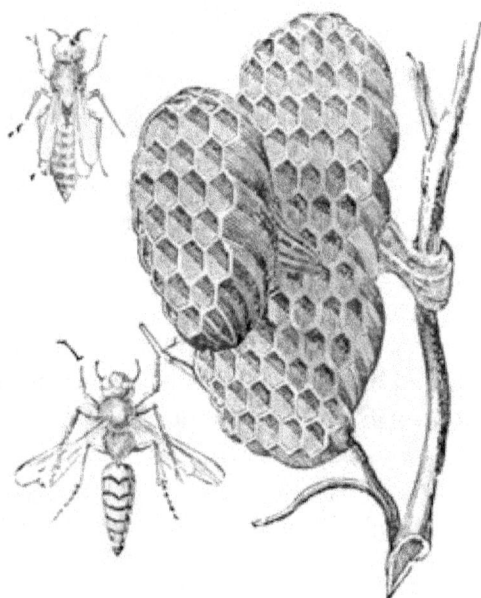

Wasps' Cells attached to a Branch.

east, and consequently is less exposed to rains, which most frequently come with southerly or westerly winds. It is another remarkable peculiarity, that, unlike the nests of other wasps, it is covered with a shining coat of varnish, to prevent moisture from soaking into the texture of the wasp's paper. The laying on this varnish, indeed, forms a considerable portion of the labour of the colony, and individuals may be seen employed for hours together spreading it on with their tongues.

There is a genus of foreign hymenoptera, called *Polistes*, which is remarkable for the building powers possessed by its members. The accompanying illustration is taken from a nest in the British Museum, and is given of the natural size. The cells are not hexagonal, like those of the *Epipone*, but are roundish in form. Those in the centre assume a roughly

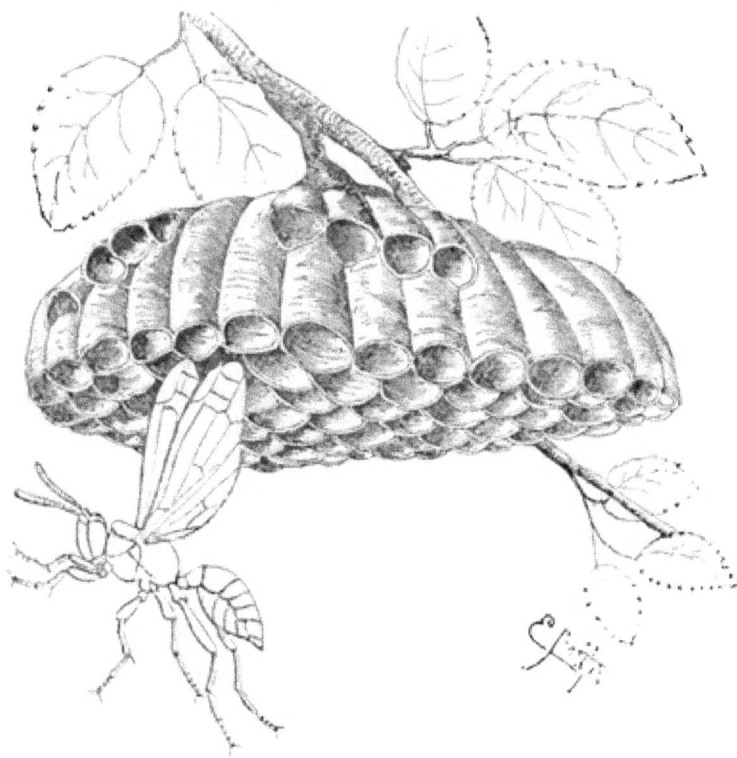

Comb of Polistes.

hexagonal form by pressure, but those which form the circumference of the cell-group are nearly round, especially on their outer sides. The cells are not of uniform width, but are narrower at the base than at the mouth, thus causing the group to assume the form which is seen in the illustration. This curious group of cells was brought from Bareilly, in the East Indies, and in the same collection there are several other specimens, varying considerably both in shape and size.]

Few circumstances are more striking, with regard to insects, as Kirby and Spence justly remark, than the great and incessant labour which maternal affection for their progeny leads them to undergo. Some of these exertions are so disproportionate to the size of the insect, that nothing short of ocular conviction could attribute them to such an agent. A wild bee, or a wasp, for instance, as we have seen, will dig a hole in a hard bank of earth some inches deep, and five or six times its own size, labouring unremittingly at this arduous task for several days in succession, and scarcely allowing itself a moment for eating or repose. It will then occupy as much time in searching for a store of food ; and no sooner is this finished, than it will set about repeating the process, and, before it dies, will have completed five or six similar cells, or even more.

We shall have occasion more particularly to dwell upon the geometrical arrangement of the cells, both of the wasp and of the social-bee, in our description of those interesting operations, which have long attracted the notice, and commanded the admiration of mathematicians and naturalists. A few observations may here be properly bestowed upon the *material* with which the wasp-family construct the interior of their nests.

The wasp is a paper-maker, and a most perfect and intelligent one. While mankind were arriving, by slow degrees, at the art of fabricating this valuable substance, the wasp was making it before their eyes, by very much the same process as that by which human hands now manufacture it with the best aid of chemistry and machinery. While some nations carved their records on wood, and stone, and brass, and leaden tablets,—others, more advanced, wrote with a style on wax,—others employed the inner bark of trees, and others the skins of animals rudely prepared,—the wasp was manufacturing a firm and durable paper. Even when the papyrus was rendered more fit, by a process of art, for the transmission of ideas in writing, the wasp was a better artisan than the Egyptians; for the early attempts at paper-making were so rude, that the substance produced was almost useless, from being extremely friable. The paper of the papyrus was formed of the leaves of the plant, dried, pressed, and polished; the wasp alone knew how to reduce vegetable fibres to a pulp, and then unite them by a size or glue, spreading the substance out into a smooth and delicate leaf. This is exactly the process of paper-making. It would seem that the wasp knows, as the modern paper-makers now know, that the fibres of rags, whether linen or cotton, are not the only materials that can be used in the formation of paper; she employs other vegetable matters, converting them into a proper consistency by her assiduous exertions. In some respects she is more skilful even than our paper-makers, for she takes care to retain her fibres of sufficient length, by which she renders her paper as strong as she requires. Many manufacturers of the present day cut their material into small bits, and thus produce a rotten article. One great distinction between good and bad paper is its toughness; and this difference is invariably produced by the fibre of which it is composed being long, and therefore tough; or short, and therefore friable.

The wasp has been labouring at her manufacture of paper from her first creation, with precisely the same instruments and the same materials; and her success has been unvarying. Her machinery is very simple, and therefore it is never out of order. She learns nothing, and she forgets nothing. Men, from time to time, lose their excellence in particular arts, and they are slow in finding out real improvements. Such improvements are often the effect of accident. Paper is now manufactured very extensively by machinery in all its stages; and thus, instead of a single sheet being made by hand, a stream of paper is poured out, which would form a roll large enough to extend round the globe, if such a length were desirable. The inventors of this machinery, Messrs. Fourdrinier, it is said, spent the enormous sum of 40,000*l.* in vain attempts to render the machine capable of determining with precision the width of the roll; and, at last, accomplished their object, at the suggestion of a bystander, by a strap revolving upon an axis, at a cost of three shillings and sixpence. Such is the difference between the workings of human knowledge and experience, and those of animal instinct. We proceed slowly and in the dark, but our course is not bounded by a narrow line, for it seems difficult to say what is the perfection of any art; animals go clearly to a given point—but they can go no further. We may, however, learn something from their perfect knowledge of what is within their range. It is not improbable that if man had attended in an earlier state of society to the labours of wasps, he would have sooner known how to make paper. We are still behind in our arts and sciences, because we have not always been observers. If we had watched the operations of insects, and the structure of insects in general, with more care, we might have been far advanced in the knowledge of many arts which are yet in their infancy, for nature has given us abundance of patterns. We have learnt to perfect some instruments of sound by examining the structure of the human ear; and the mechanism of an eye has suggested some valuable improvements in achromatic glasses.

Réaumur has given a very interesting account of the wasps of Cayenne (*Chartergus nidulans*), which hang their nests in trees.* Like the bird of Africa called the social grosbeak (*Loxia socia*), they fabricate a perfect house, capable of containing many hundreds of their community, and suspend it on high out of the reach of attack. But the Cayenne wasp is a more expert artist than the bird. He is a pasteboard-maker;—and the card with which he forms the exterior covering of his abode is so smooth, so strong, so uniform in its texture, and so white, that the most skilful manufacturer of this substance might be proud of the work. It takes ink admirably.

The nest of the pasteboard-making wasp is impervious to water. It hangs upon the branch of a tree, as represented in the engraving; and those rain-drops which penetrate through the leaves never rest upon its hard and polished

* Mémoires sur les Insectes, tom. vi.. mem. vii. See also Bonnet vol. ix.

surface. A small opening for the entrance of the insects terminates its funnel-shaped bottom. It is impossible to unite more perfectly the qualities of lightness and strength.

In the specimen from which we take our description, the length of which is nine inches, six stout circular platforms stretch internally across, like so many floors, and fixed all round to the walls of the nest. They are smooth above, with hexagonal cells on the under surface. These plat-

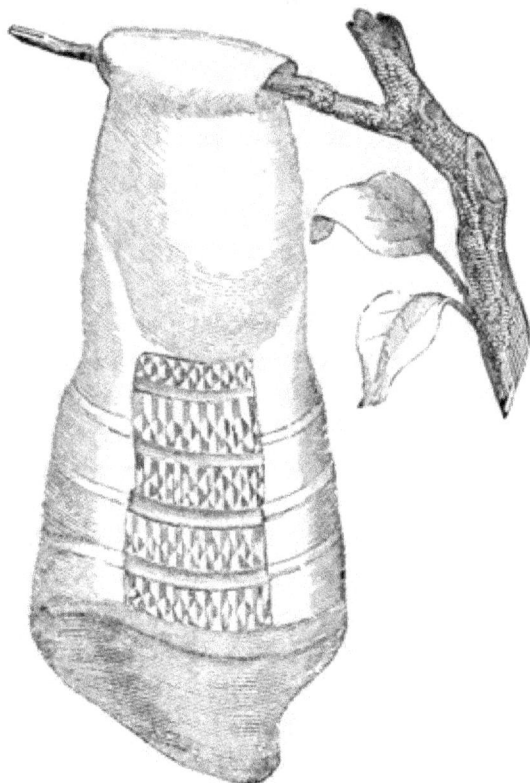

Nest of the Pasteboard-maker Wasp, with part removed to show the arrangement of the Cells.

forms are not quite flat, but rather concave above, like a watch-glass reversed; the centre of each platform is perforated for the admission of the wasps, at the extremity of a short funnel-like projection, and through this access is gained from story to story. On each platform, therefore, can the wasps walk leisurely about attending to the pupæ secured in the cells, which, with the mouths downward, cover the ceiling above their heads—the height of the latter being just convenient for their work.

[Unlike the habitations made by the British wasps, and which are vacated annually, this nest is permanent, and serves for several successive seasons. Of course, it must be enlarged continually, so as to accommodate an ever-increasing number of inhabitants. The mode of enlarging is sufficiently curious. The British wasps enlarge their nests either by making a larger covering and then removing the smaller, or by raising blisters on the outside, and eating away beneath them. But the pendulous wasp of Brazil proceeds on just the opposite principle, making new cells first, and covering them afterwards. The new tier of cells is set on the bottom of the nest, which thus becomes the floor of that tier, and a new bottom is then made beneath these new cells.]

Pendent wasps' nests of enormous size are found in Ceylon, suspended often in the talipot-tree at the height of seventy feet. The appearance of these nests thus elevated, with the larger leaves of the tree, used by the natives as umbrellas and tents, waving over them, is very singular. Though no species of European wasp is a storer of honey, yet this rule does not apply to certain species of South America. In the 'Annals and Magazine of Natural History' for June, 1841, will be found a detailed account, with a figure, of the pendent nest of a species termed by Mr. A. White *Myrapetra scutellaris.* The external case consists of stout cardboard covered with conical knobs of various sizes. The entrances are artfully protected by pent-roofs from the weather and heavy rains; and are tortuous, so as to render the ingress of a moth or other large insect difficult. Internally are fourteen combs, exclusive of a globular mass, the nucleus of several circular combs, which are succeeded by others of an arched form—that is, constituting segments of circles. Many of the uppermost combs were found to have the cells filled with honey of a brownish-red colour, but which had lost its flavour. After entering into some minute details, Mr. A.

White makes the following interesting observations:—
" Azara, in the account of his residence in various parts
of South America, mentions the fact of *several wasps* of these
countries collecting honey. The Baron Walchenaer, who
edited the French translation of this work, published in
1809, thought that the Spanish traveller, who was unskilled
in entomology, had made some mistake with regard to the
insects, and regarded the so-called *wasps* as belonging to

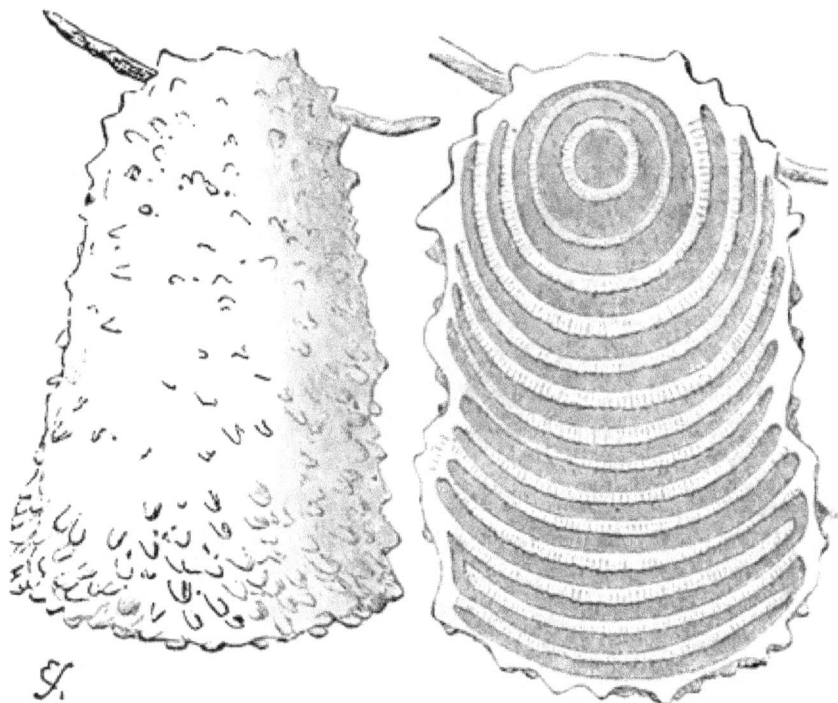

Nest of Myrapetra.

some *bee* of the genus of which *Apis amalthea* is the type
(*Melipona*). Latreille (who afterwards corrected his mistake)
also believed that they must be referred to the genera
Melipona or *Trigona*—insects which in South America take
the place of our honey-bee. These authors were afterwards
clearly convinced of the correctness of Azara's observations,
by the circumstance of M. Auguste de St. Hilaire finding
near the river Uruguay an oval grey-coloured nest of a

papery consistence, like that of the European wasps, suspended from the branches of a small shrub about a foot from the ground : he and two other attendants partook of some honey (contained in its cells) and found it of an agreeable sweetness, free from the pharmaceutic taste which so frequently accompanies European honey. He gives a detailed account of its poisonous effects on himself and his two men. Afterwards he procured specimens of the wasp, which was described by Latreille under the name of *Polistes Lecheguana*."

[The accompanying illustration shows this remarkable nest, both as it appears externally, and when divided vertically.

The material is probably the dung of the Capincha, an animal allied to the guinea-pig and the agouti. The natives, at all events, state that such is the case, and the aspect of the nest as seen through a magnifying glass carries out this assertion. The nest is hung to a branch, and is seldom more than four feet from the ground. The insect is a very little one in comparison with the size of the nest, which is sixteen inches in length, and twelve in width. The largest specimens of this insect are only one third of an inch long, while the generality scarcely exceed a quarter of an inch. Its colour is brown.

In the section is shown the very peculiar shape of the combs. At the upper part is seen the globular centre, surrounded with a comb that completely encircles it. Other combs follow in order, but are less curved as they approach the bottom of the nest. The insects obtain admission to the several tiers by means of apertures which are left between the extremities of the comb and the wall of the nest. The combs are made of the same material as the outer wall, but are very thin and paper-like. This nest may be seen in the British Museum.]

It would seem that the nest described by Mr. White agrees with that of a wasp termed *Chiguana* by Azara (or *Lecheguana*), and is very different to the slight papery nest of the *Polistes Lecheguana* of Latreille. We may add that M. Auguste de St. Hilaire speaks of two species of wasps remarkable for storing honey in South America; the honey of one is white, of the other reddish. That the habits of these honey-wasps must differ considerably from those of any of our European species we may at once admit ; perhaps in some points of their economy these insects may approach the bee.

[In the same country as is inhabited by the *Myrapetra*, and in much the same localities, is sometimes found the nest of another honey-making wasp, called *Nectarinia analis*, a small and plainly-clad insect. It is hung to the branches of low trees and underwood, and often includes both twigs and leaves in its structure. The combs of this insect are greatly curved, in order to suit the shape of the general covering, but are not arranged with that beautiful regularity which distinguishes those of the *Myrapetra*. A specimen of this nest may be seen in the British Museum, and as the outer covering has been partially taken away, the observer will be enabled to note the general form of the combs and the structure of the cells.

In the accompanying illustration are shown the habitations of two remarkable insects, both belonging to the *Hymenoptera*. Indeed, the greater number of pensile nests made by insects are formed by members of this important order ; and, if we were to exclude all the wasps, bees, and ants, we should find that we had excluded about ninety per cent. of the pensile architects.

The left-hand figure represents a nest made by a species of *Polybia*, inhabiting Brazil. It is made of a papery kind of substance, of rather slight texture, and is fixed to the stalk of a reed. The outside of the nest is seen to be marked with a series of horizontal ribs. These show the progressive stages of the nest, each rib marking a layer of paper as it was spread by the insect builder. The combs extend throughout the entire nest, the largest occupying the centre, and the smallest the ends. Each comb is firmly supported by a foot-stalk, which is fixed, not to the upper tier of cells, as is the case with the British wasps, but to the reed on which the nest is built.

Other species of *Polybia* build nests different in shape and arrangement, though still of the pensile character. One species builds a nearly globular nest, made in a rather curious manner. Carrying out still farther the principle on which

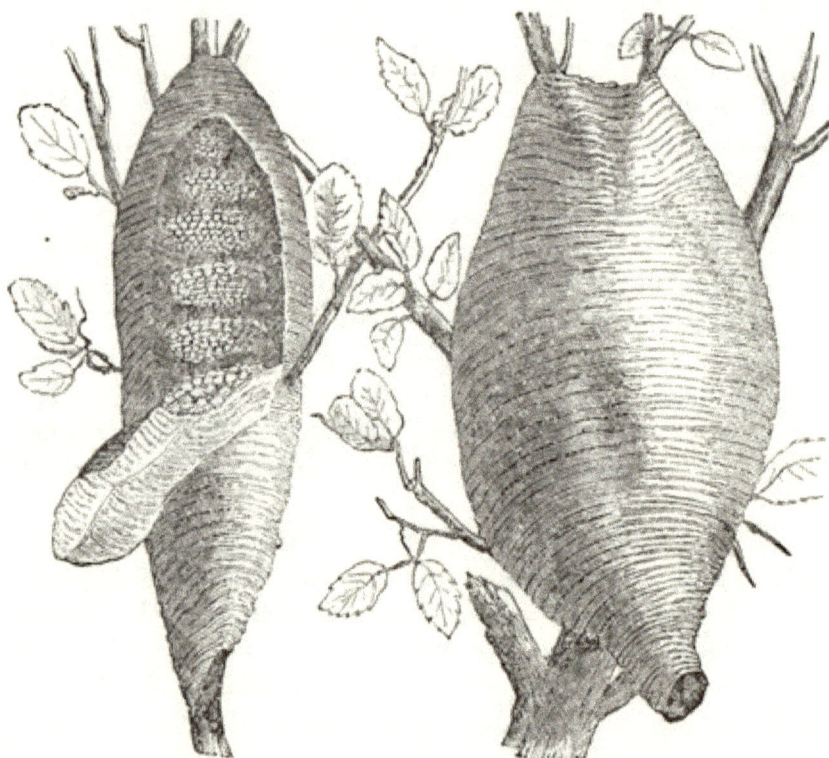

Nests of Polybia.

the cardboard wasp enlarges its nest, the *Polybia* entirely covers the outer wall with cells, and then makes a new wall over them. When a nest has reached a tolerable size, it is composed of a whole series of concentric combs, the roof of each having been originally the outer wall of the nest. There are in the British Museum some admirable specimens of these nests, in some of which the process of enlargement can be very clearly traced. Patches of new cells are seen upon the external covering, while a few breaches in the structure show the concentric combs.

One very curious point about these cells is, that they are not uniform in their direction, as is generally the case with those of social hymenoptera. The greater part, such as the various wasps, hornets, and their kin, have the mouths of the cells downwards, while the cells of the hive-bee are nearly horizontal. But the cells of this insect are arranged without the least regard to their position, all the bases pointing towards the centre of the nest, and all the mouths radiating outwards.

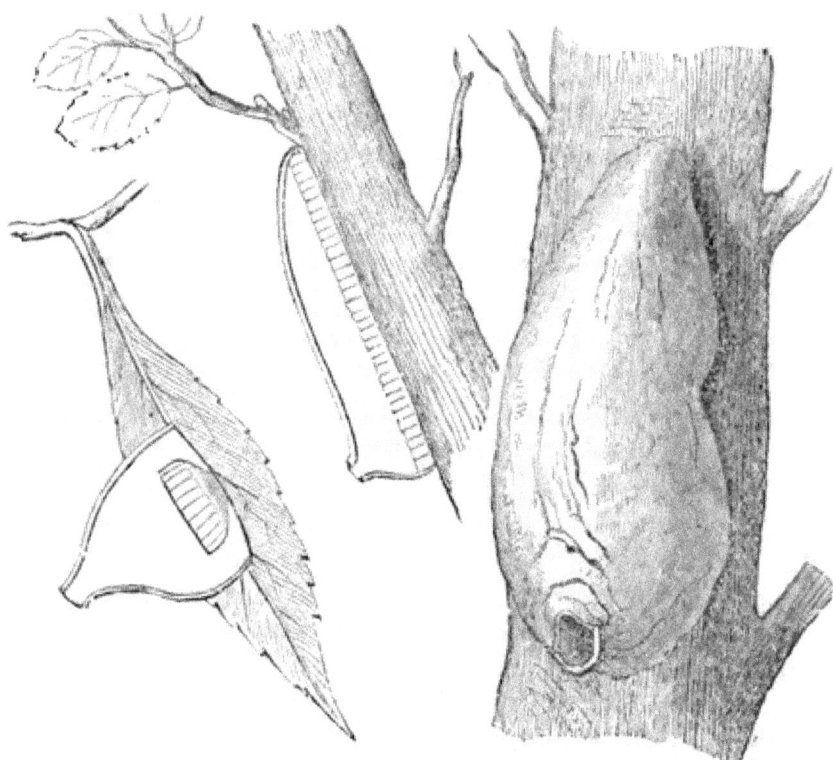

Nests of Synæca and Polybia.

There seems to be scarcely any bound to the variety which exists in the nests of the social hymenoptera. The insect which makes the nest which is represented in the illustration is a native of Brazil, and is known to entomologists as *Synæca cyanea*. The first of these names is given to it on account of its social habits, and the second, in reference to the bluish colour of its body. It is rather larger than the preceding insects, being about three quarters of an inch in length. Its wings are brown.

The shape and size of the nest are exceedingly variable, but it is almost invariably longer than wide, and is fixed to a branch or some similar object. Sometimes it attains considerable dimensions, and has been known to measure a full yard in length. Yet, however large it may be, there is only a single comb, which is set upon the side of the nest next the branch, and, in consequence, has almost all its cells placed in a horizontal direction. In the illustration, the right-hand figure represents the external appearance of the nest, and the central figure shows the manner in which the single comb is set upon the branch. The nest which occupies the left hand of the illustration is made by a species of *Polybia*, and is here given in order to show a remarkable example of similarity in the mode of building adopted by two different insects. In the one case, however, the cells are all fastened by their bases to the branch, but in the other the cells are attached to one common base which is prolonged into a footstalk.

There have been lately discovered some very remarkable social nests. Specimens of both these nests may be seen in the entomological department of the museum at Oxford.

The first is formed very much like a rather flattened Florence flask, and is hung by the neck from the branch of a tree. It is made of a strong, parchment-like substance, formed by innumerable silken threads woven and matted together into a kind of felt. When it was cut open a most singular sight was exhibited. Nearly the whole of the interior was covered with the pupæ of some butterfly, all hanging by their tails, and many of them suspended to a twig which projected downwards into the nest. Although the nest is barely eight inches in length, a great proportion of which is taken up by the neck, about one hundred pupæ were found in it. At the bottom of the nest is a small and nearly circular aperture, through which the insects could make their way as soon as they escaped from the pupal envelope, and before their wings became extended and hardened.

The butterfly which makes this singular nest is a native of Mexico, and is named *Eucheira socialis*. The colour of its wings is dark brown, with an ill-defined white band across them.

The second nest was brought from tropical Africa, and is remarkable for another peculiarity. It is shaped much like a cushion, and its measurements are, eight inches in length, five and a half in breadth, and three in depth. Instead of having only one place of exit for the inmates, it has thirteen or fourteen, all formed in the same manner. A number of short, stiff, and almost bristly threads are set round the apertures, their ends all projecting outwards, and converging to a point, where they all meet and even slightly cross each other. Owing to this structure, it is easy enough for any of the insects to pass out, as the converging hairs yield to the pressure, whereas they form an effectual barrier against any insect that wishes to creep into the nest.

The material of the nest is very strong and hard, and is formed of two layers, the inner being made of smooth brown silk, and the outer of harsher and stronger orange silk threads.]

CHAPTER V.

ARCHITECTURE OF THE HIVE-BEE.

ALTHOUGH the hive-bee (*Apis mellifica*) has engaged the attention of the curious from the earliest ages, recent discoveries prove that we are yet only beginning to arrive at a correct knowledge of its wonderful proceedings. Pliny informs us that Astromachus, of Soles, in Cilicia, devoted fifty-eight years to the study; and that Philiscus the Thracian spent his whole life in forests for the purpose of

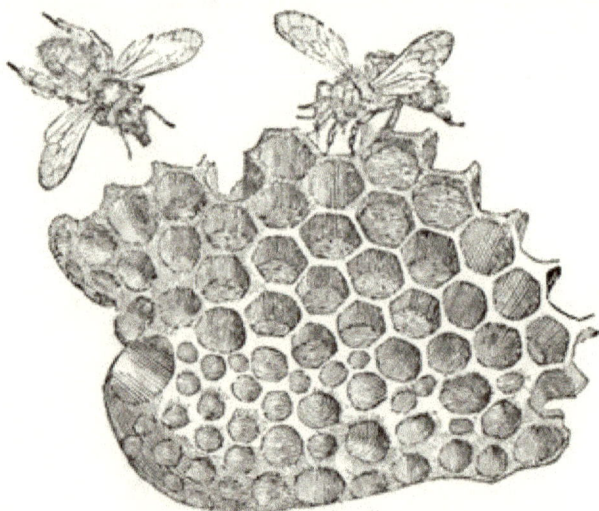

Part of a Honeycomb, and Bees at work.

observing them. But in consequence (as we may naturally infer) of the imperfect methods of research, assuming that what they did discover was known to Aristotle, Columella, and Pliny, we are justified in pronouncing the statements of these philosophers, as well as the embellished poetical pictures of Virgil, to be nothing more than conjecture, almost in every particular erroneous. It was not indeed till 1711, when glass hives were invented by Maraldi, a mathematician of Nice, that what we may call the in-door proceedings of bees could be observed. This important invention was soon afterwards taken advantage of by M. Réaumur, who laid the foundation of the more recent discoveries of John Hunter, Schirach, and the Hubers. The admirable architecture which bees exhibit in their miniature cities has, by these and other naturalists, been investigated with great care and accuracy. We shall endeavour to give as full an account of the wonderful structures as our limits will allow. In this we shall chiefly follow M. Huber the elder, whose researches appear almost miraculous when we consider that he was blind.

At the early age of seventeen this remarkable man lost his sight by *gutta serena*, the "drop serene" of our own Milton. But though cut off from the sight of Nature's works, he dedicated himself to their study. He saw them through the eyes of the admirable woman whom he married; his philosophical reasonings pointed out to her all that he wanted to ascertain; and as she reported to him from time to time the results of his ingenious experiments, he was enabled to complete, by diligent investigation, one of the most accurate and satisfactory accounts of the habits of bees which had ever been produced.

It had long been known that the bees of a hive consist of three sorts, which was ascertained by M. Réaumur to be distinguished as workers or neuters, constituting the bulk of the population; drones or males, the least numerous class; and a single female, the queen and mother of the colony. Schirach subsequently discovered the very extraordinary fact, which Huber and others have proved beyond doubt, that when a hive is accidentally deprived of a queen, the grub of a worker can be and is fed in a particular manner so as to become a queen and supply the loss.* But another

---

* It is right to remark that Huish and others have suggested that the grubs thus royalized may originally be misplaced queens; yet this admission is not necessary, since Madlle. Jurine has proved, by dissection, the workers to be imperfect females.

I

discovery of M. Huber is of more importance to the subject of architecture now before us. By minute research he ascertained that the workers which had been considered by former naturalists to be all alike, are divided into two important classes, nurse-bees and wax-makers.

The *nurse-bees* are rather smaller than the wax-workers, and even when gorged with honey their belly does not, as in the others, appear distended. Their business is to collect honey, and impart it to their companions; to feed and take care of the young grubs, and to complete the combs and cells which have been founded by the others; but they are not charged with provisioning the hive.

The *wax-workers*, on the other hand, are not only a little larger, but their stomach, when gorged with honey, is capable of considerable distension, as M. Huber proved by repeated experiments. He also ascertained that neither of the varieties can alone fulfil all the functions shared among the workers of a hive. He painted those of each class with different colours, in order to study their proceedings, and their labours were not interchanged. In another experiment, after supplying a hive deprived of a queen with brood and pollen, he saw the nurse-bees quickly occupied in the nutrition of the grubs, while those of the wax-working class neglected them. When hives are full of combs, the wax-workers disgorge their honey into the ordinary magazines, making no wax; but if they want a reservoir for its reception, and if their queen does not find cells ready made wherein to lay her eggs, they retain the honey in the stomach, and in twenty-four hours they produce wax. Then the labour of constructing combs begins.

It might perhaps be supposed that, when the country does not afford honey, the wax-workers consume the provision stored up in the hive. But they are not permitted to touch it. A portion of honey is carefully preserved, and the cells containing it are protected by a waxen covering, which is never removed except in case of extreme necessity, and when honey is not to be otherwise procured. The cells are at no time opened during summer; other reservoirs, always exposed, contribute to the daily use of the community; each bee, however, supplying itself from them with nothing but what is required for present wants. Wax-workers appear with large bellies at the entrance of their hive only when the country affords a copious collection of honey. From this it may be concluded that the production of the waxy matter depends on a concurrence of circumstances not invariably subsisting. Nurse-bees also produce wax, but in a very inferior quantity to what is elaborated by the real wax-workers. Another characteristic whereby an attentive observer can determine the moment of bees collecting sufficient honey to produce wax, is the strong odour of both these substances from the hive, which is not equally intense at any other time. From such data, it was easy for M. Huber to discover whether the bees worked in wax in his own hives, and in those of the other cultivators of the district.

There is still another sort of bee, first observed by Huber in 1809, which appear to be only casual inmates of the hive, and which are driven forth to starve, or are killed in conflict. They closely resemble the ordinary workers, but are less hairy, and of a much darker colour. These have been called *black bees*, and are supposed by Huber to be defective bees;* but Kirby and Spence conjecture that they are toil-worn superannuated workers, of no further use, and are therefore sacrificed, because burdensome to a community which tolerates no unnecessary inmates.

Preparation of Wax.

In order to build the beautiful combs, which every one must have repeatedly seen and admired, it is indispensable that the architect-bees should be provided with the materials —with the wax, in short, of which they are principally formed. Before we follow them, therefore, to the operation of building, it may be necessary to inquire how the wax itself is

---

* Huber on Bees, p. 338.

procured. Here the discoveries of recent inquirers have been little less singular and unexpected than in other departments of the history of these extraordinary insects. Now that it has been proved that wax is secreted by bees, it is not a little amusing to read the accounts given by our elder naturalists, of its being collected from flowers. Our countryman, Thorley,* appears to have been the first who suspected the true origin of wax, and Wildman (1769) seems also to have been aware of it; but Réaumur, and particularly Bonnet, though both of them in general shrewd and accurate observers, were partially deceived by appearances.

The bees, we are erroneously told, search for wax "upon all sorts of trees and plants, but especially the rocket, the simple poppy, and in general all kinds of flowers. They amass it with their hair, with which their whole body is invested. It is something pleasant to see them roll in the yellow dust which falls from the chives to the bottom of the flowers, and then return covered with the same grains; but their best method of gathering the wax, especially when it is not very plentiful. is to carry away all the little particles of it with their jaws and fore feet, to press the wax upon them into little pellets, and slide them one at a time, with their middle feet, into a socket or cavity, that opens at their hinder feet, and serves to keep the burthen fixed and steady till they return home. They are sometimes exposed to inconveniences in this work by the motion of the air, and the delicate texture of the flowers, which bend under their feet and hinder them from packing up their booty, on which occasions they fix themselves in some steady place, where they press the wax into a mass, and wind it round their legs, making frequent returns to the flowers; and when they have stocked themselves with a sufficient quantity, they immediately repair to their habitation. Two men, in the compass of a whole day, could not amass so much as two little balls of wax; and yet they are no more than the common burthen of a single bee, and the produce of one journey. Those who are employed in collecting the wax from flowers

* Melisselogia, or Female Monarchy, 8vo., Lond. 1744.

are assisted by their companions, who attend them at the door of the hive, ease them of their load at their arrival, brush their feet, and shake out the two balls of wax ; upon which the others return to the fields to gather new treasure, while those who disburthened them convey their charge to the magazine. But some bees, again, when they have brought their load home, carry it themselves to the lodge, and there deliver it, laying hold of one end by their hinder feet, and with their middle feet sliding it out of the cavity that contained it ; but this is evidently a work of supererogation which they are not obliged to perform. The packets of wax continue a few moments in the lodge, till a set of officers come, who are charged with a third commission, which is to knead this wax with their feet, and spread it out into different sheets, laid one above another. This is the unwrought wax, which is easily distinguished to be the produce of different flowers, by the variety of colours that appear on each sheet. When they afterwards come to work, they knead it over again ; they purify and whiten, and then reduce it to a uniform colour. They use this wax with a wonderful frugality ; for it is easy to observe that the whole family is conducted by prudence, and all their actions regulated by good government. Everything is granted to necessity, but nothing to superfluity ; not the least grain of wax is neglected, and if they waste it, they are frequently obliged to provide more ; at those very times when they want to get their provision of honey, they take off the wax that closed the cells, and carry it to the magazine."*

Réaumur hesitated in believing that this was a correct view of the subject, from observing the great difference between wax and pollen ; but he was inclined to think the pollen might be swallowed, partially digested, and disgorged in the form of a kind of paste. Schirach also mentions, that it was remarked by a certain Lusatian, that wax comes from the rings of the body, because, on withdrawing a bee while it is at work, and extending its body, the wax may be seen there in the form of scales.

* De la Pluche, Spectacle de la Nature, vol. i.

The celebrated John Hunter shrewdly remarked that the pellets of pollen seen on the thighs of bees are of different colours on different bees, while the shade of the new-made comb is always uniform; and therefore he concluded that pollen was not the origin of wax. Pollen also, he observed, is collected with greater avidity for old hives, where the comb is complete, than for those where it is only begun, which would hardly be the case were it the material of wax. He found that when the weather was cold and wet in June, so that a young swarm was prevented from going abroad, as much comb was constructed as had been made in an equal time when the weather was favourable and fine.

The pellets of pollen on the thighs being thence proved not to be wax, he came to the conclusion that it was an external secretion, originating between the plates of the belly. When he first observed this, he felt not a little embarrassed to explain the phenomenon, and doubted whether new plates were forming, or whether bees cast their old ones as lobsters do their shell. By melting the scales, he ascertained at least that they were wax; and his opinion was confirmed by the fact, that the scales are only to be found during the season when the combs are constructed. But he did not succeed in completing the discovery by observing the bees actually detach the scales, though he conjectured they might be taken up by others, if they were once shaken out from between the rings. *

We need not be so much surprised at mistakes committed upon this subject, when we recollect that honey itself was believed by the ancients to be an emanation of the air—a dew that descended upon flowers, as if it had a limited commission to fall only on them. The exposure and correction of error is one of the first steps to genuine knowledge; and when we are aware of the stumbling-blocks which have interrupted the progress of others, we can always travel more securely in the way of truth.

That wax is secreted is proved both by the wax-pouches

* Philosophical Trans. for 1792, p. 14?.

within the rings of the abdomen, and by actual experiment. Huber and others fed bees entirely upon honey or sugar, and, notwithstanding, wax was produced and combs formed as if they had been at liberty to select their food. "When bees were confined," says M. Huber, "for the purpose of discovering whether honey was sufficient for the production of wax, they supported their captivity patiently, and showed uncommon perseverance in rebuilding their combs as we removed them. Our experiments required the presence of grubs; honey and water had to be provided; the bees were to be supplied with combs containing brood, and at the same time it was necessary to confine them, that they might not seek pollen abroad. Having a swarm by chance, which had become useless from sterility of the queen, we devoted it for our investigation in one of my leaf-hives, which was glazed on both sides. We removed the queen, and substituted combs containing eggs and young grubs, but no cell with farina; even the smallest particle of the substance which John Hunter conjectured to be the basis of the nutriment of the young was taken away.

"Nothing remarkable occurred during the first and second day : the bees brooded over the young, and seemed to take an interest in them ; but at sunset on the third a loud noise was heard in the hive. Impatient to discover the reason, we opened a shutter, and saw all in confusion; the brood was abandoned, the workers ran in disorder over the combs, thousands rushed towards the lower part of the hive, and those about the entrance gnawed at its grating. Their design was not equivocal ; they wished to quit their prison. Some imperious necessity evidently obliged them to seek elsewhere what they could not find in the hive ; and apprehensive that they might perish if I restrained them longer from yielding to their instinct, I set them at liberty. The whole swarm escaped, but the hour being unfavourable for their collections, they flew around the hive, and did not depart far from it. Increasing darkness and the coolness of the air compelled them very soon to return. Probably these circumstances calmed their agitation; for we observed them peaceably remounting their combs; order seemed re-established, and we took advantage of this moment to close the hive.

"Next day, the 19th of July, we saw the rudiments of two royal cells, which the bees had formed on one of the brood-combs. This evening, at the same hour as on the preceding, we again heard a loud buzzing in the closed hive; agitation and disorder rose to the highest degree, and we were again obliged to let the swarm escape. The bees did not remain long absent from their habitation; they quieted and returned as before. We remarked on the 20th that the royal cells had not been continued, as would have been the case in the ordinary state of things. A great tumult took place in the evening; the bees appeared to be in a delirium; we set them at liberty, and order was restored on their return. Their captivity having endured five days, we thought it needless to protract it farther; besides, we were desirous of knowing whether the brood was in a suitable condition, and if it had made the usual progress; and we wished also to try to discover what might be the cause of the periodical agitation of the bees. M. Burnens (the assistant of Huber), having exposed the two brood-combs, the royal cells were immediately recognised; but it was obvious that they had not been enlarged. Why should they? Neither eggs, grubs, nor that kind of paste peculiar to the individuals of their species were there! The other cells were vacant likewise; no brood, not an atom of paste, was in them. Thus, the worms had died of hunger. Had we precluded the bees from all means of sustenance by removing the farina? To decide this point, it was necessary to confide other brood to the care of the same insects, now giving them abundance of pollen. They had not been enabled to make any collections while we examined their combs. On this occasion they escaped in an apartment where the windows were shut; and after substituting young worms for those they had allowed to perish, we returned them to their prison. Next day we remarked that they had resumed courage; they had consolidated the combs, and remained on the brood. They were then provided with fragments of combs, where other workers had stored up farina; and to be able to observe what they did with it, we took this substance from some of their cells, and spread it on the board of the hive. The bees soon discovered both the farina in the combs and what we had exposed to them. They crowded to the cells, and also descending to the bottom of the hives, took the pollen grain by grain in their teeth, and conveyed it to their mouths. Those that had eaten it most greedily mounted the combs before the rest, and stopping on the cells of the young worms, inserted their heads, and remained there for a certain time. M. Burnens opened one of the divisions of the hive gently, and powdered the workers, for the purpose of recognising them when they should ascend the combs. He observed them during several hours, and by this means ascertained that they took so great a quantity of pollen only to impart it to their young. Then withdrawing the portions of comb which had been placed by us on the board of the hive, we saw that the pollen had been sensibly diminished in quantity. They were returned to the bees, to augment their provision still further, for the purpose of extending the experiment. The royal, as well as several common, cells were soon closed; and, on opening the hive, all the worms were found to have prospered. Some still had their food before them; the cells of others that had spun were shut with a waxen covering.

"We witnessed these facts repeatedly, and always with equal interest. They so decisively prove the regard of the bees towards the grubs which they are intrusted with rearing, that we shall not seek for any other explanation of their conduct. Another fact, no less extraordinary, and much more difficult to be accounted for, was exhibited by bees constrained to work in wax, several times successively, from the syrup of sugar. Towards the close of the experiment they ceased to feed the young, though in the beginning those had received the usual attention. They even frequently dragged them from their cells, and carried them out of the hive."*

Mr. Wiston, of Germantown, in the United States, mentions

* Huber on Bees.

a fact conclusive on this subject. " I had," says he, " a late swarm last summer, which, in consequence of the drought, filled only one box with honey. As it was late in the season, and the food collected would not enable the bees to subsist for the winter, I shut up the hive, and gave them half-a-pint of honey every day. They immediately set to work, filled the empty cells, and then constructed new cells enough to fill another box, in which they deposited the remainder of the honey."

A more interesting proof is thus related by the same gentleman : " In the summer of 1824, I traced some wild bees, which had been feeding on the flowers in my meadow, to their home in the woods, and which I found in the body of an oak-tree, exactly fifty feet above the ground. Having caused the entrance to the hive to be closed by an expert climber, the limbs were separated in detail, until the trunk alone was left standing. To the upper extremity of this, a tackle-fall was attached so as to connect it with an adjacent tree, and, a saw being applied below, the naked trunk was cut through. When the immense weight was lowered nearly to the earth, the ropes broke, and the mass fell with a violent crash. The part of the tree which contained the hive, separated by the saw, was conveyed to my garden, and placed in a vertical position. On being released, the bees issued out by thousands, and though alarmed, soon became reconciled to the change of situation. By removing a part of the top of the block the interior of the hive was exposed to view, and the comb itself, nearly six feet in height, was observed to have fallen down two feet below the roof of the cavity. To repair the damage was the first object of the labourers : in doing which, a large part of their store of honey was expended, because it was at too late a season to obtain materials from abroad. In the following February these industrious but unfortunate insects issuing in a confused manner from the hive, fell dead in thousands around its entrance, the victims of a poverty created by their efforts to repair the ruins of their habitation."*

* American Quarterly Review for June, 1828, p. 382.

In another experiment, M. Huber confined a swarm so that they had access to nothing beside honey, and five times successively removed the combs with the precaution of preventing the escape of the bees from the apartment. On each occasion they produced new combs, which puts it beyond dispute that honey is sufficient to effect the secretion of wax without the aid of pollen. Instead of supplying the bees with honey, they were subsequently fed, exclusively, on pollen and fruit; but though they were kept in captivity for eight days under a bell-glass, with a comb containing nothing but farina, they neither made wax nor was any secreted under the rings. In another series of experiments, in which bees were fed with different sorts of sugar, it was found that nearly one-sixth of the sugar was converted into wax, dark-coloured sugar yielding more than double the quantity of refined sugar.

It may not be out of place to subjoin the few anatomical and physiological facts which have been ascertained by Huber, Maddle. Jurine, and Latreille.

The first stomach of the worker-bee, according to Latreille,[*]

Worker-bee, magnified—showing the position of the scales of Wax.

is appropriated to the reception of honey, but this is never found in the second stomach, which is surrounded with mus-

* Latreille, Mém. Acad. des Sciences, 1821.

cular rings, and from one end to the other very much resembles a cask covered with hoops. It is within these rings
that the wax is produced; but the secreting vessels for this
purpose have hitherto escaped the researches of the acutest
naturalists. Huber, however, plausibly enough conjectures
that they are contained in the internal lining of the wax-
pockets, which consists of a cellular substance reticulated
with hexagons. The wax-pockets themselves, which are
concealed by the overlapping of the rings, may be seen by
pressing the abdomen of a worker-bee so as to lengthen it,
and separate the rings further from each other. When this

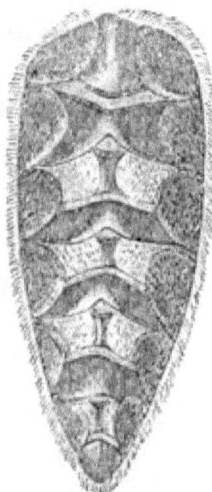

Abdomen of Wax-worker Bee.

has been done, there may be seen on each of the four intermediate hoops of the belly, and separated by what may be
called the keel (*carina*), two whitish-coloured pouches, of a
soft texture, and in the form of a trapezium. Within, the
little plates or scales of wax are produced from time to time,
and are removed and employed as we shall presently see.
We may remark, that it is chiefly the wax-workers which
produce the wax; for though the nurse-bees are furnished
with wax-pockets, they secrete it only in very small quantities; while in the queen-bee, and the males or drones, no
pockets are discoverable.

"All the scales," says Huber, "are not alike in every bee, for a difference is perceptible in consistence, shape, and thickness; some are so thin and transparent as to require a magnifier to be recognised, or we have been able to discover nothing but spiculæ similar to those of water freezing. Neither the spiculæ nor the scales rest immediately on the membrane of the pocket, a slight liquid medium is interposed, serving to lubricate the joinings of the rings, or to render the extraction of the scales easier, as otherwise they might adhere too firmly to the sides of the pockets." M. Huber has seen the scales so large as to project beyond the rings, being visible without stretching the segments, and of a whitish yellow, from greater thickness lessening their transparency. These shades of difference in the scales of various bees, their enlarged dimensions, the fluid interposed beneath them, the correspondence between the scale and the size and form of the pockets, seem to infer the oozing of this substance through the membranes whereon it is moulded. He was confirmed in this opinion by the escape of a transparent fluid on piercing the membrane, whose internal surface seemed to be applied to the soft parts of the belly. This he found coagulated in cooling, when it resembled wax, and again liquefied on exposure to heat. The scales themselves, also, melted and coagulated like wax.*

By chemical analysis, however, it appears that the wax of the rings is a more simple substance than that which composes the cells; for the latter is soluble in ether, and in spirit of turpentine, while the former is insoluble in ether, and but partially soluble in spirit of turpentine. It should seem to follow, that if the substance found lying under the rings be really the elements of wax, it undergoes some subsequent preparation after it is detached; and that the bees, in short, are capable of impregnating it with matter, imparting to it whiteness and ductility, whereas in its unprepared state it is only fusible.

* Huber on Bees. p. 325.

PROPOLIS.

WAX is not the only material employed by bees in their architecture. Beside this, they make use of a brown, odoriferous, resinous substance, called *propolis*,* more tenacious and extensible than wax, and well adapted for cementing and varnishing. It was strongly suspected by Réaumur that the bees collected the propolis from those trees which are known to produce a similar gummy resin, such as the poplar, the birch, and the willow; but he was thrown into doubt by not being able to detect the bees in the act of procuring it, and by observing them to collect it where none of those trees, nor any other of the same description, grew. His bees also refused to make use of bitumen, and other resinous substances, with which he supplied them, though Mr. Knight, as we shall afterwards see, was more successful.†

Long before the time of Réaumur, however, Mouffet, in his *Insectarum Theatrum*, quotes Cordus for the opinion that propolis is collected from the buds of trees, such as the poplar and birch; and Reim says it is collected from the pine and fir.‡ Huber at length set the question at rest; and his experiments and observations are so interesting, that we shall give them in his own words:—

"For many years," says he, "I had fruitlessly endeavoured to find them on trees producing an analogous substance, though multitudes had been seen returning laden with it.

"In July, some branches of the wild poplar, which had been cut since spring, with very large buds, full of a reddish, viscous, odoriferous matter, were brought to me, and I planted them in vessels before hives, in the way of the bees going out to forage, so that they could not be insensible of their presence. Within a quarter of an hour, they were visited by a bee, which separating the sheath of a bud with its teeth,

---

* From two Greek words προ πολις meaning *before the city*, as the substance is principally applied to the projecting parts of the hive.

† Phil. Trans. for 1807, p. 242.

‡ Schirach, Hist. des Abeilles, p. 241.

drew out threads of the viscous substance, and lodged a pellet of it in one of the baskets of its limbs ; from another bud it collected another pellet for the opposite limb, and departed to the hive. A second bee took the place of the former in a few minutes, following the same procedure. Young shoots of poplar, recently cut, did not seem to attract these insects, as their viscous matter had less consistence than the former.[*]

" Different experiments proved the identity of this substance with the propolis ; and now, having only to discover how the bees applied it to use, we peopled a hive, so prepared as to fulfil our views. The bees, building upwards, soon reached the glass above ; but, unable to quit their habitation, on account of rain, they were three weeks without bringing home propolis. Their combs remained perfectly white until the beginning of July, when the state of the atmosphere became more favourable for our observations. Serene, warm weather engaged them to forage, and they returned from the fields laden with a resinous gum, resembling a transparent jelly, and having the colour and lustre of the garnet. It was easily distinguished from the farinaceous pellets then collected by other bees. The workers bearing the propolis ran over the clusters, suspended from the roof of the hive, and rested on the rods supporting the combs, or sometimes stopped on the sides of their dwelling, in expectation of their companions coming to disencumber them of their burthen. We actually saw two or three arrive, and carry the propolis from off the limbs of each with their teeth. The upper part of the hive exhibited the most animated spectacle ; thither a multitude of bees resorted from all quarters, to engage in the predominant occupation of the collection, distribution, and application of the propolis. Some conveyed that of which they had unloaded the purveyors in their teeth, and deposited it in heaps ; others hastened, before its hardening, to spread it out like a varnish, or formed it into strings, proportioned to the interstices of the sides of the hives to be filled up.

---

[*] Kirby and Spence observed bees very busy in collecting propolis from the tacamahaca-tree (*Populus balsamifera*).—Introd., ii. 186.

Nothing could be more diversified than the operations carried on.

"The bees, apparently charged with applying the propolis within the cells, were easily distinguished from the multitude of workers, by the direction of their heads towards the horizonal pane forming the roof of the hive, and on reaching it, they deposited their burthen nearly in the middle of intervals separating the combs : then they conveyed the propolis to the real place of its destination. They suspended themselves by the claws of the hind legs to points of support, afforded by the viscosity of the propolis on the glass ; and, as it were, swinging themselves backwards and forwards, brought the heap of this substance nearer to the cells at each impulse. Here the bees employed their fore feet, which remained free, to sweep what the teeth had detached, and to unite the fragments scattered over the glass, which recovered all its transparency when the whole propolis was brought to the vicinity of the cells.

"After some of the bees had smoothed down and cleaned out the glazed cells, feeling the way with their antennæ, one desisted, and having approached a heap of propolis, drew out a thread with its teeth. This being broken off, it was taken in the claws of the fore feet, and the bee, re-entering the cell, immediately placed it in the angle of two portions that had been smoothed, in which operation the fore feet and teeth were used alternately ; but probably proving too clumsy, the thread was reduced and polished ; and we admired the accuracy with which it was adjusted when the work was completed. The insect did not stop here : returning to the cell, it prepared other parts of it to recive a second thread, for which we did not doubt that the heap would be resorted to. Contrary to our expectation, however, it availed itself of the portion of the thread cut off on the former occasion, arranged it in the appointed place, and gave it all the solidity and finish of which it was susceptible. Other bees completed the work which the first had begun : and the sides of the cells were speedily secured with threads of propolis, while some were also put on the orifices ; but we could not seize the moment when they were varnished, though it may be easily conceived how it is done." *

This is not the only use to which bees apply the propolis. They are extremely solicitous to remove such insects or foreign bodies as happen to get admission into the hive. When so light as not to exceed their powers, they first kill the insect with their stings, and then drag it out with their teeth. But it sometimes happens, as was first observed by Maraldi, and since by Réaumur and others, that an ill-fated slug creeps into the hive : this is no sooner perceived than it is attacked on all sides, and stung to death. But how are the bees to carry out so heavy a burthen? Such a labour would be in vain. To prevent the noxious smell which would arise from its putrefaction, they immediately embalm it, by covering every part of its body with propolis, through which no effluvia can escape. When a snail with a shell gets entrance, to dispose of it gives much less trouble and expense to the bees. As soon as it receives the first wound from a sting, it naturally retires within its shell. In this case, the bees, instead of pasting it all over with propolis, content themselves with gluing all round the margin of the shell, which is sufficient to render the animal for ever immovably fixed.

Mr. Knight, the learned and ingenious President of the Horticultural Society, discovered by accident an artificial substance, more attractive than any of the resins experimentally tried by Réaumur. Having caused the decorticated part of a tree to be covered with a cement composed of bees'-wax and turpentine, he observed that this was frequented by hive-bees, who, finding it to be a very good propolis ready made, detached it from the tree with their mandibles, and then, as usual, passed it from the first leg to the second, and so on. When one bee had thus collected its load, another often came behind and despoiled it of all it had collected ; a second and a third load were frequently lost in the same manner ; and yet the patient insect pursued its operations without manifesting any signs of anger.†

* Huber on Bees, p. 408.
† Philosophical Trans. for 1807, p. 242.

K

Probably the latter circumstance, at which Mr. Knight seems to have been surprised, was nothing more than an instance of the division of labour so strikingly exemplified in every part of the economy of bees.

It may not be out of place here to describe the apparatus with which the worker-bees are provided for the purpose of carrying the propolis as well as the pollen of flowers to the hive, and which has just been alluded to in the observations of Mr. Knight. The shin or middle portion of the hind pair of legs is actually formed into a triangular basket, admirably adapted to this design. The bottom of this basket is com-

Structure of the legs of the Bee, for carrying propolis and pollen, magnified.

posed of a smooth, shining, horn-like substance, hollowed out in the substance of the limb, and surrounded with a margin of strong and thickly-set bristles. Whatever materials, therefore, may be placed by the bee in the interior of this basket, are secured from falling out by the bristles around it, whose elasticity will even allow the load to be heaped beyond their points without letting it fall.

In the case of propolis, when the bee is loading her singular basket, she first kneads the piece she has detached with her mandibles, till it becomes somewhat dry and less adhesive, as otherwise it would stick to her limbs. This preliminary process sometimes occupies nearly half an hour.

She then passes it backwards by means of her feet to the cavity of her basket, giving it two or three pats to make it adhere ; and when she adds a second portion to the first, she often finds it necessary to pat it still harder. When she has procured as much as the basket will conveniently hold, she flies off with it to the hive.

THE BUILDING OF THE CELLS.

The notion commonly entertained respecting glass hives is altogether erroneous. Those who are unacquainted with bees, imagine that, by means of a glass hive, all their proceedings may be easily watched and recorded ; but it is to be remembered that bees are exceedingly averse to the intrusion of light, and their first operation in such cases is to close up every chink by which light can enter to disturb them, either by clustering together, or by a plaster composed of propolis. It consequently requires considerable management and ingenuity, even with the aid of a glass hive, to see them actually at work. M. Huber employed a hive with leaves, which opened in the manner of a book ; and for some purposes he used a glass box, inserted in the body of the hive, but easily brought into view by means of screws.

But no invention hitherto contrived is sufficient to obviate every difficulty. The bees are so eager to afford mutual assistance, and for this purpose so many of them crowd together in rapid succession, that the operations of individuals can seldom be traced. Though this crowding, however, appears to an observer to be not a little confused, it is all regulated with admirable order, as has been ascertained by Réaumur and other distinguished naturalists.

When bees begin to build the hive, they divide themselves into bands, one of which produces materials for the structure ; another works upon these, and forms them into a rough sketch of the dimensions and partitions of the cells. All this is completed by the second band, who examine and adjust the angles, remove the superfluous wax, and give the work its necessary perfection; and a third band brings provisions to the labourers, who cannot leave their work. But no distribution of food is made to those whose charge, in collecting propolis and pollen, calls them to the field, because it is supposed they will hardly forget themselves; neither is any allowance made to those who begin the architecture of the cells. Their province is very troublesome, because they are obliged to level and extend, as well as cut and adjust the wax to the dimensions required; but then they soon obtain a dismission from this labour, and retire to the fields to regale themselves with food, and wear off their fatigue with a more agreeable employment. Those who succeed them, draw their mouth, their feet, and the extremity of their body, several times over all the work, and never desist till the whole is polished and completed; and as they frequently need refreshments, and yet are not permitted to retire, there are waiters always attending, who serve them with provisions when they require them. The labourer who has an appetite, bends down his trunk before the caterer to intimate that he has an inclination to eat, upon which the other opens his bag of honey, and pours out a few drops: these may be distinctly seen rolling through the hole of his trunk, which insensibly swells in every part the liquor flows through. When this little repast is over, the labourer returns to his work, and his body and feet repeat the same motions as before.*

Before they can commence building, however, when a colony or swarm migrates from the original hive to a new situation, it is necessary first to collect propolis, with which every chink and cranny in the place where they mean to build may be carefully stopped up; and secondly, that a quantity of wax be secreted by the wax-workers to form the requisite cells. The secretion of wax, it would appear, goes on best when the bees are in a state of repose; and the wax-workers, accordingly, suspend themselves in the interior in an extended cluster, like the curtain which is composed of a series of intertwined festoons or garlands, crossing

* Spectacle de la Nature, tome i.

each other in all directions—the uppermost bee maintaining
its position by laying hold of the roof with its fore legs,
and the succeeding one by laying hold of the hind legs of
the first.

" A person," says Réaumur, " must have been born devoid
of curiosity not to take interest in the investigation of such
wonderful proceedings." Yet Réaumur himself seems not

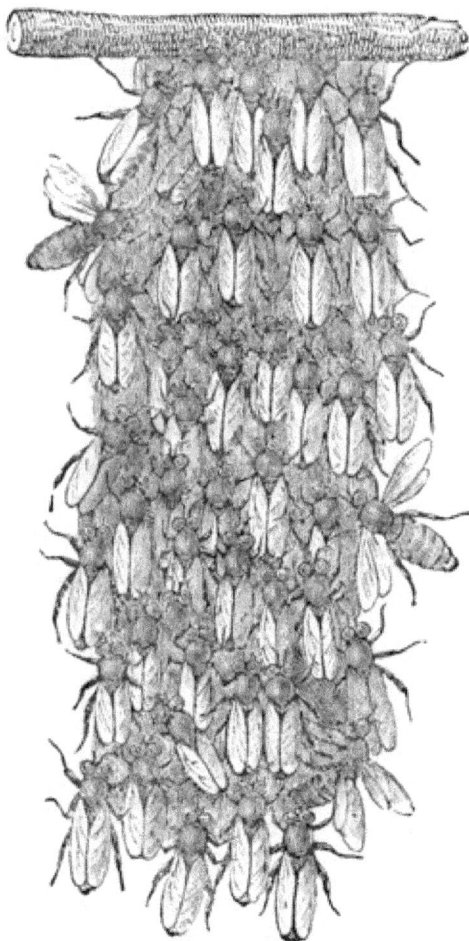

Curtain of Wax-workers secreting wax.

to have understood that the bees suspended themselves
in this manner to secrete wax, but merely, as he imagined,
to recruit themselves by rest for renewing their labours.

The bees composing the festooned curtain are individually motionless; but this curtain is, notwithstanding, kept moving by the proceedings in the interior; for the nurse-bees never form any portion of it, and continue their activity—a distinction with which Réaumur was unacquainted.

Although there are many thousand labourers in a hive, they do not commence foundations for combs in several places at once, but wait till an individual bee has selected a site, and laid the foundation of a comb, which serves as a directing mark for all that are to follow. Were we not expressly told by so accurate an observer as Huber, we might hesitate to believe that bees, though united in what appears to be an harmonious monarchy, are strangers to subordination, and subject to no discipline. Hence it is, that though many bees work on the same comb, they do not appear to be guided by any simultaneous impulse. The stimulus which moves them is successive. An individual bee commences each operation, and several others successively apply themselves to accomplish the same purpose. Each bee appears, therefore, to act individually, either as directed by the bees preceding it, or by the state of advancement in which it finds the work it has to proceed with. If there be anything like unanimous consent, it is the inaction of several thousand workers while a single individual proceeds to determine and lay down the foundation of the first comb. Réaumur regrets that, though he could by snatches detect a bee at work in founding cells or perfecting their structure, his observations were generally interrupted by the crowding of other bees between him and the little builder. He was therefore compelled rather to infer the different steps of their procedure from an examination of the cells when completed, than from actual observation. The ingenuity of Huber, even under all the disadvantages of blindness, succeeded in tracing the minutest operations of the workers from the first waxen plate of the foundation. We think the narrative of the discoverer's experiments, as given by himself, will be more interesting than any abstract of it which we could furnish :—

"Having taken a large bell-shaped glass receiver, we glued thin wooden slips to the arch at certain intervals, because the glass itself was too smooth to admit of the bees supporting themselves on it. A swarm, consisting of some thousand workers, several hundred males, and a fertile queen, was introduced, and they soon ascended to the top. Those first gaining the slips fixed themselves there by the fore-feet; others, scrambling up the sides, joined them, by holding their legs with their own, and they thus formed a kind of chain, fastened by the two ends to the upper parts of the receiver, and served as ladders or a bridge to the workers enlarging their number. The latter were united in a cluster, hanging like an inverted pyramid from the top to the bottom of the hive.

"The country then affording little honey, we provided the bees with syrup of sugar, in order to hasten their labour. They crowded to the edge of a vessel containing it; and, having satisfied themselves, returned to the group. We were now struck with the absolute repose of this hive, contrasted with the usual agitation of bees. Meanwhile, the nurse-bees alone went to forage in the country; they returned with pollen, kept guard at the entrance of the hive, cleansed it, and stopped up its edges with propolis. The wax-workers remained motionless about fifteen hours: the curtain of bees, consisting always of the same individuals, assured us that none replaced them. Some hours later, we remarked that almost all these individuals had wax scales under the rings; and next day this phenomenon was still more general. The bees forming the external layer of the cluster, having now somewhat altered their position, enabled us to see their bellies distinctly. By the projection of the wax scales, the rings seemed edged with white. The curtain of bees became rent in several places, and some commotion began to be observed in the hive.

"Convinced that the combs would originate in the centre of the swarm, our whole attention was then directed towards the roof of the glass. A worker at this time detached itself from one of the central festoons of the cluster, separated itself from the crowd, and, with its head, drove away the bees at the beginning of the row in the middle of the arch, turning round to form a space an inch or more in diameter, in which it might move freely. It then fixed itself in the centre of the space thus cleared.

" The worker now employing the pincers at the joint of one of the third pair of its limbs, seized a scale of wax projecting from a ring, and brought it forward to its mouth with the claws of its fore-legs, where it appeared in a vertical position. We remarked that, with its claws, it turned the wax in every necessary direction ; that the edge of the scale was immediately broken down, and the fragments having been accumulated in the hollow of the mandibles, issued forth

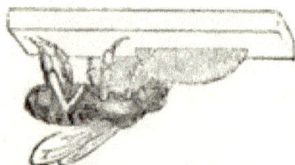

Wax-worker laying the foundation of the first Cell.

like a very narrow ribbon, impregnated with a frothy liquid by the tongue. The tongue itself assumed the most varied shapes, and executed the most complicated operations,—being sometimes flattened like a trowel, and at other times pointed like a pencil ; and, after imbuing the whole substance of the ribbon, pushed it forward again into the mandibles, whence it was drawn out a second time, but in an opposite direction.

" At length the bee applied these particles of wax to the vault of the hive, where the saliva impregnating them promoted their adhesion, and also communicated a whiteness and opacity which were wanting when the scales were detached from the rings. Doubtless this process was to give the wax that ductility and tenacity belonging to its perfect state. The bee then separated those portions not yet applied to use with its mandibles, and with the same organs afterwards arranged them at pleasure. The founder bee, a name appropriated to this worker, repeated the same operation, until all the fragments, worked up and impregnated with the fluid, were attached to the vault, when it repeated the preceding operations on the part of the scale yet kept apart, and again united to the rest what was obtained from it. A second and third scale were similarly treated by the same bee; yet the work was only sketched; for the worker did nothing but accumulate the particles of wax together. Meanwhile the founder, quitting its position, disappeared amidst its com-

Curtain of Wax-workers (see p. 132).

panions. Another, with wax under the rings, succeeded it, which suspending itself to the same spot, withdrew a scale by the pincers of the hind legs, and passing it through its mandibles, prosecuted the work; and taking care to make its deposit in a line with the former, it united their extremities. A third worker, detaching itself from the interior of the cluster, now came and reduced some of the scales to paste, and put them near the materials accumulated by its companions, but not in a straight line. Another bee, apparently sensible of the defect, removed the misplaced wax before our eyes, and carrying it to the former heap, deposited it there, exactly in the order and direction pointed out.

"From all these operations was produced a block of a rugged surface, hanging down from the arch, without any perceptible angle, or any traces of cells. It was a simple wall, or ridge, running in a straight line, and without the least inflection, two-thirds of an inch in length, above two-thirds of a cell, or two lines, high, and declining towards the extremities. We have seen other foundation walls from an inch to an inch and a half long, the form being always the same; but none ever of greater height.

"The vacuity in the centre of the cluster had permitted us to discover the first manœuvres of the bees, and the art with which they laid the foundations of their edifices. However, it was filled up too soon for our satisfaction; for workers collecting on both faces of the wall obstructed our view of their further operations."*

* Huber on Bees, p. 358.

CHAPTER VI.

THE obstruction of which M. Huber complains only operated as a stimulus to his ingenuity in contriving how he might continue his interesting observations. From the time of Pappus to the present day, mathematicians have applied the principles of geometry to explain the construction of the cells of a bee-hive; but though their extraordinary regularity, and wonderfully-selected form, had so often been investigated by men of the greatest talent, and skilled in all the refinements of science, the process by which they are constructed, involving also the causes of their regularity of form, had not been traced till M. Huber devoted himself to the inquiry.

As the wax-workers secrete only a limited quantity of wax, it is indispensably requisite that as little as possible of it should be consumed, and that none of it should be wasted. Bees, therefore, as M. Réaumur well remarks,* have to solve this difficult geometrical problem :—a quantity of wax being given, to form of it similar and equal cells of a determinate capacity, but of the largest size in proportion to the quantity of matter employed, and disposed in such a manner as to occupy the least possible space in the hive. This problem is solved by bees in all its conditions. The cylindrical form would seem to be best adapted to the shape of the insect; but had the cells been cylindrical, they could not have been applied to each other without leaving a vacant and superfluous space between every three contiguous cells. Had the cells, on the other hand, been square or triangular, they might

* Réaumur, vol. v., p. 380.

have been constructed without unnecessary vacancies; but these forms would have both required more material, and have been very unsuitable to the shape of a bee's body. The six-sided form of the cells obviates every objection; and while it fulfils the conditions of the problem, it is equally adapted with a cylinder to the shape of the bee.

M. Réaumur further remarks, that the base of each cell, instead of forming a plane, is usually composed of three pieces in the shape of the diamonds on playing cards, and placed in such a manner as to form a hollow pyramid. This structure, it may be observed, imparts a greater degree of strength, and, still keeping the solution of the problem in view, gives a great capacity with the smallest expenditure of material. This has actually, indeed, been ascertained by mathematical measurement and calculation. Maraldi, the inventor of glass hives, determined, by minutely measuring these angles, that the greater were 109° 28', and the smaller 70° 32'; and M. Réaumur, being desirous to know why these particular angles are selected, requested M. Kœnig, a skilful mathematician (without informing him of his design, or telling him of Maraldi's researches), to determine by calculation what ought to be the angle of a six-sided cell, with a concave pyramidal base, formed of three similar and equal rhomboid plates, so that the least possible matter should enter into its construction. By employing what geometricians denominate the *infinitesimal calculus*, M. Kœnig found that the angles should be 109° 26' for the greater, and 70° 34' for the smaller, or about two-sixtieths of a degree, more or less, than the actual angles made choice of by bees. The equality of inclination in the angles has also been said to facilitate the construction of the cells.

M. Huber adds to these remarks, that the cells of the first row, by which the whole comb is attached to the roof of a hive, are not like the rest; for, instead of six sides, they have only five, of which the roof forms one. The base, also, is in these different, consisting of three pieces on the face of the comb, and on the other side of two: one of these only is diamond-shaped, while the other two are of an irregular four-sided figure. This arrangement, by bringing the greatest number of points in contact with the interior surface, insures the stability of the comb.

It may, however, be said not to be quite certain, that Réaumur and others have not ascribed to bees the merit of ingenious mathematical contrivance and selection, when the construction of the cells may more probably originate in the form of their mandibles and the other instruments employed in their operations. In the case of other insects, we have, both in the preceding and subsequent pages of this volume, repeatedly noticed, that they use their bodies, or parts thereof, as the standards of measurement and modelling; and it

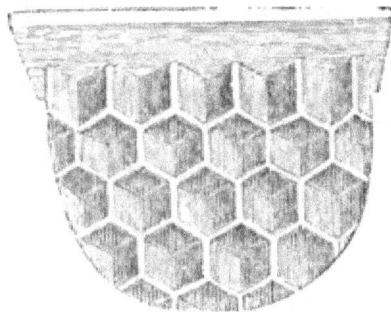

Arrangement of Cells.

is not impossible that bees may proceed on a similar principle. M. Huber replies to this objection, that bees are not provided with instruments corresponding to the angles of their cells; for there is no more resemblance between these and the form of their mandibles, than between the chisel of the sculptor and the work which he produces. The head, he thinks, does not furnish any better explanation. He admits that the antennæ are very flexible, so as to enable the insects to follow the outline of every object; but concludes that neither their structure, nor that of the limbs and mandibles, are adequate to explain the form of the cells, though all these are employed in the operations of building,—the effect, according to him, depending entirely on the object which the insect proposes.

We shall now follow M. Huber in the experiments which

he contrived, in order to observe the operations of the bees subsequent to their laying a foundation for the first cell: and we shall again quote from his own narrative :—

"It appeared to me," he says, "that the only method of isolating the architects, and bringing them individually into view, would be to induce them to change the direction of their operations and work upwards.

"I had a box made twelve inches square and nine deep, with a moveable glass lid. Combs, full of brood, honey, and pollen, were next selected from one of my leaf-hives, as containing what might interest the bees, and being cut into pieces a foot long, and four inches deep, they were arranged vertically at the bottom of the box, at the same intervals as the insects themselves usually leave between them. A small slip of wooden lath covered the upper edge of each. It was not probable that the bees would attempt to found new combs on the glass roof of the box, because its smoothness precluded the swarm from adhering to it; therefore, if disposed to build, they could do so over the slips resting on the combs, which left a vacuity five inches high above them. As we had foreseen, the swarm with which this box was peopled established itself among the combs below. We then observed the nurse-bees displaying their natural activity. They dispersed themselves throughout the hive, to feed the young grubs, to clear out their lodgment, and adapt it for their convenience. Certainly, the combs, which were roughly cut to fit the bottom of the box, and in some parts damaged, appeared to them shapeless and misplaced; for they speedily commenced their reparation. They beat down the old wax, kneaded it between their teeth, and thus formed binding materials to consolidate them. We were astonished beyond expression by such a multitude of workers employed at once in labours to which it did not appear they should have been called, at their coincidence, their zeal, and their prudence.

"But it was still more wonderful, that about half the numerous population took no part in the proceedings, remaining motionless, while the others fulfilled the functions required. The wax-workers, in a state of absolute repose, recalled our former observations. Gorged with the honey we had put within their reach, and continuing in this condition during twenty-four hours, wax was formed under their rings, and was now ready to be put in operation. To our great satisfaction, we soon saw a little foundation-wall rising on one of the slips that we had prepared to receive the superstructure. No obstacle was offered to the progress of our observations; and for the second time we beheld both the undertaking of the founder-bee, and the successive labours of several wax-workers, in forming the foundation-wall. Would that my readers could share the interest which the view of these architects inspired!

" This foundation, originally very small, was enlarged as the work required; while they excavated on one side a hollow, of about the width of a common cell, and on the opposite surface two others somewhat more elongated. The middle of the single cell corresponded exactly to the partition

Foundation-wall enlarged, and the Cells commenced.

separating the latter: the arches of these excavations, projecting by the accumulation of wax, were converted into ridges in a straight line; whence the cells of the first row were composed of five sides, considering the slip as one side, and those of the second row, of six sides.

" The interior conformation of the cavities, apparently, was derived from the position of their respective outlines. It seemed that the bees, endowed with an admirable delicacy of feeling, directed their teeth principally to the place where the wax was thickest; that is, the parts where other workers on the opposite side had accumulated it; and this explains why the bottom of the cell is excavated in an angular direction behind the projection on the sides of which the sides of the corresponding cells are to rise. The largest of the excavations, which was opposite to three others, was divided into three parts, while the excavations of the first row on the other face, applied against this one, were composed of only two.

" In consequence of the manner in which the excavations were opposed to each other, those of the second row, and all subsequent, partially applied to three cavities, were composed of three equal diamond-shaped lozenges. I may here remark, that each part of the labour of bees appears the natural result of what has preceded it: therefore, chance has no share in these admirable combinations.

" A foundation-wall rose above the slip like a minute vertical partition, five or six lines long, two lines high, but only half a line in thickness; the edge circular, and the surface rough. Quitting the cluster among the combs, a nurse-bee mounted the slip, turned around the block, and visiting both sides, began to work actively in the middle. It removed as much wax with its teeth as might equal the diameter of a common cell; and after kneading and moistening the particles, deposited them on the edge of the excavation. This insect having laboured some seconds, retired, and was soon replaced by another; a third continued the work, raising the margin of the edges, now projecting from the cavity, and with assistance of its teeth and feet fixing the particles, so as to give these edges a straighter form. More than twenty bees successively participated in the same work; and when the cavity was little above a line and a half in height, though equalling a cell in width, a bee left the swarm, and after encircling the block, commenced its operations on the opposite face, where yet untouched. But its teeth acting only on one half of this side, the hollow which it formed was opposite to only one of the slight prominences bordering the first cavity. Nearly at the same time another worker began on the right of the face that had been untouched, wherein both were occupied in forming cavities which may be designed the second and third; and they also were replaced by substitutes. These two latter cavities were separated only by the common margin, framed of particles of wax withdrawn from them; which margin corresponded with the centre of the cavity on the opposite surface. The foundation-wall itself was still of insufficient dimensions to admit the full diameter of a cell : but while the excavations were deepened, wax-workers, extracting their scales of wax, applied them in enlarging its circumference ; so that it rose nearly two lines further around the circular arch. The nurse-bees, which appeared more especially charged with sculpturing the cells, being then enabled to continue their outlines, prolonged the cavities, and heightened their margins on the new addition of wax.

" The arch, formed by the edge of each of these cavities, was next divided as by two equal chords, in the line of which the bees formed stages or projecting borders, or margins meeting at an obtuse angle : the cavities now had four margins, two lateral and perpendicular to the supporting slip, and two oblique, which were shorter.

" Meantime, it became more difficult to follow the operations of the bees, from their frequently interposing their heads between the eye of the observer and the bottom of the cell; but the partition, whereon their teeth laboured, had become so transparent as to expose what passed on the other side.

" The cavities of which we speak formed the bottom of the first three cells ; and while the bees engaged were advancing them to perfection, other workers commenced sketching a second row of cells above the first, and partly behind those in front—for, in general, their labour proceeds by combination. We cannot say, 'When bees have finished this cell, they will begin new ones;' but, 'while particular workers advance a certain portion, we are certain that others will carry on the adjacent cells.' Further, the work begun on one face of the comb is already the commencement of that which is to follow on the reverse. All this depends on a reciprocal relation, or a mutual connexion of the parts, rendering the whole subservient to each other. It is undoubted, therefore, that slight irregularities on the front will affect the form of the cells on the back of the comb."*

* Huber on Bees, p. 368.

L

When they have in this manner worked the bottoms of the first row of cells into the required forms, some of the nurse-bees finish them by imparting a sort of polish, while others proceed to cut out the rudiments of a second row from a fresh wall of wax which has been built in the meanwhile by the wax-workers, and also on the opposite side of this wall; for a comb of cells is always double, being arranged in two layers, placed end to end. The cells of this second row are engrafted on the borders of cavities hollowed out in the wall, being founded by the nurse-bees, bringing the contour of all the bottoms, which is at first unequal, to the same level; and this level is kept uniform in the margins of the cells till they are completed. At first sight nothing appears more simple than adding wax to the margins; but from the inequalities occasioned by the shape of the bottom, the bees must accumulate wax on the depressions, in order to bring them to a level. It follows accordingly that the surface of a new comb is not quite flat, there being a progressive slope produced as the work proceeds, and the comb being therefore in the form of a lens, the thickness decreasing towards the edge, and the last-formed cells being shallower or shorter than those preceding them. So long as there is room for the enlargement of the comb, this thinning of its edge may be remarked; but as soon as the space within the hive prevents its enlargement, the cells are made equal, and two flat and level surfaces are produced.

M. Huber observed, that while sketching the bottom of a cell, before there was any upright margin on the reverse, their pressure on the still soft and flexible wax gave rise to a projection, which sometimes caused a breach of the partition. This, however, was soon repaired, but a slight prominence always remained on the opposite surface, to the right and left of which they placed themselves to begin a new excavation; and they heaped up part of the materials between the two flutings formed by their labour. The ridge thus formed becomes a guide to the direction which the bees are to follow for their vertical furrow of the front cell.

We have already seen that the first cell determines the

place of all that succeed it, and two of these are never, in ordinary circumstances, begun in different parts of the hive at the same time, as is alleged by some early writers. When some rows of cells, however, have been completed in the first comb, two other foundation-walls are begun, one on each side of it, at the exact distance of one-third of an inch, which is sufficient to allow two bees employed on the opposite cells to pass each other without jostling. These new walls are also parallel to the former; and two more are afterwards begun exterior to the second, and at the same parallel distance. The combs are uniformly enlarged, and lengthened in a progression proportioned to the priority of their origin; the middle comb being always advanced beyond the two adjoining ones by several rows of cells, and these again beyond the ones exterior to them. Did the bees lay the foundations of all their combs at the same time, they would not find it easy to preserve parallelism and an equality in their distances. It may be remarked further, that beside the vacancies of half an inch between the cells, which form what we call the highways of the community, the combs are pierced in several places with holes which serve as postern-gates for easy communication from one to another, to prevent loss of time in going round. The equal distance between the combs is of more importance to the welfare of the hive than might at first appear; for were they too distant, the bees would be so scattered and dispersed, that they could not reciprocally communicate the heat indispensable for hatching the eggs and rearing the young. If the combs, on the other hand, were closer, the bees could not traverse the intervals with the freedom necessary to facilitate the work of the hive. On the approach of winter, they sometimes elongate the cells which contain honey, and thus contract the intervals between the combs. But this expedient is in preparation for a season when it is important to have copious magazines, and when, their activity being relaxed, it is unnecessary for their communications to be so spacious and free. On the return of spring, the bees hasten to contract the elongated cells, that they may become fit for receiving the eggs which the queen is about to deposit, and in this manner they re-establish the regular distance.*

We are indebted to the late Dr. Barclay of Edinburgh, well known as an excellent anatomist, for the discovery that each cell in a honeycomb is not simply composed of one wall, but consists of two. We shall give the account of his discovery in his own words:—

"Having inquired of several naturalists whether or not they knew any author who had mentioned that the partitions between the cells of the honeycomb were double, and whether or not they had ever remarked such a structure themselves, and they having answered in the negative, I now take the liberty of presenting to the Society pieces of honeycomb, in which the young bees had been reared, upon breaking which, it will be clearly seen that the partitions between different cells, at the sides and the base, are all *double;* or, in other words, that each cell is a distinct, separate, and in some measure an independent structure, agglutinated only to the neighbouring cells; and that when the agglutinating substance is destroyed, each cell may be entirely separated from the rest.

"I have also some specimens of the cells formed by wasps, which show that the partitions between them are also double, and that the agglutinating substance between them is more easily destroyed than that between the cells of the bee."†

IRREGULARITIES IN THEIR WORKMANSHIP.

Though bees, however, work with great uniformity when circumstances favour their operations, they may be compelled to vary their proceedings. M. Huber made several ingenious experiments of this kind. The following, mentioned by Dr. Bevan, was accidental, and occurred to his friend Mr. Walond. "Inspecting his bee-boxes at the end of October, 1817, he perceived that a centre comb, burthened with honey, had separated from its attachments, and was

---

* Huber on Bees, p. 220.
† Memoirs of the Wernerian Nat. Hist. Soc., vol. ii. p. 260.

leaning against another comb so as to prevent the passage
of the bees between them. This accident excited great
activity in the colony; but its nature could not be ascertained at the time. At the end of a week, the weather
being cold, and the bees clustered together Mr. Walond
observed, through the window of the box, that they had
constructed two horizontal pillars betwixt the combs alluded
to, and had removed so much of the honey and wax from
the top of each as to allow the passage of a bee : in about
ten days more there was an uninterrupted thoroughfare ; the
detached comb at its upper part had been secured by a
strong barrier, and fastened to the window with the spare
wax. This being accomplished, the bees removed the horizontal pillars first constructed, as being of no further use."*

A similar anecdote is told by M. Huber. " During the
winter," says he, "a comb in one of my bell-glass hives,
having been originally insecure, fell down, but preserved its
position parallel to the rest. The bees were unable to fill
up the vacuity left above it, because they do not build combs
of old wax, and none new could be then obtained. At a
more favourable season they would have engrafted a new
comb on the old one ; but now their provision of honey could
not be spared for the elaboration of this substance, which
induced them to insure the stability of the comb by another
process.

" Crowds of bees taking wax from the lower part of other
combs, and even gnawing it from the surface of the orifices
of the deepest cells, they constructed so many irregular
pillars, joists, or buttresses, between the sides of the fallen
comb, and others on the glass of the hive. All these were
artificially adapted to localities. Neither did they confine
themselves to repairing the accidents which their works had
sustained. They seemed to profit by the warning to guard
against a similar casualty.

" The remaining combs were not displaced ; therefore,
while solidly adhering by the base, we were greatly surprised
to see the bees strengthen their principal fixtures with old

* Bevan on Bees, p. 326.

wax. They rendered them much thicker than before, and fabricated a number of new connections, to unite them more firmly to each other and to the sides of their dwelling. All this passed in the middle of January, a time that these insects commonly keep in the upper part of their hive, and when work is no longer seasonable."*

M. Huber the younger shrewdly remarks, that the tendency to symmetry observable in the architecture of bees does not hold so much in small details as in the whole work, because they are sometimes obliged to adapt themselves to particular localities. One irregularity leads on to another, and it commonly arises from mere accident, or from design on the part of the proprietor of the bees. By allowing, for instance, too little interval between the spars for receiving the foundation of the combs, the structure has been continued in a particular direction. The bees did not at first appear to be sensible of the defect, though they afterwards began to suspect their error, and were then observed to change their line of work till they gained the customary distance. The cells having been by this change of direction in some degree curved, the new ones which were commenced on each side of it, by being built everywhere parallel to it, partook of the same curvature. But the bees did not relish such approaches to the "line of beauty," and exerted themselves to bring their buildings again into the regular form.

In consequence of several irregularities which they wished to correct, the younger Huber has seen bees depart from their usual practice, and at once lay on a spar two foundation-walls not in the same line. They could consequently neither be enlarged without obstructing both, nor from their position could the edges unite, had they been prolonged. The little architects, however, had recourse to a very ingenious contrivance: they curved the edges of the two combs, and brought them to unite so neatly that they could be both prolonged in the same line with ease; and when carried to some little distance, their surface became quite uniform and level.

* Huber on Bees, p. 416.

" Having seen bees," says the elder Huber, "work both
up and down, I wished to try to investigate whether we could
compel them to construct their combs in any other direction.
We endeavoured to puzzle them with a hive glazed above
and below, so that they had no place of support but the
upright sides of their dwelling; but, betaking themselves to
the upper angle, they built combs perpendicular to one of
these sides, and as regularly as those which they usually
build under a horizontal surface. The foundations were laid
on a place which does not serve naturally for the base, yet,
except in the difference of direction, the first row of cells
resembled those in ordinary hives, the others being distributed on both faces, while the bottoms alternately corresponded with the same symmetry. I put the bees to a still
greater trial. As they now testified their inclination to
carry their combs, by the shortest way, to the opposite side
of the hive (for they prefer uniting them to wood, or a
surface rougher than glass), I covered it with a pane.
Whenever this smooth and slippery substance was interposed
between them and the wood, they departed from the straight
line hitherto followed, and bent the structure of their comb
at a right angle to what was already made, so that the pro-
longation of the extremity might reach another side of the
hive, which had been left free.

" Varying this experiment in several ways, I saw the bees
constantly change the direction of their combs, when I
presented to them a surface too smooth to admit of their
clustering on it. They always sought the wooden sides. I
thus compelled them to curve the combs in the strangest
shapes, by placing a pane at a certain distance from their
edges. These results indicate a degree of instinct truly
wonderful. They denote even more than instinct: for glass
is not a substance against which bees can be warned by
nature. In trees, their natural abode, there is nothing that
resembles it, or with the same polish. The most singular
part of their proceeding is changing the direction of the
work before arriving at the surface of the glass, and while
yet at a distance suitable for doing so. Do they anticipate

the inconvenience which would attend any other mode of
building? No less curious is the plan adopted by the bee
for producing an angle in the combs: the wonted fashion of
their work, and the dimensions of the cells, must be altered.
Therefore, the cells on the upper or convex side of the
combs are enlarged; they are constructed of three or four
times the width of those on the opposite surface. How can
so many insects, occupied at once on the edges of the combs,
concur in giving them a common curvature from one extremity
to the other? How do they resolve on establishing cells so
small on one side, while dimensions so enlarged are bestowed
on those of the other? And is it not still more singular,
that they have the art of making a correspondence between
cells of such reciprocal discrepance? The bottom being
common to both, the tubes alone assume a taper form.
Perhaps no other insect has afforded a more decisive proof of
the resources of instinct, when compelled to deviate from the
ordinary course.

"But let us study them in their natural state, and there
we shall find that the diameter of their cells must be adapted
to the individuals which shall be bred in them. The cells
of males have the same figure, the same number of lozenges
and sides, as those of workers, and angles of the same size.
Their diameter is $3\frac{1}{2}$ lines, while those of workers are
only $2\frac{2}{5}$.

"It is rarely that the cells of males occupy the higher
part of the combs. They are generally in the middle or on
the sides, where they are not isolated. The manner in
which they are surrounded by other cells alone can explain
how the transition in size is effected. When the cells of
males are to be fabricated under those of workers, the bees
make several rows of intermediate cells, whose diameter
augments progressively, until gaining that proportion proper
to the cells required; and in returning to those of workers,
a lowering is observed in a manner corresponding.

"Bees, in preparing the cells of males, previously establish a block or lump of wax on the edge of their comb,
thicker than is usually employed for those of workers. It is

also made higher, otherwise the same order and symmetry could not be preserved on a larger scale.

" Several naturalists notice the irregularities in the cells of bees as so many defects. What would have been their astonishment had they observed that part of them are the result of calculation? Had they followed the imperfection of their organs, some other means of compensating them would have been granted to the insects. It is much more surprising that they know how to quit the ordinary route, when circumstances demand the construction of enlarged cells ; and, after building thirty or forty rows of them, to return to the proper proportions from which they have departed by successive reductions. Bees also augment the dimensions of their cells when there is an opportunity for a great collection.of honey. Not only are they then constructed of a diameter much exceeding that of the common cells, but they are elongated throughout the whole space admitting it. A great portion of irregular comb contains cells an inch, or even an inch and a half, in depth.

" Bees, on the contrary, sometimes are induced to shorten their cells. When wishing to prolong an old comb, whose cells have received their full dimensions, they gradually reduce the thickness of its edges, by gnawing down the sides of the cells, until they restore it to its original lenti-cular form. They add a waxen block around the whole circumference, and on the edge of the comb construct pyra-midal bottoms, such as those fabricated on ordinary occa-sions. It is a certain fact, that a comb never is extended in any direction unless the bees have thinned the edges, which are diminished throughout a sufficient space to remove any angular projection.

" The law which obliges these insects partly to demolish the cells on the edges of the comb before enlarging it, un-questionably demands more profound investigation. How can we account for instinct leading them to undo what they have executed with the utmost care ? The wonted regular gradation which may be necessary for new cells, subsists among those adjoining the edges of a comb recently constructed. But afterwards, when those on the edge are deepened like the cells of the rest of the surface, the bees no longer preserve the decreasing gradation which is seen in the new combs." *

THE FINISHING OF THE CELLS.

While the cells are building they appear to be of a dull white colour, soft, even, though not smooth, and translucent; but in a few days they become tinged with yellow, particularly on the interior surface; and their edges, from being thin, uniform, and yielding, become thicker, less regular, more heavy, and so firm that they will bend rather than break. New combs break on the slightest touch. There is also a glutinous substance observable around the orifices of the yellow cells, of reddish colour, unctuous, and odoriferous. Threads of the same substance are applied all around the interior of the cells, and at the summit of their angles, as if it were for the purpose of binding and strengthening the walls. These yellow cells also require a much higher temperature of water to melt them than the white ones.

It appeared evident, therefore, that another substance, different from wax, had been employed in varnishing the orifices, and strengthening the interior of the cells. M. Huber, by numerous experiments, ascertained the resinous threads lining the cells, as well as the resinous substance around their orifice, to be propolis; for he traced them, as we mentioned in our account of propolis, from the poplar buds where they collected it, and saw them apply it to the cells; but the yellow colour is not imparted by propolis, to which it bears no analogy. We are, indeed, by no means certain what it is, though it was proved by experiment not to arise from the heat of the hives, nor from emanations of honey, nor from particles of pollen. Perhaps it may be ascribed to the bees rubbing their teeth, feet, and other parts of their body, on the surfaces where they seem to rest; or to their tongue (haustellum) sweeping from right to left like a fine

* Huber on Bees, p. 391.

pliant pencil, when it appears to leave some sprinkling of a transparent liquid.

Besides painting and varnishing their cells in this manner, they take care to strengthen the weaker part of their edifice by means of a mortar composed of propolis and wax, and named *pissoceros* * by the ancients who first observed it, though Réaumur was somewhat doubtful respecting the existence of such a composition. We are indebted to the shrewd observations of Huber for a reconcilement of the Roman and the French naturalists. The details which he has given of his discovery are perhaps the most interesting in his delightful book.

"Soon," he says, "after some new combs had been finished in a hive, manifest disorder and agitation prevailed among the bees. They seemed to attack their own works. The primitive cells, whose structure we had admired, were scarcely recognizable. Thick and massive walls, heavy, shapeless pillars, were substituted for the slight partitions previously built with such regularity. The substance had changed along with the form, being composed apparently of wax and propolis. From the perseverance of the workers in their devastation, we suspected that they proposed some useful alteration of their edifices; and our attention was directed to the cells least injured. Several were yet untouched; but the bees soon rushed precipitately on them, destroyed the tubes, broke down the wax, and threw all the fragments about. But we remarked that the bottom of the cells of the first row were spared; neither were the corresponding parts on both faces of the comb demolished at the same time. The bees laboured at them alternately, leaving some of the natural supports, otherwise the comb would have fallen down, which was not their object: they wished, on the contrary, to provide it a more solid base, and to secure its union to the vault of the hive, with a substance whose adhesive properties infinitely surpassed those of wax. The propolis employed on this occasion had been deposited in a mass over a cleft of the hive, and had hardened in

* From two Greek words, signifying *pitch* and *wax*.

drying, which probably rendered it more suitable for the purpose. But the bees experienced some difficulty in making any impression on it ; and we thought, as also had appeared to M. de Réaumur, that they softened it with the same frothy matter from the tongue which they use to render wax more ductile.

" We very distinctly observed the bees mixing fragments of old wax with the propolis, kneading the two substances together to incorporate them ; and the compound was employed in rebuilding the cells that had been destroyed. But they did not now follow their ordinary rules of architecture. for they were occupied by the solidity of their edifices alone. Night intervening, suspended our observations, but next morning confirmed what we had seen.

" We find, therefore, that there is an epoch in the labour of bees, when the upper foundation of their combs is constructed simply of wax, as Réaumur believed ; and that, after all the requisite conditions have been attained, it is converted to a mixture of wax and propolis, as remarked by Pliny so many ages before us. Thus is the apparent contradiction between these two great naturalists explained. But this is not the utmost extent of the foresight of these insects. When they have plenty of wax, they make their combs the full breadth of the hive, and solder them to the glass or wooden sides, by structures more or less approaching the form of cells, as circumstances admit. But should the supply of wax fail before they have been able to give sufficient diameter to the combs whose edges are rounded, large intervals remain between them and the upright sides of the hive, and they are fixed only at the top. Therefore, did not the bees provide against it, by constructing great pieces of wax mixed with propolis, in the intervals, they might be borne down by the weight of the honey. These pieces are of irregular shape, strangely hollowed out, and their cavities void of symmetry."*

It is remarked by the lively Abbé la Pluche, that the foundations of our houses sink with the earth on which they

* Huber on Bees, p. 415.

are built, the walls begin to stoop by degrees, they nod with age, and bend from their perpendicular;—lodgers damage everything, and time is continually introducing some new decay. The mansions of the bees, on the contrary, grow stronger the oftener they change inhabitants. Every bee-grub, before its metamorphosis into a nymph, fastens its skin to the partitions of its cell, but in such a manner as to make it correspond with the lines of the angle, and without in the least disturbing the regularity of the figure. During summer, accordingly, the same lodging may serve for three or four grubs in succession; and in the ensuing season it may accommodate an equal number. Each grub never fails to fortify the panels of its chamber by arraying them with its spoils, and the contiguous cells receive a similar augmentation from its brethren.* Réaumur found as many as seven or eight of these skins spread over one another: so that all the cells being incrusted with six or seven coverings, well dried and cemented with propolis, the whole fabric daily acquires a new degree of solidity.

It is obvious, however, that by a repetition of this process the cell might be rendered too contracted; but in such a case the bees know well how to proceed, by turning the cells to other uses, such as magazines for bee-bread and honey. It has been remarked, however, that in the hive of a new swarm, during the months of July and August, there are fewer small bees or nurse-bees than in one that has been tenanted four or five years. The workers, indeed, clean out the cell the moment that a young bee leaves its cocoon, but they never detach the silky film which it has previously spun on the walls of its cell. But though honey is deposited after the young leave the cells, the reverse also happens; and accordingly, when bees are bred in contracted cells, they are by necessity smaller, and constitute, in fact, the important class of nurse-bees.

We are not disposed, however, to go quite so far as an American periodical writer, who says, "Thus we see that the contraction of the cell may diminish the size of a bee,

* Spectacle de la Nature, vol. i.

even to the extinction of life, just as the contraction of a Chinese shoe reduces the foot even to uselessness."* We know, on the contrary, that the queen-bee will not deposit eggs in a cell either too small or too large for the proper rearing of the young. In the case of large cells, M. Huber took advantage of a queen that was busy depositing the eggs of workers to remove all the common cells adapted for their reception, and left only the larger cells appropriated for males. As this was done in June, when bees are most active, he expected that they would have immediately repaired the breaches he had made; but to his great surprise they did not, make the slightest movement for that purpose. In the meanwhile the queen, being oppressed by her eggs, was obliged to drop them about at random, preferring this to depositing them in the male cells, which she knew to be too large. At length she did deposit six eggs in the large cells, which were hatched as usual three days after. The nurse-bees, however, seemed to be aware that they could not be reared there, and though they supplied them with food, did not attend to them regularly. M. Huber found that they had been all removed from the cells during the night, and the business both of laying and nursing was at a complete stand for twelve days, when he supplied them again with a comb of small cells, which the queen almost immediately filled with eggs, and in some cells she laid five or six.

[The accompanying illustration exhibits these three kinds of bees, namely, the Queen, the Drone, and the Worker, together with the cells which they respectively inhabit. Fig. 1 shows the queen-bee as she appears when in command of a hive. When she first issues from the royal cell, she is much smaller in the body, and an inexperienced observer might have some difficulty in distinguishing her from an ordinary worker. But any one who has been accustomed to bees can pick her out as soon as his eyes rest upon her. Her body is rather larger and narrower than those of the workers, and the wings are shorter in proportion, slightly crossing at the tips when she is at rest. Fig. 2

* North American Rev., Oct. 1828, p. 355.

represents the common worker-bee, which, as has already been mentioned, is simply an undeveloped female. Fig. 3 is the male or drone-bee, which is easily distinguishable, even by a novice. He is larger, stouter, and heavier built

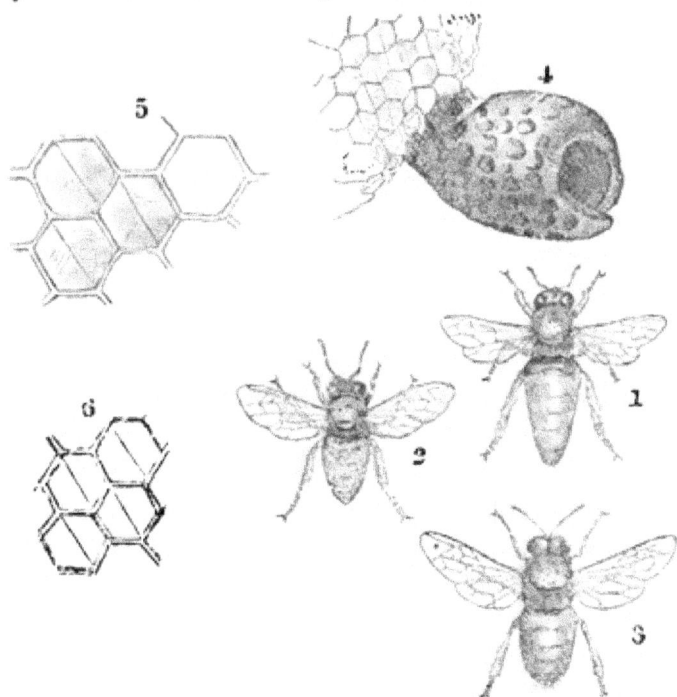

than the female; his eyes are so enormous that they seem to occupy nearly the entire head, and he has some well-defined tufts of hair on the end of the abdomen. He can even be detected by the ear, as he flies, the deep droning hum being quite unlike the fussy, business-like sound produced by the worker. Fig. 4 represents one of the royal cells, a little reduced in size. In making this cell, the bees lose sight of their habitual economy of wax, and use enough material for fifty ordinary cells. It is probable that the great size of the cell enables the inclosed insect to expand, and so to be capable of becoming the mother as well as the ruler of her subjects. The royal cell is always placed at the edge of a comb, so as not to interfere with the other cells, which contain honey, bee-bread, and grubs; and in each hive there are generally several of these cells in different stages of structure. Figs. 5 and 6 represent the proportionate sizes of the cells which contain the drone and worker bees.]

The architecture of the hive, which we have thus detailed, is that of bees receiving the aid of human care, and having external coverings of a convenient form, prepared for their reception. In this country bees are not found in a wild state; though it is not uncommon for swarms to stray from their proprietors. But these stray swarms do not spread colonies through our woods, as they are said to do in America. In the remoter parts of that continent there are no wild bees. They precede civilization; and thus when the Indians observe a swarm they say, "The white man is coming." There is evidence of bees having abounded in these islands, in the earlier periods of our history; and Ireland is particularly mentioned by the Venerable Bede as being "rich in milk and honey."[*] The hive-bee has formed an object of economical culture in Europe at least for two thousand years; and Varro describes the sort of hives used in his time, 1870 years ago. We are not aware, however, that it is now to be found wild in the milder clime of Southern Europe, any more than it is in our own island.

The wild bees of Palestine principally hived in rocks. "He made him," says Moses, "to suck honey out of the rock."[†] "With honey out of the rock," says the Psalmist, "should I have satisfied thee."[‡] In the caves of Salsette and Elephanta, at the present day, they hive in the clefts of the rocks, and the recesses among the fissures, in such numbers as to become very troublesome to visitors. Their nests hang in innumerable clusters.[§]

We are told of a little black stingless bee found in the island of Guadaloupe, which hives in hollow trees or in the cavities of rocks by the sea-side, and lays "up honey

---

[*] "Hibernia dives lactis ac *mellis* insula."—Beda, Hist. Eccles. i. 7.
[†] Deut. xxxii. 13.   [‡] Psalm lxxxi. 16.
[§] Forbes, Orien. Mem. i.

in cells about the size and shape of pigeons' eggs. These cells are of a black or deep-violet colour, and so joined together as to leave no space between them. They hang in clusters almost like a bunch of grapes."* The following are mentioned by Lindley as indigenous to Brazil. "On an excursion towards Upper Tapagippe," says he, "and skirting the dreary woods which extend to the interior, I observed the trees more loaded with bees' nests than even in the neighbourhood of Porto Seguro. They consist of a ponderous shell of clay, cemented similarly to martins' nests, swelling from high trees about a foot thick, and forming an oval mass full two feet in diameter. When broken, the wax is arranged as in our hives, and the honey abundant."†

Captain Basil Hall found in South America the hive of a honey-bee very different from the Brazilian, but nearly allied to, if not the same as, that of Guadaloupe. "The hive we saw opened," he says, "was only partly filled, which enabled us to see the economy of the interior to more advantage. The honey is not contained in the elegant hexagonal cells of our hives, but in wax bags, not quite so large as an egg. These bags or bladders are hung round the sides of the hive, and appear about half full; the quantity being probably just as great as the strength of the wax will bear without tearing. Those near the bottom, being better supported, are more filled than the upper ones. In the centre of the lower part of the hive we observed an irregularly-shaped mass of comb, furnished with cells like those of our bees, all containing young ones in such an advanced state, that, when we broke the comb, and let them out, they flew merrily away."

Clavigero, in his 'History of Mexico,' evidently describing the same species of bee, says it abounds in Yucatan, and makes the honey of Estabentum, the finest in the world, and which is taken every two months. He mentioned another species of bee, smaller in size, and also without a sting, which forms its nest of the shape of a sugar-loaf, and as large

* Amer. Q. Rev., iii. p. 383.
† Roy. Mil. Chron. quoted by Kirby and Spence.

M

or larger.   These are suspended from trees, particularly from
the oak, and are much more populous than our common
hives.

Wild honey-bees of some species appear also to abound in
Africa.   Mr. Park, in his second volume of travels, tells us
that some of his associates imprudently attempted to rob a
numerous hive of its honey, when the exasperated bees,
rushing out to defend their property, attacked their assail-
ants with great fury, and quickly compelled the whole company to fly.

At the Cape of Good Hope the bees themselves must be
less formidable, or more easily managed, as their hives are
sought for with avidity.   Nature has there provided man
with a singular and very efficient assistant in a bird, most appropriately named the honey-guide (*Indicator major*, VEILLOT;
Cuculus indicator, LATHAM).   The honey-guide, it is said, so
far from being alarmed at the presence of man, appears
anxious to court his acquaintance, and flits from tree to
tree with an expressive note of invitation, the meaning of
which is both well known to the colonists and the Hottentots.
A person thus invited by the honey-guide seldom refuses to
follow it onward till it stops, as it is certain to do, at some
hollow tree containing a bee-hive, usually well stored with
honey and wax.   It may be that the bird finds itself inadequate to the attack of a legion of bees, or to penetrate
into the interior of the hive, and is thence led to invite an
agent more powerful than itself.   The person invited, indeed,
always leaves the bird a share of the spoil, as it would be
considered sacrilege to rob it of its due, or in any way to
injure so sacred a bird.

Useful, however, as is the honey-guide, it must always be
carefully watched, and the traveller must not follow it without keeping his eyes well open.   For although, as a general
fact, the bird will lead its followers to honey, it has a strange
habit of leading them to the spot where lies hidden some
dangerous animal.   Sometimes it brings them to a rhinoceros,
wallowing in a mud pool.   Sometimes it directs them to a
solitary buffalo, one of the most dangerous animals that

Southern Africa produces, and one which the natives fear but little less than the lion itself. And more than once the too-confiding traveller has followed the honey-guide, and been led to a spot where was lying one of the venomous serpents.

The Americans, who have not the African honey-guide, employ several well-known methods to track bees to their hives. One of the most common though ingenious modes is to place a piece of bee-bread on a flat surface, a tile for instance, surrounding it with a circle of wet white paint. The bee, whose habit it is always to alight on the edge of any plane, has to travel through the paint to reach the bee-bread. When, therefore, she flies off, the observer can track her by the white on her body. The same operation is repeated at another place, at some distance from the first, and at right angles to the bee-line just ascertained. The position of the hive is easily determined, for it lies in the angle made by the intersection of the bee-lines. Another method is described in the 'Philosophical Transactions for 1721.' The bee-hunter decoys, by a bait of honey, some of the bees into his trap, and when he has secured as many as he judges will suit his purpose, he encloses one in a tube, and, letting it fly, marks its course by a pocket-compass. Departing to some distance, he liberates another, observes its course, and in this manner determines the position of the hive, upon the principle already detailed. These methods of bee-hunting depend upon the insect's habit of always flying in a right line to its home. Those who have read Cooper's tale of the 'Prairie' must well remember the character of the bee-hunter, and the expression of "lining a bee to its hive."

In reading these and similar accounts of the bees of distant parts of the world, we must not conclude that the descriptions refer to the same species as the common honey-bee. There are numerous species of social bees, which, while they differ in many circumstances, agree in the practice of storing up honey, in the same way as we have numerous species of the mason-bee and of the humble-bee.

CHAPTER VII.

THE operations of an insect in boring into a leaf or a bud to form a lodgment for its eggs appear very simple. The tools, however, by which these effects are performed are very complicated and curious. In the case of gall-flies (*Cynips*), the operation itself is not so remarkable as its subsequent chemical effects. These effects are so different from any others that may be classed under the head of Insect Architecture, that we shall reserve them for the latter part of this volume, although, with reference to the use of galls, the protection of eggs and larvæ, they ought to find a place here. We shall, however, at present confine ourselves to those which simply excavate a nest, without producing a tumour.

The first of these insects which we shall mention is celebrated for its song, by the ancient Greek poets, under the name of *Tettix*. The Romans called it *Cicada*, which we sometimes, but erroneously, translate " grasshopper ;" for the grasshoppers belong to an entirely different order of insects. We shall, therefore, take the liberty of calling the Cicadæ *Tree-hoppers*, to which the cuckoo-spit insect (*Tettigonia spumaria*, OLIV.) is allied ; but there is only one of the true Cicadæ hitherto ascertained to be British, namely, the *Cicada hæmatodes* (LINN.), which was discovered in the New Forest, Hampshire, by Mr. Daniel Bydder.

M. Réaumur was exceedingly anxious to study the economy of these insects ; but they not being indigenous in the neighbourhood of Paris, he commissioned his friends to send him some from more southern latitudes, and he procured in this way specimens not only from the South of France and from Italy, but also from Egypt. From these specimens he has given the best account of them yet published ; for though, as he tells us, he had never had the pleasure of seeing one of them alive, the most interesting parts of their structure can be studied as well in dead as in living specimens. We ourselves possess several specimens from New Holland, upon which we have verified some of the more interesting observations of Réaumur.

Virgil tells us that in his time "the cicadæ burst the very shrubs with their querulous music;"* but we may well suppose that he was altogether unacquainted with the singular instrument by means of which they can, not poetically, but actually, cut grooves in the branches they select for depositing their eggs. It is the male, as in the case of birds, which fills the woods with his song; while the female, though mute, is no less interesting to the naturalist on account of her curious ovipositor. This instrument, like all those with which insects are furnished by nature for cutting, notching, or piercing, is composed of a horny substance, and is also considerably larger than the size of the tree-hopper would proportionally indicate. It can on this account be partially examined without a microscope, being, in some of the larger species, no less than five lines† in length.

The ovipositor, or auger (*tarière*), as Réaumur calls it, is lodged in a sheath which lies in a groove of the terminating ring of the belly. It requires only a very slight pressure to cause the instrument to protrude from its sheath, when it appears to the naked eye to be of equal thickness throughout, except at the point, where it is somewhat enlarged and angular, and on both sides finely indented with teeth. A more minute examination of the sheath demonstrates that it is composed of two horny pieces slightly curved, and ending in the form of an elongated spoon, the concave part of which is adapted to receive the convex end of the ovipositor.

When the protruded instrument is further examined with a microscope, the denticulations, nine in number on each side, appear strong, and arranged with great symmetry, increasing in fineness towards the point, where there are three or four

* " Cantu querulæ rumpent arbusta cicadæ."—Georg. iii. 328
† A line is about the twelfth part of an inch.

very small ones, beside the nine that are more obvious. The magnifier also shows that the instrument itself, which appeared simple to the naked eye, is, in fact, composed of three different pieces; two exterior armed with the teeth before mentioned, denominated by Réaumur files (*limes*), and another pointed like a lancet, and not denticulated. The denticulated pieces, moreover, are capable of being moved forwards and backwards, while the centre one remains stationary; and as this motion is effected by pressing a pin or the blade of a knife over the muscles on either side at the origin of the ovipositor, it may be presumed that those muscles are destined for producing similar movements when the insect requires them. By means of a finely-pointed pin carefully introduced between the pieces, and pushed very gently downwards, they may be, with no great difficulty, separated in their whole extent.

The contrivance by which those three pieces are held united, while at the same time the two files can be easily put in motion, is similar to those of our own mechanical inventions, with this difference, that no human workman could construct an instrument of this description so small, fine, exquisitely polished, and fitting so exactly. We should have been apt to form the grooves in the central piece, whereas they are scooped out in the handles of the files, and play upon two projecting ridges in the central piece, by which means this is rendered stronger. M. Réaumur discovered that the best manner of showing the play of this extraordinary instrument is to cut it off with a pair of scissors near its origin, and then, taking it between the thumb and the finger at the point of section, work it gently to put the files in motion.

Beside the muscles necessary for the movement of the files, the handle of each is terminated by a curve of the same hard horny substance as itself, which not only furnishes the muscles with a sort of lever, but serves to press, as with a spring, the two files close to the central piece, as is shown in the lower figure.

M. Pontedera, who studied the economy of the tree-hoppers

with some care, was anxious to see the insect itself make use of the ovipositor in forming grooves in wood, but found that it was so shy and easily alarmed, that it took to flight whenever he approached; a circumstance of which Réaumur takes

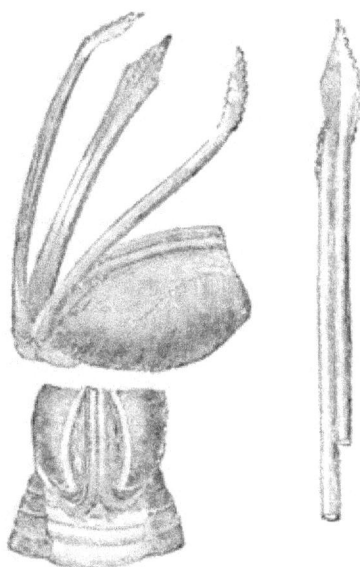

Ovipositors, with files, of Tree-hopper, magnified.

advantage, to soothe his regret that the insects were not indigenous in his neighbourhood. But of their workmanship, when completed, he had several specimens sent to him from Provence and Languedoc by the Marquis de Caumont.

The gall-flies, when about to deposit their eggs, select growing plants and trees; but the tree-hoppers, on the contrary, make choice of dead, dried branches, for the mother

Excavations for eggs of Tree-hopper, with the chip-lids raised.

seems to be aware that moisture would injure her progeny. The branch, commonly a small one, in which eggs have been deposited, may be recognised by being covered with little oblong elevations caused by small splinters of the wood, detached at one end, but left fixed at the other, by the insect. These elevations are for the most part in a line, rarely in a double line, nearly at equal distances from each other, and form a lid to a cavity in the wood about four lines in length, containing from four to ten eggs. It is to be remarked that the insect always selects a branch of such dimensions that it can get at the pith, not because the pith is more easily bored, for it does not penetrate into it all, but to form a warm and safe bed for the eggs. M. Pontedera says, that when the eggs have been deposited, the insect closes the mouth of the hole with a gum capable of protecting them from the weather; but M. Réaumur thinks this only a fancy, as, out of a great number which he examined, he could discover nothing of the kind. Neither is such a protection wanted; for the woody splinters above mentioned furnish a very good covering.

The grubs hatch from these eggs (of which, M. Pontedera says, one female will deposit from five to seven hundred), issue from the same holes through which the eggs have been introduced, and betake themselves to the ground to feed on the roots of plants. They are not transformed into chrysalides, but into active nymphs, remarkable for their fore limbs, which are thick, strong, and furnished with prongs for digging; and when we are told, by Dr. Le Ferve, that they make their way easily into hard stiff clay, to the depth of two or three feet, we perceive how necessary to them such a conformation must be.

SAW-FLIES.

An instrument for cutting grooves in wood, still more ingeniously contrived than that of the tree-hopper, was first observed by Vallisnieri, an eminent Italian naturalist, in a four-winged fly, most appropriately denominated by M. Réaumur the *saw-fly* (*Tenthredo*), of which many sorts are indigenous to Great Britain. The grubs from which these flies originate are indeed but too well known, as they frequently strip our rose, gooseberry, raspberry, and red currant trees of their leaves, and are no less destructive to birch, alder, and willows; while turnips and wheat suffer still more seriously by their ravages. These grubs may readily be distinguished from the caterpillars of moths and butterflies by having from sixteen to twenty-eight feet, by which they usually hang to the leaf they feed on, while they coil up the hinder part of their body in a spiral ring. The perfect flies are distinguished by four transparent wings; and some of the most common have a flat body of a yellow or orange colour, while the head and shoulders are black.

In order to see the ovipositor, to which we shall for the present turn our chief attention, a female saw-fly must be taken, and her belly gently pressed, when a narrow slit will be observed to open at some distance from the apex, and a short, pointed, and somewhat curved body, of a brown colour

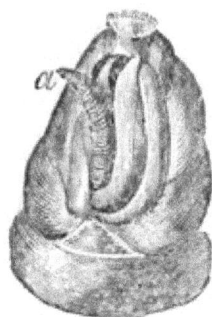

a, Ovipositor of Saw-fly, protruded from its sheath, magnified.

and horny substance, will be protruded. The curved plates which form the sides of the slit are the termination of the sheath, in which the instrument lies concealed till it is wanted by the insect. The appearance of this instrument, however, and its singular structure, cannot be well understood without the aid of a microscope.

The instrument thus brought into view is a very finely-contrived saw, made of a horny substance, and adapted for penetrating branches and other parts of plants where the eggs are to be deposited. The ovipositor-saw of the insect is much more complicated than any of those employed by our carpenters. The teeth of our saws are formed in a line, but in such a manner as to cut in two lines parallel to, and at a

small distance-from, each other. This is effected by slightly
bending the points of the alternate teeth right and left, so
that one-half of the whole teeth stand a little to the right,
and the other half a little to the left. The distance of the
two parallel lines thus formed is called the *course* of the saw,
and it is only the portion of wood which lies in the course
that is cut into saw-dust by the action of the instrument. It
will follow that in proportion to the thinness of a saw there
will be the less destruction of wood which may be sawed.
When cabinet-makers have to divide valuable wood into very

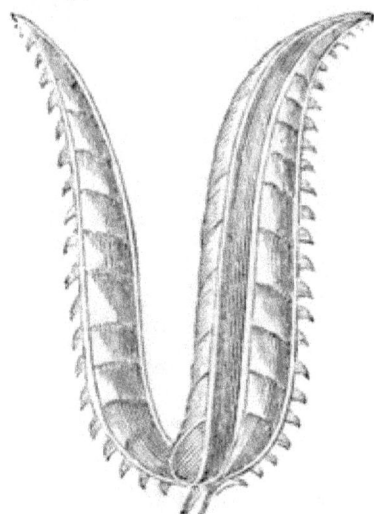

Ovipositor-saw of Saw-fly, with rasps shown in the cross lines.

thin leaves, they accordingly employ saws with a narrow
course, while sawyers who cut planks use one with a broad
course. The ovipositor-saw being extremely fine, does not
require the teeth to diverge much; but from the manner in
which they operate, it is requisite that they should not stand,
like those of our saws, in a straight line. The greater
portion of the edge of the instrument, on the contrary, is
towards the point somewhat concave, similar to a scythe,
while towards the base it becomes a little convex, the whole
edge being nearly the shape of an Italic *f*.

The ovipositor-saw of the fly is put in motion in the same
way as a carpenter's hand-saw, supposing the tendons attached

to its base to form the handle, and the muscles which put it in motion to be the hand of the carpenter. But the carpenter can only work one saw at a time, whereas each of these flies is furnished with two, equal and similar, which it works at the same time—one being advanced and the other retracted alternately. The secret, indeed, of working more saws than one at once is not unknown to our mechanics; for two or three are sometimes fixed in the same frame. These, however, not only all move upwards and downwards simultaneously, but cut the wood in different places; while the two saws of the ovipositor work in the same cut, and consequently, though the teeth are extremely fine, the effect is similar to a saw with a wide set.

It is important, seeing that the ovipositor-saws are so fine, that they be not bent or separated while in operation—and this, also, nature has provided for, by lodging the backs of the saws in a groove, formed by two membranous plates, similar to the structure of a clasp-knife. These plates are thickest at the base, becoming gradually thinner as they approach the point, which the form of the saws requires. According to Vallisnieri, it is not the only use of this apparatus to form a back for the saws, he having discovered, between the component membranes, two canals, which he supposes are employed to conduct the eggs of the insect into the grooves which it has hollowed out for them.

The teeth of a carpenter's saw, it may be remarked, are simple, whereas the teeth of the ovipositor-saw are themselves

Portion of a Saw-Fly's comb-toothed rasp, and saw.

denticulated with fine teeth. The latter, also, combines at the same time the properties of a saw and of a rasp or file. So far as we are aware, these two properties have never been combined in any of the tools of our carpenters. The rasping part of the ovipositor, however, is not constructed like our rasps, with short teeth thickly studded together, but has teeth almost as long as those of the saw, and placed contiguous to them, on the back of the instrument, resembling in their form and setting the teeth of a comb, as may be seen in the figure. Of course, such observations are conducted with the aid of a microscope.

When a female saw-fly has selected the branch of a rose-tree, or any other, in which to deposit her eggs, she may be seen bending the end of her belly inwards, in form of a crescent, and protruding her saw, at the same time, to penetrate the bark or wood. She maintains this recurved position so long as she works in deepening the groove; but when she has attained the depth required, she unbends her body into a straight line, and in this position works upon the place lengthways, by applying the saw more horizontally. When she has rendered the groove as large as she wishes, the motion of the tendons ceases, and an egg is placed in the cavity. The saw is then withdrawn into the sheath for about two-thirds of its length, and at the same moment a sort of frothy liquid, similar to a lather made with soap, is dropped over the egg, either for the purpose of gluing it in its place or sheathing it from the action of the juices of the tree. She proceeds in the same manner in sawing out a second groove, and so on in succession, till she has deposited all her eggs, sometimes to the number of twenty-four. The grooves are usually placed in a line, at a small distance from one another, on the same branch; but sometimes the mother-fly shifts to another, or to a different part of the branch, when she is either scared or finds it unsuitable. She commonly, also, takes more than one day to the work, notwithstanding the superiority of her tools. Réaumur has seen a saw-fly make six grooves in succession, which occupied her about ten hours and a half.

The grooves, when finished, have externally little elevation above the level of the bark, appearing like the puncture of a lancet in the human skin; but in the course of a day or two the part becomes first brown and then black, while it also becomes more and more elevated. This increased elevation is not owing to the growth of the bark, the fibres of which, indeed, have been destroyed by the ovipositor-saw, but to the actual growth of the egg; for when a new-laid egg of the saw-fly is compared with one which has been several days enclosed in the groove, the latter will be found to be very considerably the larger. This growth of the egg is contrary to the analogy observable in the eggs of birds, and even of most other insects; but it has its advantages. As it continues to increase, it raises the bark more and more, and consequently widens, at the same time, the slit at the entrance; so that, when the grub is hatched, it finds a passage ready for its exit. The mother-fly seems to be aware of this growth of her eggs, for she takes care to deposit them at such distances as may prevent their disturbing one another by their development.

Another species of saw-fly, with a yellow body and deep violet-coloured wings, which also selects the rose-tree, deposits her eggs in a different manner. Instead of making a groove for each egg, like the preceding, she forms a large single groove, sufficient for about two dozen eggs. These eggs are all arranged in pairs, forming two straight lines parallel to the sides of the branch. The eggs, however, though thus deposited in a common groove, are carefully kept

Nest of eggs of Saw-fly, in rose-tree.

each in its place; for a ridge of the wood is left to prevent those on the right from touching those on the left—and not only so, but between each egg of a row a thin partition of wood is left, forming a shallow cell.

The edges of this groove, it will be obvious, must be

farther apart than those which only contain a single egg,
and, in fact, the whole is open to inspection ; but the eggs
are kept from falling out, both by the frothy glue before
mentioned, and by the walls of the cells containing them.
They were observed also, by Vallisnieri, to increase in size
like the preceding.

[In the middle of summer, plenty of these grooves may be
seen, by looking at the under lid of leaf-stalks or delicate
young twigs. Row upon row of the grooves are sometimes
found, so the all-destructive power of the insects must
indeed be great. The larvæ, when full fed, dispose of themselves in various ways. Those of the gooseberry-fly, for
example (*Nematus Ribesii*), after they have stripped the bush
of its leaves, either seek the ground or remain on the
branches, and spin a series of cocoons, attaching them to each

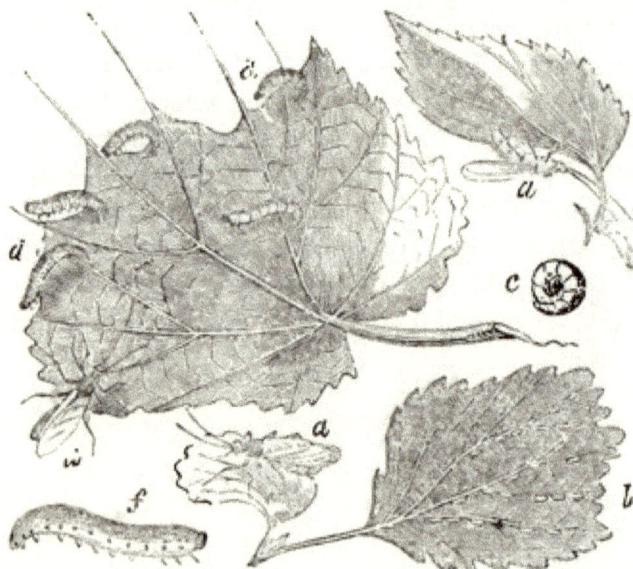

a a a, Saw-fly of the gooseberry (*Nematus Ribesii*, STEPHENS). *b*, its eggs on the
nervures of a leaf. *d d*, the caterpillars eating. *c*, one rolled up. *f*, one extended.

branch by their ends. Those, therefore, who wish to destroy
these little pests, must know both localities of the cocoons,
or they will allow one half to escape while destroying the
other.]

This insect has a flat yellow body and four pellucid wings, the two outer ones marked with brown on the edge. In April it issues from the pupa, which has lain under ground from the preceding September. The female of the gooseberry saw-fly does not, like some of the family, cut a groove in the branch to deposit her eggs;—"of what use, then," asks Réaumur, "is her ovipositor-saw?" In order to satisfy himself on this point, he introduced a pair of the flies under a bell-glass along with a branch bent from a red-currrant bush, that he might watch the process. The female immediately perambulated the leaves in search of a place suited to her purpose, and passing under a leaf began to lay, depositing six eggs within a quarter of an hour. Each time she placed herself as if she wished to cut into the leaf with her saw; but, upon taking out the leaf, the eggs appeared rather projecting than lodged in its substance. The cater-pillars are hatched in two or three weeks; and they feed in company till after midsummer, frequently stripping both the leaves and fruit of an extensive plantation. The caterpillar has six legs and sixteen prolegs, and is of a green colour mixed with yellow, and covered with minute black dots raised like shagreen. In its last skin it loses the black dots and becomes smooth and yellowish white. The Caledonian Horticultural Society have published a number of plans for destroying these caterpillars.

[Another remarkable mode of disposing of the pupa is shown in the accompanying illustration; it represents the nest of an exotic saw-fly, named *Deilocenes Ellisii*. In this instance, the numerous larvæ unite in spinning for them-selves a common envelope of considerable strength; it is seen as it appears when attached to the branch of a tree. The material of which it is composed is the tough silken fibre spun by the larvæ of so many insects, which may be seen in perfection in the cocoons of the Microgaster. Two species of this curious group will be described in a future page.

[By the side of the branch is seen a diagram of the same nest, as it would appear in section. The irregularly angular cells are seen in the centre, and around them is the common envelope composed of fibres. As may be seen from the upper figure, as soon as the insects have attained their perfect form, they gnaw their way out of the cell and the covering also. The insect is shown as it appears when flying.

[We will conclude this chapter by a few remarks upon some exotic insects, whose nests are not only remarkable in their form, but are valuable to the entomologist in affording grounds for the reception or rejection of certain familiar theories upon the subject of this volume—Insect Architecture. Several of these nests are of comparatively late discovery, and are therefore found in this work.

[The curious series of cells shown in the left-hand figure is made by a hymenopterous insect belonging to the genus Icaria, and the specimens from which the drawing was taken may be seen in the British Museum. They are made of a paper-like substance, much resembling in look the material of which the common wasp builds its cells, but as they are exposed to the air, they are necessarily tougher and stronger than ordinary wasp cells, which are shielded from the elements. The insects belonging to this genus make nests

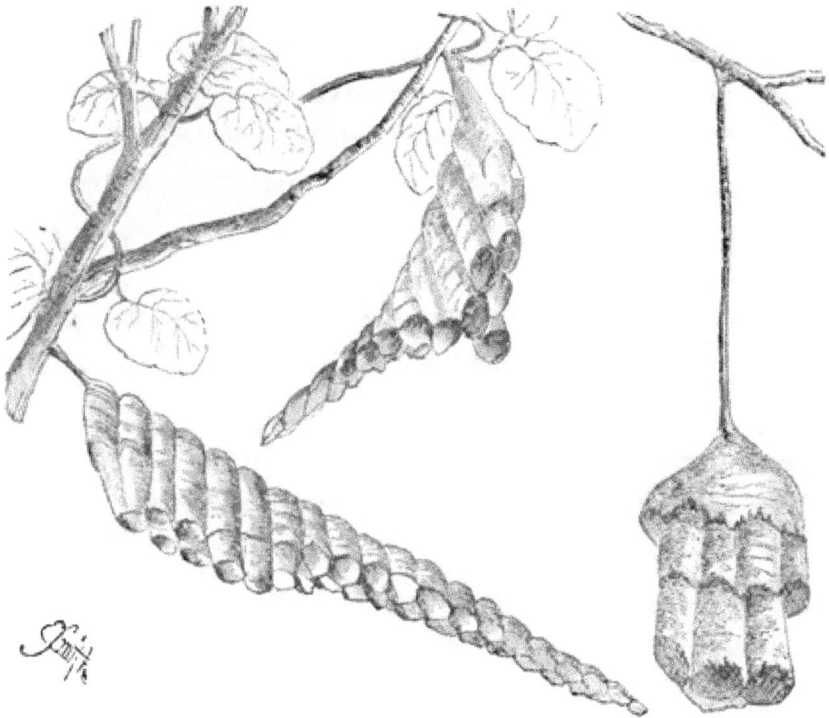

of very diverse forms, some of which are stuck on leaves in a most curious manner, reminding the observer of the parasitic mollercoids that cover the stems and fronds of large seaweeds. Others, however, are not dependent upon leaves for their support, but stand out boldly from the branches to which they are fixed, supported entirely by a foot-stalk composed of the same material as the cells, though necessarily of a harder and more compact substance.

[As many of these nests have been found in India, it is easy to trace the manner in which they were made. The mother insect began by kneading woody fibre into a paste, and making the footstalk of the future nest. One end of this footstalk is attached very strongly to the branch, and to the other end is fastened the first cell. As soon as the Icaria has made the first beginning of the cell, and raised—or rather lowered—the walls to a fourth or so of their complete dimensions, she inserts an egg into the yet imperfect cell, and adds to the walls while the egg is being hatched. Her next duty is, to add a second cell, and this is quickly followed by a third, all these cells being fastened to each other on three or four of their sides, leaving the others free and unattached. It is evident that by this mode of construction the cells nearest the branch must be the longest, because they are begun the soonest, and this will always be found to be the case.

[Now, there is a point respecting which the attention of the reader must be specially solicited. On looking at the cells, he will see that they are partly cylindrical and partly angular, and may perhaps think that this fact goes towards proving that the hexagonal shape of bee cells is owing to mutual pressure, the outer sides of the cells being rounded, while the inner are angular. But, there are other cells in existence, built by allied insects, and formed in an analogous manner, and which are either angular or cylindrical, exactly according to the instinctive powers of the insect which built them.

[On the right hand of the Icarian nest may be seen a singular-looking structure pendent at the end of a long footstalk. This is the nest of an insect called *Mischocyttarus labiatus*, one of the Polistidæ. In this case, the cells are built so as to be defended from the rain by a sort of penthouse, over which all the raindrops would run, and so fall harmless to the ground. The cells of this insect are soft in texture, and are more cylindrical than angular, the angles being but very slightly marked.

[Here, however, is the nest of an insect called *Raphigaster Guiniensis*, which is built in a manner similar to that of the Icaria, the cells being closely in contact with each other. The material of which they are made is peculiarly soft, something like very thin and flimsy grey paper. Consequently, they must press strongly upon each other, and we might reasonably expect to find that their angles are well

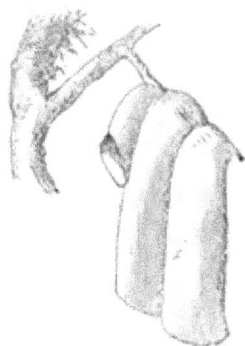

and boldly developed. But, instead of that, we find that they have no angles at all, but remain smooth and rounded throughout their length.

[Perhaps the most powerful argument against the equal pressure theory is to be found in the nest of a species of Icaria, which is shown in the accompanying illustration.

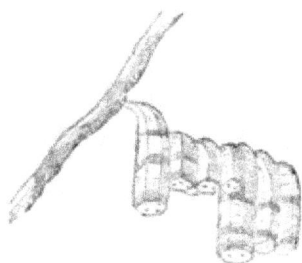

[As may be seen by reference to the illustration, the material of which they are made is so soft, that they bend over by their own weight, and therefore we might expect to find that they would follow the shape of the Raphigaster

and the Mischocyttarus.   But, we find that all the cells are
boldly angular, and that the angles are just as sharp on the
exterior of each cell as on the sides which cement the cells
together.   It is clear that the bold lines and decided angles
of these cells cannot have been produced mechanically, and
that they must have been intentionally formed by the
insect architect.

[One single cell, such as is here shown, is sufficient to
overthrow the theory of " equal pressure," by which insects
were deprived of all mechanical skill, and supposed to labour
like so many animated machines, without caring or knowing
anything about the work on which they were engaged.
According to the equal pressure theory, each of these cells
would have required six similar cells around it before it could
have assumed the hexagonal form, and yet we find that a cell
which is only connected with its neighbour by one side, has
its other five sides angular, and with the angles boldly defined.]

CHAPTER VIII.

LEAF-ROLLING CATERPILLARS.

THE labours of those insect-architects, which we have endeavoured to describe in the preceding pages, have been chiefly those of mothers to form a secure nest for their eggs, and the young hatched from them, during the first stage of their existence. But a much more numerous and not less ingenious class of architects may be found among the newly-hatched insects themselves, who, untaught by experience, and altogether unassisted by previous example, manifest the most marvellous skill in the construction of tents, houses, galleries, covert-ways, fortifications, and even cities, not to speak of subterranean caverns and subaqueous apartments, which no human art could rival.

The caterpillars, which are familiarly termed leaf-rollers, are perfect hermits. Each lives in a cell, which it begins to construct almost immediately after it is hatched; and the little structure is at once a house which protects the caterpillar from its enemies, and a store of food for its subsistence, while it remains shut up in its prison. But the insect only devours the inner folds. The art which these caterpillars exercise, although called into action but once, perhaps, in their lives, is perfect. They accomplish their purpose with a mechanical skill, which is remarkable for its simplicity and unerring success. The art of rolling leaves into a secure and immovable cell may not appear very difficult: nor would it be so if the caterpillars had fingers, or any parts which were equivalent to those delicate and admirable natural instruments with which man accomplishes his most elaborate works. And yet the human fingers could not roll a rocket-case of paper more regularly than the caterpillar rolls his house of leaves. A leaf is not a very easy substance to roll. In some trees it is very brittle. It has also a natural elasticity,—a disposition to spring back if it be bent,—which is caused by the continuity of its threads, or nervures. This elasticity is speedily overcome by the ingenuity with which the caterpillar works; and the leaf is thus retained in its artificial position for many weeks, under every variety of temperature. We will examine, in detail, how these little leaf-rollers accomplish their task.

One of the most common as well as the most simple fabrics constructed by caterpillars, may be discovered during summer on almost every kind of bush and tree. We shall take as examples those which are found on the lilac and on the oak.

A small but very pretty chocolate-coloured moth, abundant

Lilac-tree Moth.　(*Lozotænia rileana*, STEPHENS?)

in every garden, but not readily seen, from its frequently alighting on the ground, which is so nearly of its own colour,

Nest of a Lilac-leaf Roller.

deposits its eggs on the leaves of the currant, the lilac, and of some other trees, appropriating a leaf to each egg. As soon as the caterpillar is hatched, it begins to secure itself from birds and predatory insects by rolling up the lilac leaf into the form of a gallery, where it may feed in safety. We have repeatedly seen one of them when just escaped from the egg, and only a few lines long, fix several silk threads from one edge of a leaf to the other, or from the edge to the mid-rib; then going to the middle of the space, he shortened the threads by bending them with his feet, and consequently pulled the edges of the leaves into a circular form; and he retained them in that position by gluing down each thread as he shortened it. In their younger state, those caterpillars seldom roll more than a small portion of the leaf; but, when

Another nest of Lilac-leaf Roller.

farther advanced, they unite the two edges together in their whole extent, with the exception of a small opening at one end, by which an exit may be made in case of need.

Another species of caterpillar, closely allied to this, rolls up the lilac leaves in a different form, beginning at the end of a leaf, and fixing and pulling its threads till it gets it nearly into the shape of a scroll of parchment. To retain this form more securely, it is not contented, like the former insect, with threads fixed on the inside of the leaf; but has also recourse to a few cables which it weaves on the outside.

Another species of moth, allied to the two preceding, is of

Small green Oak-moth. (*Tortrix viridana.*)

a pretty green colour, and lays its eggs upon the leaves of the oak. This caterpillar folds them up in a similar manner, but with this difference, that it works on the under surface of the leaf, pulling the edge downwards and backwards, instead of forwards and upwards. This species is very

Nests of Oak-leaf-rolling Caterpillars.

abundant, and may readily be found as soon as the leaves expand. In June, when the perfect insect has appeared, by beating a branch of an oak, a whole shower of these pretty green moths may be shook into the air.

Among the leaf-rolling caterpillars, there is a small dark-brown one, with a black head and six feet, very common in gardens, on the currant-bush, or the leaves of the rose-tree (*Lozotænia rosana*, STEPHENS). It is exceedingly destructive to the flower-buds. The eggs are deposited in the summer, and probably also in the autumn or in spring, in little oval or circular patches of a green colour. The grub makes its appearance with the first opening of the leaves, of whose structure in the half-expanded state it takes advantage to construct its summer tent. It is not, like some of the other leaf-rollers, contented with a single leaf, but weaves together as many as there are in the bud where it may chance to have been hatched, binding their discs so firmly with silk, that all the force of the ascending sap, and the increasing growth of the leaves, cannot break through; a farther expansion is of course prevented. The little inhabitant in the meanwhile banquets securely on the partitions of its tent, eating door-ways from one apartment into another, through which it can escape in case of danger or disturbance.

The leaflets of the rose, it may be remarked, expand in nearly the same manner as a fan, and the operations of this ingenious little insect retain them in the form of a fan nearly shut. Sometimes, however, it is not contented with one bundle of leaflets, but by means of its silken cords unites all which spring from the same bud into a rain-proof canopy, under the protection of which it can feast on the flower-bud, and prevent it from ever blowing.

In the instance of the currant-leaves, the proceedings of the grub are the same; but it cannot unite the plaits so smoothly as in the case of the rose leaflets, and it requires more labour, also, as the nervures, being stiff, demand a greater effort to bend them. When all the exertions of the insect prove unavailing in its endeavours to draw the edges of a leaf together, it bends them inwards as far as it can, and weaves a close web of silk over the open space between.

This is well exemplified in one of the commonest of our leaf-rolling caterpillars, which may be found as early as February on the leaves of the nettle and the white archangel (*Lamium album*). It is of a light dirty-green colour, spotted with black, and covered with a few hairs. In its young state it confines itself to the bosom of a small leaf, near the insertion of the leaf-stalk, partly bending the edges inwards, and covering in the interval with a silken curtain. As this sort of covering is not sufficient for concealment when the animal advances in growth, it abandons the base of the leaf for the middle, where it doubles up one side in a very secure and ingenious manner.

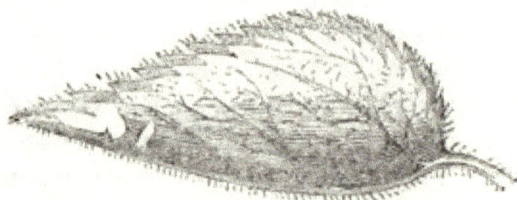

Nest of the Nettle-leaf-rolling Caterpillar.

We have watched this little architect begin and finish his tent upon a nettle in our study, the whole operation taking more than half an hour. (J. R.) He began by walking over the plant in all directions, examining the leaves severally, as if to ascertain which was best fitted for his purpose by being pliable, and bending with the weight of his body. Having found one to his mind, he placed himself along the mid-rib, to the edge of which he secured himself firmly with the pro-legs of his tail; then stretching his head to the edge of the leaf, he fixed a series of parallel cables between it and the mid-rib, with another series crossing these at an acute angle. The position in which he worked was most remarkable, for he did not, as might have been supposed, spin his cables with his face to the leaf, but throwing himself on his back, which was turned towards the leaf, he hung with his whole weight by his first-made cables. This, by drawing them into the form of a curve, shortened them, and consequently pulled the edge of the leaf down towards the mid-rib. The weight of his body was not, however, the only power which he employed ; for, using the terminal pro-legs as a point of support, he exerted the whole muscles of his body to shorten his threads, and pull down the edge of the leaf. When he had drawn the threads as tight as he could, he held them till he spun fresh ones of sufficient strength to retain the leaf in the bent position into which he had pulled it. He then left the first series to hang loose while he shortened the fresh-spun ones as before. This process was continued till he had worked down about an inch and a half of the leaf, as much as he deemed sufficient for his habitation. This was the first part of the architecture.

By the time he had worked to the end of the fold, he had brought the edge of the leaf to touch the mid-rib ; but it was only held in this position by a few of the last-spun threads, for all the first-spun ones hung loose within. Apparently aware of this, the insect protruded more than half of its body through the small aperture left at the end, and spun several bundles of threads on the outside precisely similar to those ropes of a tent which extend beyond the canvas, and are pegged into the ground. Unwilling to trust the exposure of his whole body on the outside, lest he should be seized by the first sand-wasp (*odynerus*) or sparrow which might descry him, he now withdrew to complete the internal portion of his dwelling, where the threads were hanging loose and disorderly. For this purpose he turned his head about, and proceeded precisely as he had done at the beginning of his task, but taking care to spin his new threads so as to leave the loose ones on the outside, and make his apartment smooth and neat. When he again reached the opposite end, he constructed there also a similar series of cables on the outside, and then withdrew to give some final touches to the interior.

It is said by Kirby and Spence,* that when these leaf-rolling insects find that the larger nervures of the leaves are so strong as to prevent them from bending, they " weaken it by gnawing it here and there half through." We have never

* Introd., vol. i, p. 457.

observed the circumstance, though we have witnessed the
process in some hundreds of instances; and we doubt the
statement, from the careful survey which the insect makes
of the capabilities of the leaf before the operation is begun.
If she found upon examination that a leaf would not bend,
she would reject it, as we have often seen happen, and pass to
another. (J. R.)

A species of leaf-roller, of the most diminutive size,
merits particular mention, although it is not remarkable
in colour or figure. It is without hair, of a greenish-white,
and has all the vivacity of the other leaf-rollers. Sorrel is
the plant on which it feeds; and the manner in which it
rolls a portion of the leaf is very ingenious.

The structure which it contrives is a sort of conical
pyramid, composed of five or six folds lapped round each
other. From the position of this little cone the caterpillar

Leaf-rolling Caterpillars of the Sorrel.

has other labours to perform, beside that of rolling the leaf.
It first cuts across the leaf, its teeth acting as a pair of
scissors; but it does not entirely detach this segment. It
rolls it up very gradually, by attaching threads of silk to
the plane surface of the leaf, as we have before seen , and
then, having cut in a different direction, sets the cone upright,

by weaving other threads, attached to the centre of the roll
and the plane of the leaf, upon which it throws the weight
of its body. This, it will be readily seen, is a somewhat
complicated effort of mechanical skill. It has been minutely
described by M. Réaumur; but the preceding representation
will perhaps make the process clearer than a more detailed
account.

This caterpillar, like those of which we have already
spoken, devours all the interior of the roll. It weaves,
also, in the interior, a small and thin cocoon of white silk,
the tissue of which is made compact and close. It is then
transformed into a chrysalis.

The caterpillars of two of our largest and handsomest
butterflies, the painted lady (*Cynthia cardui,* STEPHENS),
and the admiral, or *Alderman* of the London fly-fanciers

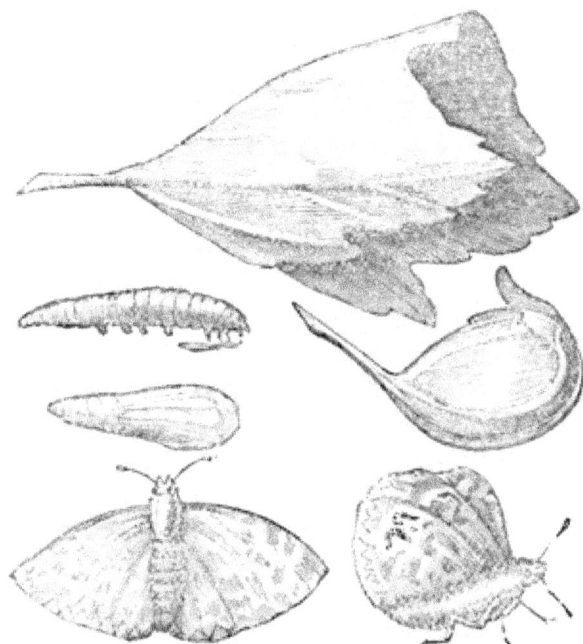

Nests of the Hesperia malvæ, with Caterpillar, Chrysalis, and Butterflies.

(*Vanessa atalanta*), are also leaf-rollers. The first selects
the leaves of the great spear-thistle, and sometimes those
of the stemless or star-thistle, which might be supposed

rather difficult to bend; but the caterpillar is four times as large and strong as those which we have been hitherto describing. In some seasons it is plentiful; in others it is rarely to be met with: but the admiral is seldom scarce in any part of the country; and by examining the leaves of nettles which appear folded edge to edge, in July and August, the caterpillar may be readily found.

Another butterfly (*Hesperia malvæ*) is met with on dry banks where mallows grow, in May, or even earlier, and also in August, but is not indigenous. The caterpillar, which is grey, with a black head, and four sulphur-coloured spots on the neck, folds around it the leaves of the mallow, upon which it feeds. There is nothing, however, peculiarly different in its proceedings from those above described; but the care with which it selects and rolls up one of the smaller leaves, when it is about to be transformed into a chrysalis, is worthy of remark; it joins it, indeed, so completely round and round, that it has somewhat the resemblance of an egg. Within this green cell it lies secure, till the time arrives when it is ready to burst its cerements, and trust to the quickness of its wings for protection against its enemies.

Among the nests of caterpillars which roll up *parcels* of leaves, we know none so well contrived as those which are found upon willows and a species of osier. The long and narrow leaves of these plants are naturally adapted to be adjusted parallel to each other; for this is the direction which they have at the end of each stalk, when they are not entirely developed. One kind of small smooth caterpillar (*Tortrix chlorana*), with sixteen feet, the under part of which is brown, and streaked with white, fastens these leaves together, and makes them up into parcels. There is nothing particularly striking in the mechanical manner in which it constructs them. It does precisely what we should do in a similar case: it winds a thread round those leaves which must be kept together, from a little above their termination to a very short distance from their extreme point; and as it finds the leaves almost constantly lying near each other, it has little difficulty in bringing them together, as is shown in the following cut, *a*.

The prettiest of these parcels are those which are made upon a kind of osier, the borders of whose leaves sometimes form columnar bundles before they become developed. A section of these leaves has the appearance of filigree-work (see *b*).

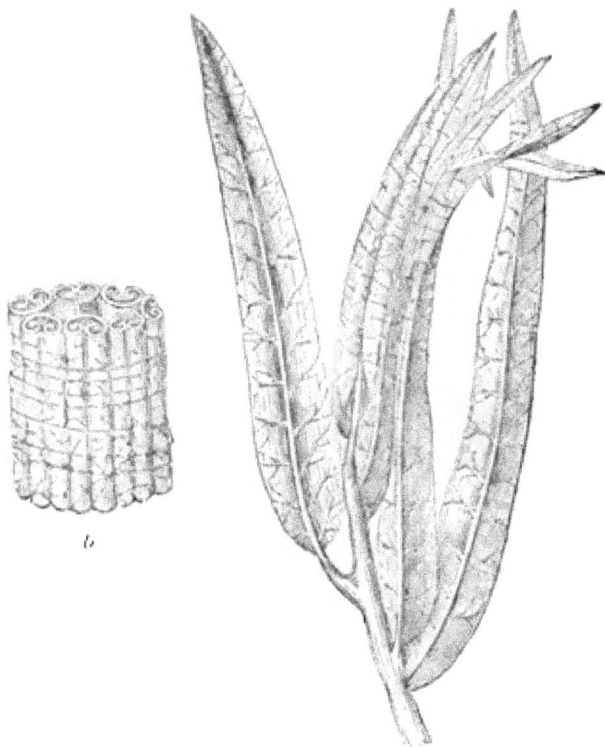

Nest of Willow-leaf Roller.

A caterpillar which feeds upon the willow, and whose singular attitudes have obtained for it the trivial name of *Ziczac*, also constructs for itself an arbour of the leaves, by drawing them together in an ingenious manner. M. Roesel * has given a tolerable representation of this nest, and of the caterpillar. The caterpillar is found in June;

* Roesel, cl. ii., Pap. Nocturn., tab. xx. fig. 1, 2, 3, 4, 5, 6.

and the moth (*Notodonta ziczac*) from May to July in the
following year (see cut, p. 151).

Beside those caterpillars which live solitary in the folds
of a leaf, there are others which associate, employing their
united powers to draw the leaves of the plants they feed
upon into a covering for their common protection. Among
these we may mention the caterpillar of a small butterfly,
the plantain or Glanville fritillary (*Melitea cinxia*), which is
very scarce in this country.

Although a colony of these caterpillars is not numerous,
seldom amounting to a hundred individuals, the place which

Ziczac Caterpillar and Nest.

they have selected is not hard to discover. Their abode
may be seen in the meadow in form of a tuft of herbage
covered with a white web, which may readily be mistaken,
at first view, for that of a spider, but closer inspection soon
corrects this notion. It is, in fact, a sort of common tent,

in which the whole brood lives, eats, and undergoes the usual transformations. The shape of this tent, for the most part, approaches the pyramidal, though that depends much upon the natural growth of the herbage which composes it. The interior is divided into compartments formed by the union of several small tents, as it were, to which others have been from time to time added according to the necessities of the community.

When they have devoured all the leaves, or at least those which are most tender and succulent, they abandon their first camp, and construct another contiguous to it under a tuft of fresh leaves. Several of these encampments may sometimes be seen within the distance of a foot or two, when they can find plantain (*Plantago lanceolata*) fit for their purpose; but though they prefer this plant, they content themselves with grass if it is not to be procured.

When they are about to cast their skins, but particularly when they perceive the approach of winter, they construct a more durable apartment in the interior of their principal tent. The ordinary web is thin and semi-transparent, permitting the leaves to be seen through it; but their winter canvas, if we may call it so, is thick, strong, and quite opaque, forming a sort of circular hall without any partition, where the whole community lie coiled up and huddled together.

Early in spring they issue forth in search of fresh food, and again construct tents to protect them from cold and rain, and from the mid-day sun.

M. Réaumur found upon trial, that it was not only the caterpillars hatched from the eggs of the same mother which would unite in constructing the common tent; for different broods, when put together, worked in the same social and harmonious manner. We ourselves ascertained, during the present summer (1829), that this principle of sociality is not confined to the same species, nor even to the same genus. The experiment which we tried was to confine two broods of different species to the same branch, by placing it in a glass of water to prevent their escape.

The caterpillars which we experimented on were several broods of the brown-tail moth (*Porthesia auriflua*) and the lackey (*Clisiocampa neustria*). These we found to work with as much industry and harmony in constructing the common tent as if they had been at liberty on their native trees; and when the lackeys encountered the brown-tails they manifested no alarm nor uneasiness, but passed over the backs of one another, as if they had made only a portion of the branch. In none of their operations did they seem to be subject to any discipline, each individual appearing to work, in perfecting the structure, from individual instinct, in the same manner as was remarked by M. Huber in the case of the hive-bees. In making such experiments, it is obvious that the species of caterpillars experimented with must feed upon the same sort of plant.* (J. R.)

The design of the caterpillars in rolling up the leaves is not only to conceal themselves from birds and predatory insects, but also to protect themselves from the cuckoo-flies, which lie in wait in every quarter to deposit their eggs in their bodies, that their progeny may devour them. Their mode of concealment, however, though it appear to be cunningly contrived and skilfully executed, is not always successful, their enemies often discovering their hiding-place. We happened to see a remarkable instance of this last summer (1828), in the case of one of the lilac caterpillars which had changed into a chrysalis within the closely-folded leaf. A small ichneumon, aware it should seem of the very spot where the chrysalis lay within this leaf, was seen boring through it with her ovipositor, and introducing her eggs through the punctures thus made into the body of the dormant insect. We allowed her to lay all her eggs, about six in number, and then put the leaf under an inverted glass. In a few days the eggs of the cuckoo-fly were hatched, the grubs devoured the lilac chrysalis, and finally changed into pupæ in a case of yellow silk, and into perfect insects like their parents. (J. R.)

* See p. 100.

CHAPTER IX.

THE habitations of the insects which we have just described consist of growing leaves, bent, rolled, or pressed together, and fixed in their positions by silken threads. But there are other habitations of a similar kind, which are constructed by cutting out and detaching a whole leaf, or a portion of a leaf. We have already seen how dexterously the upholsterer-bees cut out small parts of leaves and petals with their mandibles, and fit them into their cells. Some of the caterpillars do not exhibit quite so much neatness and elegance as the leaf-cutting bees, though their structures answer all the purposes intended; but there are others, as we shall presently see, that far excel the bees, at least in the delicate minutiæ of their workmanship. We shall first advert to those structures which are the most simple.

Not far from Longchamps, in a road through the Bois de Boulogne, is a large marsh, which M. Réaumur never observed to be in a dry state even during summer. This marsh is surrounded with very lofty oaks, and abounds with pondweed, the water-plant named by botanists *potamogeton*. The shining leaves of this plant, which are as large as those of the laurel or orange-tree, but thicker and more fleshy, are spread upon the surface of the water. Having pulled up several of these about the middle of June, M. Réaumur observed, beneath one of the first which he examined, an elevation of an oval shape, which was formed out of a leaf of the same plant. He carefully examined it, and discovered that threads of silk were attached to this elevation. Breaking the threads, he raised up one of the ends, and saw a cavity. in which a caterpillar (*Hydrocampa potamogeta*) was lodged.

An indefatigable observer, such as M. Réaumur, would naturally follow up this discovery; and he has accordingly given us a memoir of the pondweed tent-maker, distinguished by his usual minute accuracy.

In order to make a new habitation, the caterpillar fastens itself on the under side of a leaf of the *Potamogeton*. With its mandibles it pierces some part of this leaf, and afterwards gradually gnaws a curved line, marking the form of the piece which it wishes to detach. When the caterpillar has cut off, as from a piece of cloth, a patch of leaf of the size and shape suited to its purpose, it is provided with half of the materials requisite for making a tent. It takes hold of this piece by its mandibles, and conveys it to the situation on the under side of its own or another leaf, whichever is found most appropriate. It is there disposed in such a manner that the under part of the patch—the side which was the under part of the entire leaf—is turned towards the under part of the new leaf, so that the inner walls of the cell or tent are always made by the under part of two portions of leaf. The leaves of the potamogeton are a little concave on the under side; and thus the caterpillar produces a hollow cell, though the rims are united.

The caterpillar secures the leaf in its position by threads of white silk. It then weaves in the cavity a cocoon, which is somewhat thin, but of very close tissue. There it shuts itself up, to appear again only in the form of the perfect insect, and is soon transformed into a chrysalis. In this cocoon of silk no point touches the water; whilst the tent of leaves, lined with silk, has been constructed underneath the water. This fact proves that the caterpillar has a particular art by which it repels the water from between the leaves.

When the caterpillar, which has thus conveyed and disposed a patch of leaf against another leaf, is not ready to be transformed into a chrysalis, it applies itself to make a tent or habitation which it may carry everywhere about with it. It begins by slightly fixing the piece against the perfect leaf, leaving intervals all round, between the piece and leaf, at which it may project its head. The piece which it has fixed serves as a model for cutting out a similar piece in the other leaf. The caterpillar puts them accurately together, except at one end of the oval, where an opening is left for the insect to project its head through. When the caterpillar is inclined to change its situation, it draws itself forward by means of its scaly limbs, riveted upon the leaf. The membraneous limbs, which are riveted against the inner sides of the tents, oblige it to follow the anterior part of the body, as it advances. The caterpillar, also, puts its head out of the tent every time it desires to eat.

There is found on the common chickweed (*Stellaria media*), towards the end of July, a middle-sized smooth green caterpillar, having three brown spots bordered with white on the back, and six legs and ten pro-legs, whose architecture is worthy of observation. When it is about to go into chrysalis, towards the beginning of August, it gnaws off, one by one, a number of the leaves and smaller twigs of the chickweed, and adjusts them into an oval cocoon, somewhat rough and unfinished externally, but smooth, uniform, and finely tapestried with white silk within. Here it undergoes its transformation securely, and, when the period of its pupa trance has expired in the following July, it makes its exit in the form of a yellowish moth, with several brown spots above, and a brown band on each of its four wings below. It is also furnished with a sort of tail.

On the cypress-spurge (*Euphorbia cyparissias*), a native woodland plant, but not of very common occurrence, may be found, towards the end of October, a caterpillar of a middle size, sparely tufted with hair, and striped with black, white, red, and brown. The leaves of the plant, which are in the form of short narrow blades of grass, are made choice of by the caterpillar to construct its cocoon, which it does with great neatness and regularity, the end of each leaf, after it has been detached from the plant, being fixed to the stem, and the other leaves placed parallel, as they are successively added. The other ends of all these are bent inwards, so as to form a uniformly rounded oblong figure, somewhat larger at one end than at the other.

A caterpillar which builds a very similar cocoon to the last-mentioned may be found upon a more common plant—the yellow snap-dragon or toad-flax (*Antirrhinum linaria*)—which is to be seen in almost every hedge. It is somewhat shaped like a leech, is of a middle size, and the prevailing colour pearl-grey, but striped with yellow and black. It spins up about the beginning of September, forming the outer coating of pieces of detached leaves of the plant, and sometimes of whole leaves placed longitudinally, the whole disposed with great symmetry and neatness. The moth appears in the following June.

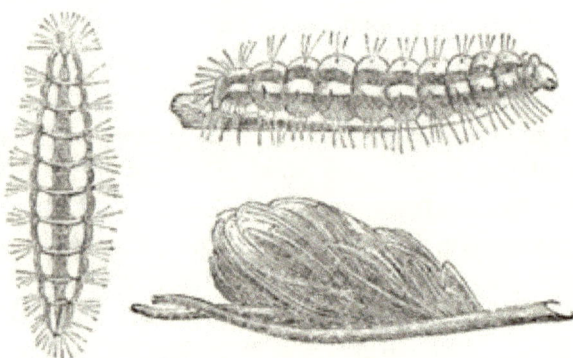

Cypress-Spurge Caterpillar – (*Acronycta Euphrasiæ*)—with a Cocoon, on a branch.

It is worthy of remark, as one of the most striking instances of instinctive foresight, that the caterpillars which build structures of this substantial description are destined to lie much longer in their chrysalis trance than those which spin merely a flimsy web of silk. For the most part, indeed, the latter undergo their final transformation in a few weeks; while the former continue entranced the larger portion of a year, appearing in the perfect state the summer after their architectural labours have been completed. (J. R.) This is a remarkable example of the instinct which leads these little creatures to act as if under the dictates of prudence, and with a perfect knowledge of the time, be it long or short, which will elapse before the last change of the pupa takes place. That the caterpillar, while weaving its cocoon and preparing to assume the pupa state, exercises any reflective faculties, or is aware of what is about to occur relative to its own self, we cannot admit. It enters upon a work of which it has had no previous experience, and which is performed, as far as contingencies allow, in the same manner by every caterpillar of the same species. Its labours, its mode of carrying them on, and the very time in which they are to be commenced, are all pre-appointed; and an instinctive impulse urges and guides; and with this instinct its organic endowments are in precise harmony; nor does instinct ever impel to labours for which an animal is not provided. "The same wisdom," says Bonnet, "which has constructed and arranged with so much art the various organs of animals, and has made them concur towards one determined end, has also provided that the different operations which are the natural results of the economy of the animal should concur towards the same end. The creature is directed towards his object by an invisible hand; he executes with precision, and by one effort, those works which we so much admire; he appears to act as if he reasoned, to return to his labour at the proper time, to change his scheme in case of need. But in all this he only obeys the secret influence which drives him on. He is but an instrument which cannot judge of each action, but is wound up by that adorable Intelligence, which has traced out for every insect its proper labours, as he has traced the orbit of each planet. When, therefore, I see an insect working at the construction of a nest, or a cocoon, I am impressed with respect, because it seems to me that I am at a spectacle where the Supreme Artist is hid behind the curtain."*

There is a small sort of caterpillar which may be found on old walls, feeding upon minute mosses and lichens, the proceedings of which are well worthy of attention. They are similar, in appearance and size, to the caterpillar of the small cabbage-butterfly (*Pontia rapæ*), and are smooth and bluish. The material which they use in building their cocoons is composed of the leaves and branchlets of green

* Contemplation de la Nature, part xv. chap. 38.

moss, which they cut into suitable pieces, detaching at the same time along with them a portion of the earth in which they grow. They arrange these upon the walls of their building, with the moss on the outside, and the earth on the inside, making a sort of vault of the tiny bits of green moss turf, dug from the surface of the wall. So neatly, also, are the several pieces joined, that the whole might well be supposed to be a patch of moss which had grown in form of an oval tuft, a little more elevated than the rest growing on the wall. When these caterpillars are shut up in a box with some moss, without earth, they construct with it cells in form of a hollow ball, very prettily plaited and interwoven.

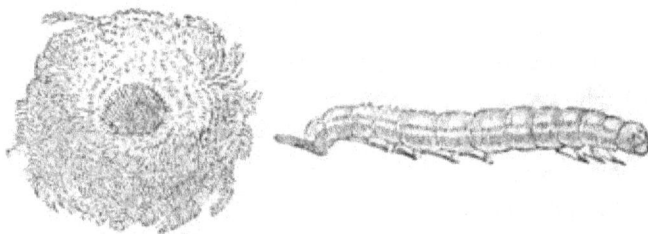

Moss-Cell of small Caterpillar (*Bryophila perla?*)

In May last (1829), we found on the walls of Greenwich Park a great number of caterpillars, whose manners bore some resemblance to those of the grub described by M. Réaumur. (J. R.) They were of middle size, with a dull-orange stripe along the back; the head and sides of the body black, and the belly greenish. Their abodes were constructed with ingenuity and care. A caterpillar of this sort appears to choose either a part where the mortar contains a cavity, or it digs one suited to its design. Over the opening of the hollow in the mortar it builds an arched wall, so as to form a chamber considerably larger than is usual with other architect caterpillars. It selects grains of mortar, brick, or lichen, fixing them, by means of silk, firmly into the structure. As some of these vaulted walls were from an inch to an inch and a half long, and about a third of an inch wide and deep, it may be well imagined that it would require no little industry and labour to complete the work; yet it does not demand more than a few hours for the insect to raise it from the foundation. Like all other insect architects, this caterpillar uses its own body for a measuring-rule, and partly for a mould, or rather a block or centre to shape the walls by, curving itself round and round concentrically with the arch which it is building.

We afterwards found one of these caterpillars, which had dug a cell in one of the softest of the bricks, covering itself on the outside with an arched wall of brick-dust, cemented with silk. As this brick was of a bright-red colour, we were thereby able to ascertain that there was **not a particle** of lichen employed in the structure.

The neatness mentioned by Réaumur, as remarkable in his moss-building caterpillars, is equally observable in that which we have just described; for, on looking at the surface of the wall, it would be impossible for a person unacquainted with those structures to detect where they were placed, as they are usually, on the outside, level with the adjoining brick-work; and it is only when they are opened by the entomologist, that the little architect is perceived lying snug in his chamber. If a portion of the wall be thus broken down, the caterpillar immediately commences repairing the breach, by piecing in bits of mortar and fragments of lichen, till we can scarcely distinguish the new portion from the old.

CHAPTER X.

CADDIS-WORMS AND CARPENTER-CATERPILLARS.

THERE is a very interesting class of grubs which live under water, where they construct for themselves moveable tents of various materials as their habits direct them, or as the substances they require can be conveniently procured. Among the materials used by these singular grubs, well-known to fishermen by the name of *caddis-worms*, and to naturalists as the *larvæ* of the four-winged flies in the order *Trichoptera* of Kirby and Spence, we may mention sand, stones, shells, wood, and leaves, which are skilfully joined and strongly cemented. One of these grubs forms a pretty case of leaves glued together longitudinally, but leaving an aperture sufficiently large for the inhabitant to put out its head and shoulders when it wishes to look about for food.

Leaf Nest of Caddis-Worm.

Another employs pieces of reed cut into convenient lengths, or of grass, straw, wood, &c., carefully joining and cementing each piece to its fellow as the work proceeds; and he frequently finishes the whole by adding a broad piece longer

Reed Nest of Cadd s-Worm.

than the rest to shade his door-way overhead, so that he may not be seen from above. A more laborious structure is reared by the grub of a beautiful caddis-fly (*Phryganea*), which weaves together a group of the leaves of aquatic plants into a roundish ball, and in the interior of this forms a cell for its abode. The following figure from Roesel will give a more precise notion of this structure than a lengthened description.

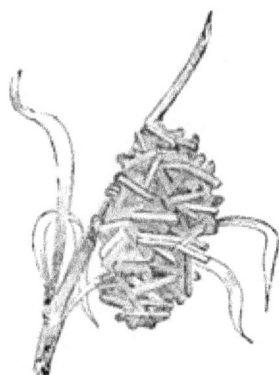

Another of these aquatic architects makes choice of the tiny shells of young fresh-water mussels and snails (*Planorbis*), to form a moveable grotto ; and as these little shells are for the most part inhabited, he keeps the poor animals close

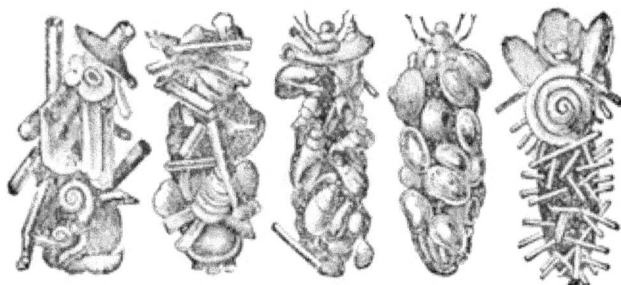

Shell Nests of Caddis-Worms.

prisoners, and drags them without mercy along with him. These grotto-building grubs are by no means uncommon in ponds ; and in chalk districts, such as the country about Woolwich and Gravesend, they are very abundant.

One of the most surprising instances of their skill occurs in the structures of which small stones are the principal material. The problem is to make a tube about the width

of the hollow of a wheat-straw or a crow-quill, and equally
smooth and uniform. Now the materials being small stones
full of angles and irregularities, the difficulty of performing
this problem will appear to be considerable, if not insur-
mountable : yet the little architects, by patiently examining
their stones and turning them round on every side, never
fail to accomplish their plans. This, however, is only part

Stone Nest of Caddis-Worm.

of the problem, which is complicated with another condition,
and which we have not found recorded by former observers,
namely, that the under-surface shall be flat and smooth,
without any projecting angles which might impede its progress when dragged along the bottom of the rivulet where it
resides. The selection of the stones, indeed, may be accounted for, from this species living in streams where, but
for the weight of its house, it would to a certainty be swept
away. For this purpose, it is probable that the grub makes
choice of larger stones than it might otherwise want ; and
therefore also it is that we frequently find a case composed
of very small stones and sand, to which, when nearly

Sand Nest balanced with a Stone.

finished, a large stone is added by way of ballast. In other
instances, when the materials are found to possess too great
specific gravity, a bit of light wood, or a hollow straw, is
added to buoy up the case.

Nest of Caddis-Worm balanced with Straws.

It is worthy of remark, that the cement, used in all these

cases, is superior to pozzolana * in standing water, in which it is indissoluble. The grubs themselves are also admirably adapted for their mode of life, the portion of their bodies which is always enclosed in the case being soft like a meal-worm, or garden-caterpillar, while the head and shoulders, which are for the most part projected beyond the door-way in search of food, are firm, hard, and consequently less liable to injury than the protected portion, should it chance to be exposed.

We have repeatedly tried experiments with the inhabitants of those aquatic tents, to ascertain their mode of building. We have deprived them of their little houses, and furnished them with materials for constructing new ones, watching their proceedings from their laying the first stone or shell of the structure. They work at the commencement in a very clumsy manner, attaching a great number of chips to whatever materials may be within their reach with loose threads of silk, and many of these they never use at all in their perfect building. They act, indeed, much like an unskilful workman trying his hand before committing himself upon an intended work of difficult execution. Their main intention is, however, to have abundance of materials within reach: for after their dwelling is fairly begun, they shut themselves up in it, and do not again protrude more than half of their body to procure materials; and even when they have dragged a stone, a shell, or a chip of reed within building reach, they have often to reject it as unfit. (J. R.)

[We have here some examples of the latter kind of nest, *i. e.*, those habitations which are made of stones and shells. Beginning at the upper left-hand figure, we find one that is made of moderately-sized stones cemented together in a way that reminds the observer of the manner in which a builder forms irregular stones into a wall. Next to it is another, in which the stones are larger and narrower, and are arranged much as some of the caddis-worms arrange pieces of stick and straw.

* A cement prepared of volcanic earth, or lava.

In the second, and on the left-hand side, is a very long
and simple tube, made of a grass stem, and balanced by three
little sticks attached to its centre. The next figure represents
a number of sand-tubes attached to each other. These are
built up laboriously of single particles of sand, and are
remarkable for their peculiar horn-like shape, the tube
having the same regular curve as the horn of an ox or
antelope, and tapering gradually from the base to the top.
A somewhat similar tube, but of larger size, is shown in the
right-hand figure.

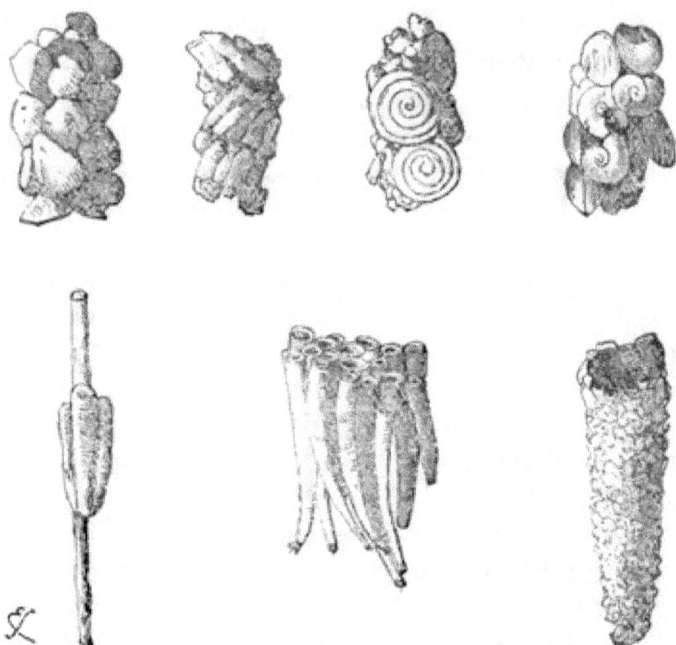

Any one who wishes to see one of these creatures rebuild
its house can do so by carefully removing it from its tube,
and supplying it with fresh material. Very great care must
be taken in the removal, as the grub is easily damaged, and
it holds so tightly to the tube with a pair of pincers at
the end of its body, that it must rather be coaxed than
driven out.

If desirable, they can be made to build their new houses

of most singular materials. A lady, Miss Smee, was very successful in a series of experiments which she made with these insects, forcing them to make tubes of different colours and patterns, by supplying them with coloured sand, pieces of stained glass of various hues, gold dust, and similar materials. Although there was scarcely any material which they would not use, they seemed to consider a certain amount of angularity as essential, and rejected any object, such as a bead, of which the surface was perfectly rounded, while they would accept the same, if it were broken or indented.

When the caddis-grub has ceased from feeding, and is about to pass into the perfect stage, it spins over the mouth of the tube a strong silken web. This web is made in quite a pretty pattern, and being woven with rather wide meshes, it allows the water to flow through the tube while it prevents any aquatic foes from penetrating and destroying the pupa.

The remaining figures of the illustration represent tubes, around which are built a quantity of small shells. Generally, stones are mixed with the shells; but in some cases, shells seem to be almost the only material.]

CARPENTER-CATERPILLARS.

Insects, though sometimes actuated by an instinct apparently blind, unintelligent, or unknown to themselves, manifest in other instances a remarkable adaptation of means to ends. We have it in our power to exemplify this in a striking manner by the proceedings of the caterpillar of a goat-moth (*Cossus ligniperda*) which we kept till it underwent its final change.

This caterpillar, which abounds in Kent and many other parts of the island, feeds on the wood of willows, oaks, poplars, and other trees, in which it eats extensive galleries; but it is not contented with the protection afforded by these galleries during the colder months of winter, before the arrival of which it scoops out a hollow in the tree, if it do not find one ready prepared, sufficiently large to contain its body in a bent or somewhat coiled-up position. On sawing off a portion of an old poplar in the winter of 1827, we found such a cell with a caterpillar coiled up in it.

Caterpillar of Goat-Moth in a Willow Tree.

It had not, however, been contented with the bare walls of the retreat which it had hewn out of the tree, for it had lined it with a fabric as thick as coarse broadcloth, and equally warm, composed of the raspings of the wood scooped

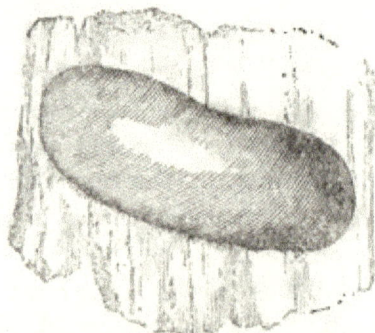

Winter Nest of the Goat-Caterpillar.

out of the cell, united with the strong silk which every species of caterpillar can spin. In this snug retreat our caterpillar, if it had not been disturbed, would have spent the winter without eating; but upon being removed into a warm room and placed under a glass along with some pieces of wood, which it might eat if so inclined, it was roused for a time from its dormant state, and began to move about. It was not long, however, in constructing a new cell for itself, no less ingenious than the former. It either could not gnaw into the fir plank, where it was now placed with a glass above it, or it did not choose to do so; for it left it untouched, and made it the basis of the edifice it began to construct. It formed, in fact, a covering for itself precisely like the one from which we had dislodged it,— composed of raspings of wood detached for the purpose from what had been given it as food, the largest piece of which was employed as a substantial covering and protection for the whole. It remained in this retreat, motionless, and without food, till revived by the warmth of the ensuing spring, when it gnawed its way out, and began to eat voraciously, to make up for its long fast.

These caterpillars are three years in arriving at their final change into the winged state; but as the one just mentioned was nearly full grown, it began, in the month of May, to prepare a cell, in which it might undergo its metamorphosis. Whether it had actually improved its skill in architecture by its previous experience we will not undertake to say, but its second cell was greatly superior to the first. In the first there was only one large piece of

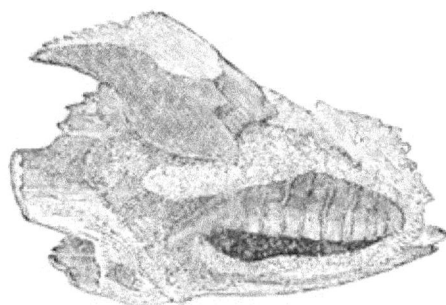

Nest of Goat-Moth.—Figured from specimen, and raised to show the Pupa.

wood employed; in the second, two pieces were placed in such a manner as to support each other, and beneath the angle thus formed an oblong structure was made, composed, as before, of wood-raspings and silk, but much stronger in texture than the winter cell. In a few weeks (four, if we recollect aright) the moth came forth.   (J. R.)

[I have now before me a series of three cocoons, made by one caterpillar of the goat-moth, showing its increase in size during the three years that it remained in the larval state. They were found in an old willow tree, and occupied different parts of the same burrow.   The ravages which a goat-moth caterpillar can make in a tree are almost incredible to those who have not seen the long and tortuous burrows which the insect will construct, burrows which at first are small and insignificant, but which afterwards become large enough to admit a man's finger.

[Sometimes the tunnel runs just under the bark, and sometimes it goes straight towards the centre of the tree; and no small labour is required before it can be fully traced. Still, the result is worth the labour, for it is most interesting to trace the creature through its whole existence, from the tiny hole which it made soon after its exit from the egg, to the large aperture through which it emerged as a moth.   The whole of the tunnel is strongly imbued with the peculiar and unpleasant odour which has given to the goat-moth its popular name; and the scent is so persistent, that it adheres to the fingers which have touched the sides of the tunnel, and can scarcely be removed even by repeated washings.

[The moth itself is a well-known insect, though rarely seen except by night. It is large, brown, round bodied; the wings are covered with a soft and downy clothing, which strongly reminds the observer of the plumage of an owl.]

A wood-boring caterpillar, of a species of moth much rarer than the preceding (*Ægeria asiliformis*, STEPHENS), exhibits great ingenuity in constructing a cell for its metamorphosis. We observed above a dozen of them during this summer (1829) in the trunk of a poplar, one side of which had been stripped of its bark. It was this portion of the trunk which all the caterpillars selected for their final retreat, not one having been observed where the tree was covered with bark. The ingenuity of the little architect consisted in scooping its cell almost to the very surface of the wood, leaving only an exterior covering of unbroken wood, as thin as writing-paper. Previous, therefore, to the chrysalis making its way through this feeble barrier, it could not have been suspected that an insect was lodged under the smooth wood. We observed more than one of these in the act of breaking through this covering, within which there is, besides, a round moveable lid of a sort of brown wax. (J. R.)

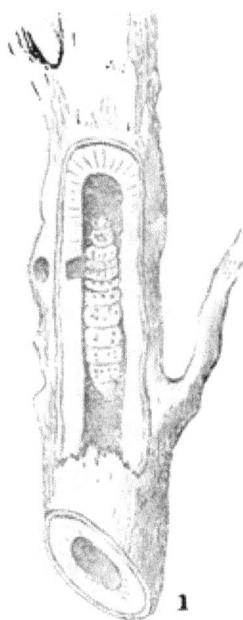

Larva of Ægeria.

Another architect caterpillar, frequently to be met with in July on the leaves of the willow and the poplar, is, in the fly-state, called the puss-moth (*Cerura vinula*). The caterpillar is produced from brown-coloured shining eggs, about the size of a pin's head, which are deposited—one, two, or more together—on the upper surface of a leaf. In the course of six or eight weeks (during which time it casts its skin thrice) it arrives at its full growth, when it

Eggs of the Puss-Moth.

is about as thick, and nearly as long, as a man's thumb, and begins to prepare a structure in which the pupa may sleep securely during the winter. As we have, oftener than once, seen this little architect at work, from the foundation till the completion of its edifice, we are thereby enabled to give the details of the process.

The puss, it may be remarked, does not depend for protection on the hole of a tree, or the shelter of an overhanging branch, but upon the solidity and strength of the fabric which it rears. The material it commonly uses is the bark of the tree upon which the cell is constructed; but when this cannot be procured, it is contented to employ whatever analogous materials may be within reach. One which we had shut up in a box substituted the marble paper it was lined with for bark, which it could not procure.* With

* It is justly remarked by Réaumur, that when caterpillars are left at liberty among their native plants, it is only by lucky chance they can be observed building their cocoons, because the greater number abandon the plants upon which they have been feeding, to spin up in places at some distance. In order to see their operations, they must be kept in confinement, particularly in boxes with glazed doors, where they may be always under the eye of the naturalist. In such circumstances, however, we may be ignorant what building materials we ought to provide them with for their structures. A red caterpillar, with a few tufts of hair, which Réaumur found in July feeding upon the flower bunches of the nettle, and refusing to touch the leaves, began in a few days to prepare its cocoon, by gnawing the paper lid of the box in which it was placed. This, of course, was a material which it could not have procured in the fields, but it was the nearest in properties that it could procure; for, though it had the leaves and stems of nettles, it never used a single fragment of either. When Réaumur found that it was likely to gnaw through the paper

silk it first wove a thin web round the edges of the place
which it marked out for its edifice, then it ran several
threads in a spare manner from side to side, and from end
to end, but very irregularly in point of arrangement; these
were intended for the skeleton or frame-work of the building When this outline was finished, the next step was to
strengthen each thread of silk by adding several (sometimes

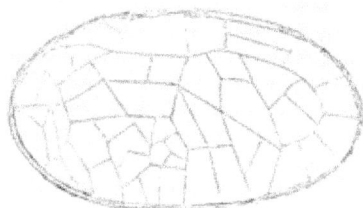

Rudiments of the Cell of the Puss-Moth.

six or eight) parallel ones, all of which were then glued
together into a single thread, by the insect running its
mandibles, charged with gluten, along the line. The
meshes, or spaces, which were thus widened by the com-
pression of the parallel threads, were immediately filled
up with fresh threads, till at length only very small spaces
were left. It was in this stage of the operation that the
paper came into requisition, small portions of it being
gnawed off the box and glued into the meshes. It was not,
however, into the meshes only that the bits of paper were
inserted ; for the whole fabric was in the end thickly
studded over with them. In about half a day from the
first thread of the frame-work being spun the building was
completed. It was at first, however, rather soft, and
yielded to slight pressure with the finger ; but as soon as
it became thoroughly dry, it was so hard that it could with

lid of the box, and might effect its escape, he furnished it with bits of rumpled
paper, fixed to the lid by means of a pin ; and these it chopped down into such
pieces as it judged convenient for its structure, which it took a day to complete.
The moth appeared four weeks after, of a brownish-black colour, mottled with
white, or rather grey, in the manner of lace.
  Bonnet also mentions more than one instance in which he observed caterpillars making use of paper, when they could not procure other materials.

difficulty be penetrated with the point of a penknife.
(J. R.)

[One puss-moth larva, which I reared, made its nest in a
rather curious manner. After it had ceased feeding it had
been placed on a marble mantelpiece under a glass tumbler,

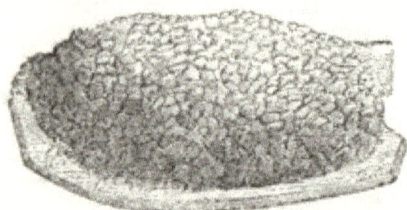

Cell built by the Larva of the Puss-Moth.

as a temporary residence until a more appropriate dwelling
could be found for it. But its instincts urged it to make
its nest without delay, and it accordingly set to work, and
spun itself up in a cocoon composed entirely of its own
silk, neither the glass tumbler or the mantelpiece affording
it any material with which to harden the walls of its
dwelling.

[Consequently, the texture of the cocoon was of a rather
singular nature. The silken threads had been fused together so as to form a translucent cocoon, looking as if it
had been made of gelatine, and being nearly equally transparent, the chrysalis being plainly visible through its walls.
The cocoon was thin and elastic, as if it had been made of
very thin horn; and it was so tightly fixed to the mantelpiece as well as to the tumbler, that it could not be removed
without damage. The moth suffered no injury from the
privation which the larva had to undergo.

[The cocoons of the puss-moth are to be found upon the
trunks of trees, but they are so rough, and so greatly resembling the bark, with which, indeed, their walls are
strengthened, that an inexperienced eye would fail to detect
them. Even when they have been pointed out to a novice in
practical entomology, he has failed to find them again whenever his eye has been taken off their rugged outlines.]

A question will here suggest itself to the curious inquirer, how the moth, which is not, like the caterpillar, furnished with mandibles for gnawing, can find its way through so hard a wall. To resolve this question, it is asserted by recent naturalists (see Kirby and Spence, vol. iii. p. 15) that the moth is furnished with a peculiar acid for dissolving itself a passage. We have a specimen of the case of a puss-moth, in which, notwithstanding its strength, one of the ichneumons had contrived to deposit its eggs. In the beginning of summer, when we expected the moth to appear, and felt anxious to observe the recorded effects of the acid, we were astonished to find a large orange cuckoo-fly make its escape; while another, which attempted to follow, stuck by the way and died. On detaching the cell from the box, we found several others, which had not been able to get out, and had died in their cocoons. (J. R.)

Ichneumon (*Ophion luteum*), figured from the one mentioned.

Among the carpenter-grubs may be mentioned that of the purple capricorn-beetle (*Callidium violaceum*), of which the Rev. Mr. Kirby has given an interesting account in the fifth volume of the 'Linnæan Transactions.' This insect feeds principally on fir timber which has been felled some time without having had the bark stripped off; but it is often found on other wood. Though occasionally taken in this kingdom, it is supposed not to have been originally a native. The circumstance of this destructive little animal attacking only such timber as had not been stripped of its bark ought to be attended to by all persons who have any concern in this article; for the bark is a temptation not only to this, but to various other insects; and much of the injury done in timber might be prevented, if the trees were all barked as soon as they were felled. The female is furnished, at the posterior extremity of her body, with a

flat retractile tube, which she inserts between the bark
and the wood, to the depth of about a quarter of an inch,
and there deposits a single egg. By stripping off the bark,
it is easy to trace the whole progress of the grub, from the
spot where it is hatched, to that where it attains its full
size. It first proceeds in a serpentine direction, filling the
space which it leaves with its excrement, resembling sawdust, and so stopping all ingress to enemies from without.
When it has arrived at its utmost dimensions, it does not
confine itself to one direction, but works in a kind of
labyrinth, eating backwards and forwards, which gives the
wood under the bark a very irregular surface : by this
means its paths are rendered of considerable width. The
bed of its paths exhibits, when closely examined, a curious
appearance, occasioned by the gnawings of its jaws, which
excavate an infinity of little ramified canals. When the
insect is about to assume its chrysalis state, it bores down
obliquely into the solid wood, to the depth sometimes of
three inches, and seldom if ever less than two, forming
holes nearly semi-cylindrical, and of exactly the form of
the grub which inhabits them. At first sight one would
wonder how so small and seemingly so weak an animal
could have strength to excavate so deep a mine ; but when
we examine its jaws, our wonder ceases. These are large,
thick, and solid sections of a cone divided longitudinally,
which, in the act of chewing, apply to each other the
whole of their interior plane surface, so that they grind
the insect's food like a pair of millstones. Some of the
grubs are hatched in October ; and it is supposed that
about the beginning of March they assume their chrysalis
state. At the place in the bark opposite to the hole from
whence they descended into the wood, the perfect insects
gnaw their way out, which generally takes place betwixt
the middle of May and the middle of June. These insects
are supposed only to fly in the night, but during the day
they may generally be found resting on the wood from
which they were disclosed. The grubs are destitute of
feet, pale, folded, somewhat hairy, convex above, and

divided into thirteen segments. Their head is large and convex.*

It would not be easy to find a more striking example of ingenuity than occurs in a small caterpillar which may be found in May, on the oak, and is supposed by Kirby and Spence to be that of the *Pyralis strigulalis.* It is of a whitish-yellow colour, tinged with a shade of carnation, and studded with tufts of red hairs on each segment, and two brown spots behind the head. It has fourteen feet, and the upper part of its body is much flatter than is common in caterpillars. When this ingenious little insect begins to form its cell, it selects a smooth young branch of the oak, near an offgoing of the branchlets whose angle may afford it some protection. It then measures out, with its body for a rule, the space destined for its structure, the

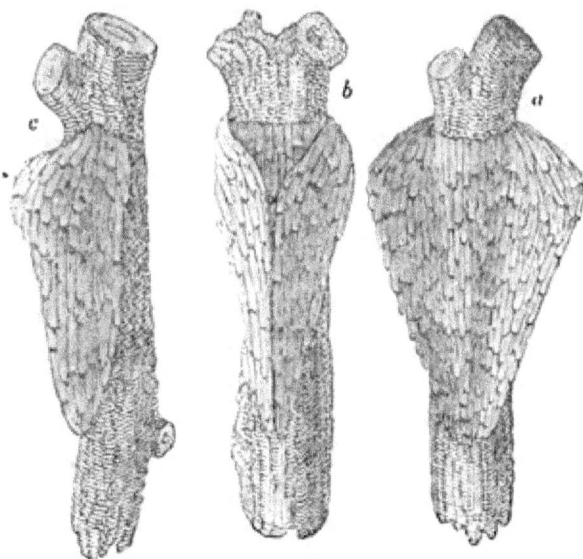

Magnified Cells of Pyralis strigulalis?

a. The walls before they are joined. b. Walls joined, but not closed at top. c. Side view of structure complete.

basement of which is of a triangular form, with the apex at the lower end. The building itself is composed of small, rectangular, strap-shaped pieces of the outer bark of the branch cut out from the immediate vicinity; the insect

* Kirby, in ' Linn. Trans.,' vol. v. p. 246, and Introd. n.

indeed never travels further for materials than the length of its own body. Upon the two longest sides of the triangular base it builds uniform walls, also of a triangular shape, and both gradually diverging from each other as they increase in height. These are formed with so much mathematical precision, that they fit exactly when they are afterwards brought into contact. As soon as the little architect has completed these walls, which resemble very much the feathers of an arrow, it proceeds to draw them together in a manner similar to that which the leaf-rolling caterpillars employ in constructing their abodes, by pulling them with silken cords till they bend and converge. Even when the two longest sides are thus joined, there is an opening left at the upper end, which is united in a similar manner. When the whole is finished, it requires close inspection to distinguish it from the branch, being formed of the same materials, and having consequently the same colour and gloss. Concealment, indeed, may be supposed, with some justice, to be the final object of the insect in producing this appearance, the same principle being extensively exemplified in numerous other instances.

CHAPTER XI.

EARTH-MASON CATERPILLARS.

MANY species of caterpillars are not only skilful in concealing themselves in their cocoons, but also in the concealment of the cocoon itself; so that even when that is large, as in the instance of the death's-head hawk-moth (*Acherontia atropos*), it is almost impossible to find it. We allude to the numerous class of caterpillars which, previous to their changing into the pupa state, bury themselves in the earth. This circumstance would not be surprising, were it confined to those which are but too well known in gardens, from their feeding upon and destroying the roots of lettuce, chicory, and other plants, as they pass a considerable portion of their lives under ground; nor is it surprising that those which retire under ground during the day, and come abroad to feed in the night, should form their cocoons where they have been in the habit of concealing themselves. But it is very singular and unexpected, that caterpillars which pass the whole of their life on plants and even on trees, should afterwards bury themselves in the earth. Yet, the fact is, that perhaps a greater number make their cocoons under than above ground, particularly those which are not clothed with hair.

Some of those caterpillars which go into the ground previous to their change make no cocoon at all, but are contented with a rude masonry of earth as a nest for their pupæ: into the details of their operations it will not be so necessary for us to go, as into those which exhibit more ingenuity and care. When one of the latter is dug up it has the appearance of nothing more than a small clod of earth, of a roundish or oblong shape, but, generally, by no means uniform. The interior, however, when it is laid open, always exhibits a cavity, smooth, polished, and regular, in which the cocoon or the chrysalis lies secure (Fig. B, p. 221). The polish of the interior is precisely such as might be given to soft earth by moistening and kneading it with great care. But beside this, it is usually lined with a tapestry of silk, more or less thick, though this cannot always be discovered without the aid of a magnifying glass. This species of caterpillars, as soon as they have completed their growth, go into the earth, scoop out, as the cossus does in wood, a hollow cell of an oblong form, and line it with pellets of earth, from the size of a grain of sand to that of a pea—united, by silk or gluten, into a fabric more or less compact, according to the species, but all of them fitted for protecting the inhabitant, during its winter sleep, against cold and moisture.

Outside view of Nests of Earth-mason Caterpillars.

One of the examples of this occurs in the ghost-moth (*Hepialus humuli*), which, before it retires into the earth, feeds upon the roots of the hop or the burdock. Like other insects which construct cells under ground, it lines the cemented earthen walls of its cell with a smooth tapestry of silk, as closely woven as the web of the house-spider.

Inaccurate observers have inferred that these earthen structures were formed by a very rude and unskilful process —the caterpillar, according to them, doing nothing more than roll itself round, while the mould adhered to the gluey perspiration with which they describe its body to be covered. This is a process as far from the truth as Aristotle's account of the spider spinning its web from wool taken from its body. Did the caterpillar do nothing more than roll itself in the earth, the cavity would be a long tube fitted exactly to its body (Fig. c) : it is essentially different.

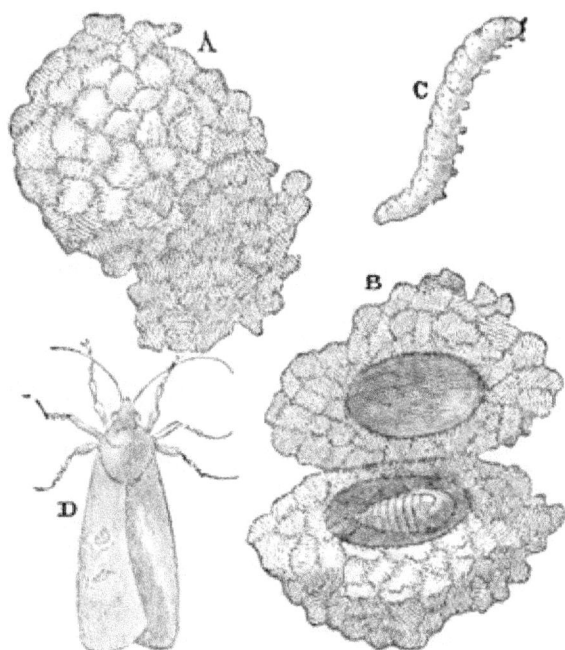

Nests, &c., of an Earth-mason Caterpillar.

It does not indeed require very minute observation to perceive that every grain of earth in the structure is united to the contiguous grains by threads of silk ; and that consequently, instead of the whole having been done at once, it must have required very considerable time and labour. This construction is rendered more obvious by throwing one of these earthen cases into water, which dissolves the earth, but does not act on the silk which binds it together. To understand how this is performed, it may not be uninteresting to follow the little mason from the beginning of his task.

When one of these burrowing caterpillars has done feeding, it enters the earth to the depth of several inches, till it finds mould fit for its purpose. Having nowhere to throw the earth which it may dig out, the only means in its power of forming a cavity is to press it with its body; and, by turning round and round for this purpose, an oblong hollow is soon made. But were it left in this state, as Réaumur well remarks, though the vault might endure the requisite time by the viscosity of the earth alone, were no change to take place in its humidity, yet, as a great number are wanted to hold out for six, eight, and ten months, they require to be substantially built; a mere lining of silk, therefore, would not be sufficient, and it becomes necessary to have the walls bound with silk to some thickness.

When a caterpillar cannot find earth sufficiently moist to bear kneading into the requisite consistence, it has the means of moistening it with a fluid which it ejects for the purpose; and as soon as it has thus prepared a small pellet of earth, it fits it into the wall of the vault, and secures it with silk. As the little mason, however, always works on the *inside* of the building, it does not, at first view, appear in what manner it can procure materials for making one or two additional walls on the inside of the one first built. As the process takes place under ground, it is not easy to discover the particulars, for the caterpillars will not work in glazed boxes. The difficulty was completely overcome by M. Réaumur, in the instance of the caterpillar of the water-betony moth (*Cucullia scrophulariæ*, SCHRANK), which he permitted to construct the greater part of its under-ground building, and then dug it up and broke a portion off from the end, leaving about a third part of the whole to be re-built. Those who are unacquainted with the instinct of insects might have supposed that, being disturbed by the demolition of its walls, it would have left off work; but the stimulus of providing for the great change is so powerful, that scarcely any disturbance will interrupt a caterpillar in this species of labour.

The little builder accordingly was not long in recommencing its task for the purpose of repairing the disorder, which it accomplished in about four hours. At first it protruded its body almost entirely beyond the breach which had been made, to reconnoitre the exterior for building materials.

Earth was put within its reach, of the same kind as it had previously used, and it was not long in selecting a grain adapted to its purpose, which it fitted into the wall and secured with silk. It first enlarged the outside of the wall

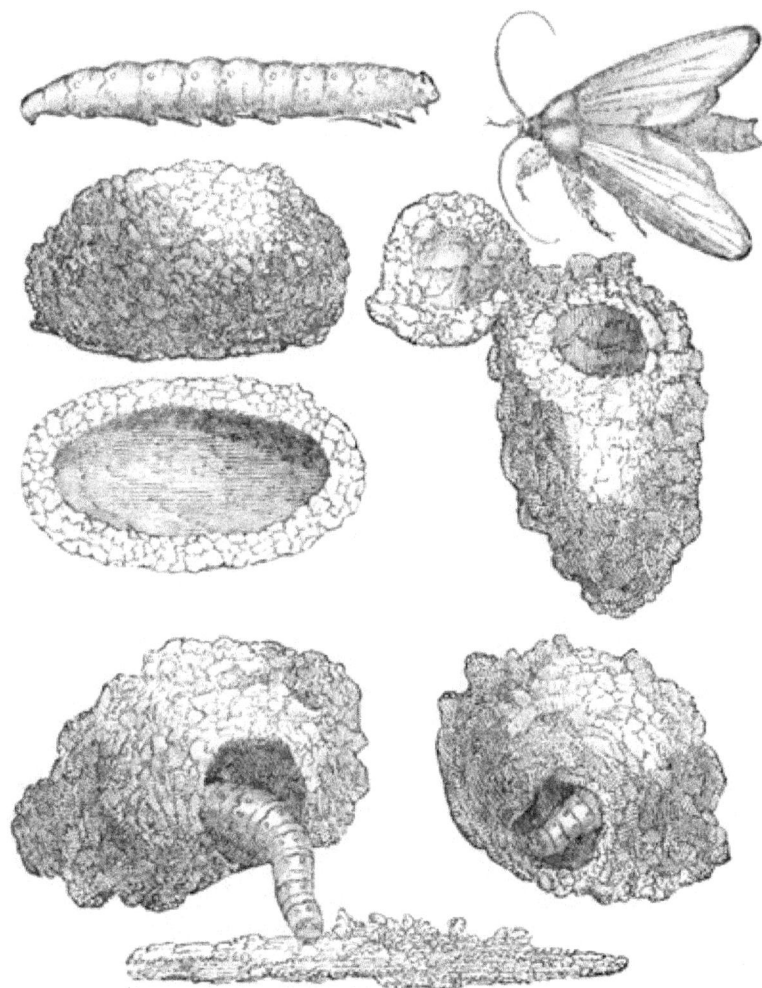

Earth-mason Caterpillars' Nests, with the perfect Moth, &c.

by the larger and coarser grains, and then selected finer for the interior. But before it closed the aperture, it collected a quantity of earth on the inside, wove a pretty thick net-work tapestry of silk over the part which remained open, and into the meshes of this, by pushing and pressing, it thrust grains of earth, securing them with silk till the whole was rendered opaque; and the further operations of the insect could no longer be watched, except that it was observed to keep in motion, finishing, no doubt, the silken tapestry of the interior of its little chamber. When it was completed, M. Réaumur ascertained that the portion of the structure which had been built under his eye was equally thick and compact with the other, which had been done under ground.

The grubs of several of the numerous species of may-fly (*Ephemera*) excavate burrows for themselves in soft earth, on the banks of rivers and canals, under the level of the water, an operation well described by Scopoli, Swammerdam, and Réaumur. The excavations are always proportioned to the size of the inhabitant; and consequently, when it is young and small, the hole is proportionally small, though, with respect to extent, it is always at least double the length of its body. The hole, being under the level of the river, is always filled with water, so that the grub swims in its native element, and while it is secure from being preyed upon by fishes, it has its own food within easy reach. It feeds, in fact, if we may judge from its *egesta*, upon the slime or moistened clay with which its hole is lined.

In the bank of the stream at Lee, in Kent, we had occasion to take up an old willow stump, which, previous to its being driven into the bank, had been perforated in numerous places by the caterpillar of the goat-moth (*Cossus ligniperda*). From having been driven amongst the moist clay, these perforations became filled with it, and the grubs of the ephemerae found them very suitable for their habitation : for the wood supplied a more secure protection than if their galleries had been excavated in the clay. In these holes of the wood we found several empty, and some in which were full-grown grubs. (J. R.)

The architecture of the grub of a pretty genus of beetles, known to entomologists by the name of *Cicindela*, is peculiarly interesting. It was first made known by the eminent

French naturalists, Geoffroy, Desmarest, and Latreille. This grub, which may be met with during spring, and also in summer and autumn, in sandy places, is long, cylindric, soft whitish, and furnished with six scaly brown feet. The

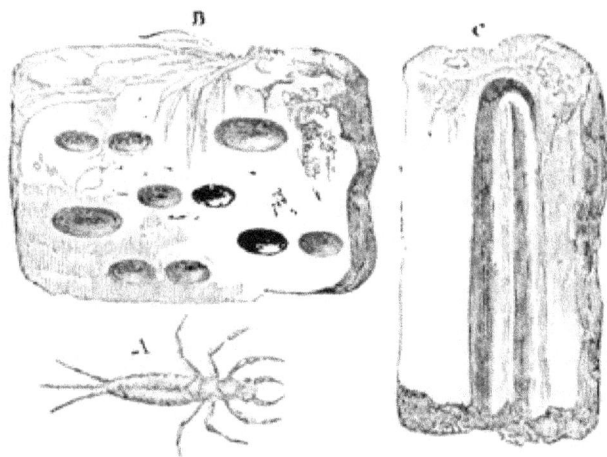

Nest of the Grubs of Ephemeræ.
A. The Grub. B, Perforations in a river bank. C, One laid open to show the parallel structure.

head is of a square form, with six or eight eyes, and very large in proportion to the body. They have strong jaws, and on the eighth joint of the body there are two fleshy tubercles, thickly clothed with reddish hairs, and armed

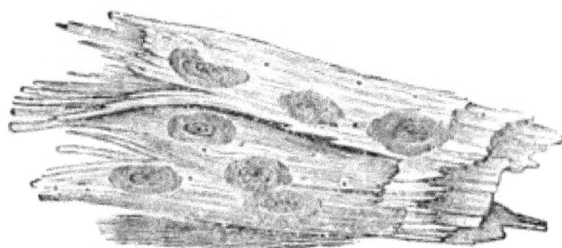

Nests of Ephemeræ in holes of Cossus.

with a recurved horny spine, the whole giving to the grub the form of the letter Z.

With their jaws and feet they dig into the earth to the depth of eighteen inches, forming a cylindrical cavity of greater diameter than their body, and furnished with a perpendicular entrance. In constructing this, the grub first clears away the particles of earth and sand by placing them on its broad trapezoidal head, and carrying the load in this manner beyond the area of the excavation. When it gets deeper down, it climbs gradually up to the surface with similar loads by means of the tubercles on its back, above described. This process is a work of considerable time and difficulty, and in carrying its loads the insect has often to rest by the way to recover strength for a renewed exertion. Not unfrequently, it finds the soil so ill adapted to its operations, that it abandons the task altogether, and begins anew in another situation. When it has succeeded in forming a complete den, it fixes itself at the entrance by the hooks of its tubercles, which are admirably adapted for the purpose, forming a fulcrum or support, while the broad plate on the top of the head exactly fits the aperture of the excavation, and is on a level with the soil. In this position the grub remains immovable, with jaws expanded, and ready to seize and devour every insect which may wander within its reach, particularly the smaller beetles ; and its voracity is so great, that it does not spare even its own species. It precipitates its prey into the excavation, and in case of danger it retires to the bottom of its den, a circumstance which renders it not a little difficult to discover the grub. The method adopted by the French naturalists was to introduce a straw or pliant twig into the hole, while they dug away, by degrees and with great care, the earth around it, and usually found the grub at the bottom of the cell, resting in a zig-zag position like one of the caterpillars of the geometric moths.

When it is about to undergo its transformation into a pupa, it carefully closes the mouth of the den, and retires to the bottom in security.

It does not appear that the grub of the genus *Cincindela* uses the excavation just described for the purpose of a trap or pitfall, any further than that it can more effectually secure its prey by tumbling them down into it ; but there are other species of grubs which construct pitfalls for the express purpose of traps. Among these is the larva of a fly (*Rhagio vermileo*), not unlike the common flesh maggot. The den which it constructs is in the form of a funnel, the sides of which are composed of sand or loose earth. It forms this pitfall of considerable depth, by throwing out the earth obliquely on all sides; and when its trap is finished, it stretches itself along the bottom, remaining stiff and motionless, like a piece of wood. The last segment of the body is bent at an angle with the rest, so as to form a strong point of support in the struggles which it must often have to encounter with vigorous prey. The instant that an insect tumbles into the pitfall, the grub pounces upon it, writhes itself round it like a serpent, transfixes it with its jaws, and sucks its juices at its ease. Should the prey by any chance escape, the grub hurls up jets of sand and earth, with astonishing rapidity and force, and not unfrequently succeeds in again precipitating it to the bottom of its trap.

The Ant-Lion.

The observations of the continental naturalists have made known to us a pitfall constructed by an insect, the details of whose operations are exceedingly curious; we refer to the grub of the ant-lion (*Myrmeleon formicarius*), which, though marked by Dr. Turton and Mr. Stewart as British, has not (at least of late years) been found in this country. As it is not, however, uncommon in France and Switzerland, it is probable it may yet be discovered in some spot hitherto unexplored, and if so, it will well reward the search of the curious.

The ant-lion grub being of a grey colour, and having its body composed of rings, is not unlike a wood-louse (*Oniscus*), though it is larger, more triangular, has only six legs, and most formidable jaws, in form of a reaping-hook, or a pair of calliper compasses. These jaws, however, are not for masticating, but are perforated and tubular, for the purpose of sucking the juices of ants, upon which it feeds. Vallisnieri was therefore mistaken, as Réaumur well remarks, when he

supposed that he had discovered its mouth. Its habits require that it should walk backwards, and this is the only
species of locomotion which it can perform. Even this sort
of motion it executes very slowly; and were it not for the
ingenuity of its stratagems, it would fare but sparingly,
since its chief food consists of ants, whose activity and swift-
ness of foot would otherwise render it impossible for it to
make a single capture. Nature, however, in this, as in
nearly every other case, has given a compensating power to
the individual animal, to balance its privations. The ant-
lion is slow, but it is extremely sagacious; it cannot follow
its prey, but it can entrap it.

The snare which the grub of the ant-lion employs consists
of a funnel-shaped excavation formed in loose sand, at the

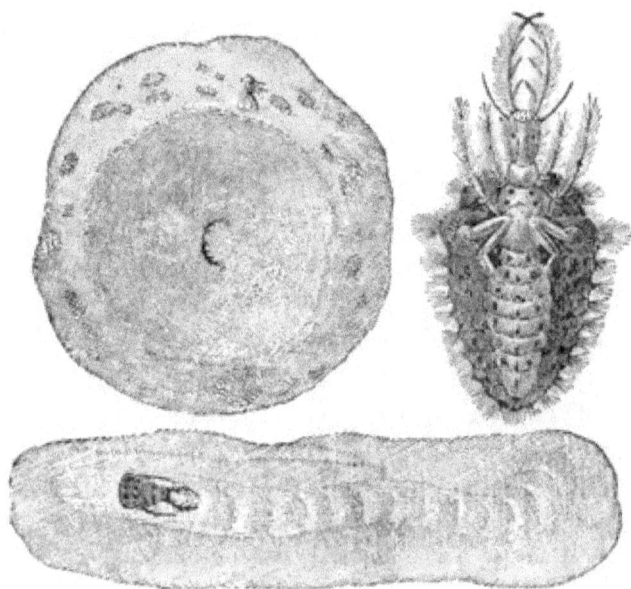

Grub of the Ant-Lion, magnified, with one perfect Trap, and another begun.

bottom of which it lies in wait for the ants that chance to
stumble over the margin, and cannot, from the looseness of
the walls, gain a sufficient footing to effect their escape.

By shutting up one of these grubs in a box with loose
sand, it has been repeatedly observed constructing its trap

of various dimensions, from one to nearly three inches in diameter, according to circumstances.

In the 'Magazine of Natural History,' 1838, p. 601, Mr. Westwood gives a very interesting account of the mode in which the ant-lion proceeds in the excavation of its pitfall, as witnessed by himself in specimens procured in the Parc de Belle Vue, near Paris, where, at the foot of a very high sand-bank, these pits were numerous, and of various sizes, but none exceeded an inch and a half or two inches in diameter, and two-thirds of an inch deep. "The ant-lions were of various sizes, corresponding to the size of their retreats. I brought many of them to Paris, placing several together in a box filled with sand. They, however, destroyed one another whilst shut up in these boxes; and I only succeeded in bringing three of them alive to England, one of which almost immediately afterwards (on the 23rd of July) enclosed itself in a globular cocoon of fine sand. The other two afforded me many opportunities of observing their proceedings. They were unable to walk forwards,—an anomalous circumstance, and not often met with in animals furnished with well-developed legs. It is generally backwards, working in a spiral direction, that the creature moves, pushing itself backwards and downwards at the same time, the head being carried horizontally, and the back much arched, so that the extremity of the body is forced into the sand. In this manner it proceeds backwards (to use an Hibernianism), forming little mole-hills in the sand. But it does not appear to me that this retrograde motion has anything to do with the actual formation of the cell, since, as soon as it has fixed upon a spot for its retreat, it commences throwing up the sand with the back of its head, jerking the sand either behind its back or on one or the other side. It shuts its long jaws, forming them into a kind of shovel, the sharp edges of which it thrusts laterally into the sand on each side of its head, and thereby contrives to lodge a quantity of the sand upon the head as well as the jaws. The motion is in fact something like that of the head of a goat, especially when butting sideways in play. In this manner it contrives to throw away the sand, and by degrees to make a hole entirely with its head, the fore legs not affording the slightest assistance in the operation. During this performance the head only is exposed, the insect having previously pushed itself beneath the surface of the sand; but when it has made the hole sufficiently deep, it withdraws the head also, leaving only the jaws exposed, which are spread open in a line, and laid on the sand so as to be scarcely visible. If alarmed, the insect immediately takes a step backwards, withdrawing the jaws; but when an insect falls into the hole, the jaws are instinctively and instantaneously closed, and the insect seized by the leg, wing, or body, just as it may chance to fall within the reach of the ant-lion's jaws. If, however, the insect be not seized, but attempts to escape, no matter in what direction, the ant-lion immediately begins twisting its head about, and shovelling up the sand with the greatest agility, jerking it about on each side and backwards, but never forwards, as misrepresented in some figures, until the hole is made so much deeper, and such a disturbance caused in the sides of the hole, that the insect is almost sure to be brought down to the bottom, when it is seized by the ant-lion, which immediately endeavours to draw it beneath the sand; and if it be very boisterous, the ant-lion beats it about, holding it firmly by the jaws until it is too weak for further resistance. Hence, as the head of the ant-lion is immersed in the sand, it is evident that the accounts given in popular works of the instinct by which it throws the sand in the direction of the escaping prey are not quite correct. The act of throwing up the sand, when an insect has fallen into the pit and attempts to escape, has evidently for its chief object that of making the pit deeper and more conical, and therefore more difficult of ascent."

It is by the action of the hinder pair of its legs that the ant-lion drags itself backwards, the other four pair being extended trailing after it, and leaving an impression on the surface of the fine sand over which it has passed; and when burrowing its way beneath the surface of the sand, it proceeds by short steps backwards. A portion of sand at each step is thrown on the head, owing to the hump-like form of the back : this is immediately jerked away, the body at the same time advancing another step in its backward and spiral motion. Where it rests, a little hillock of sand is raised by the body of the ant-lion underneath ; while its jaws emerge and spread flat on the surface. It now probably commences its pitfall, the mode of excavating which we have given in detail. From the spiral course described by the ant-lion in its backward progress appears to have arisen the idea of its tracing out a circle as the outline of its pitfall— as would an architect or engineer ; but whence sprang the often-repeated statement, that the ant-lion loads its head with sand by means of one of its legs, that nearest the centre of the circle, we cannot conjecture. Nor do we know how, as it works entirely buried with the exception of the head, the ant-lion can act when it meets with a stone or other obstacle, as M. Bonnet states he has repeatedly witnessed. He observes that if the stone be small, it can manage to jerk it out in the same manner as the sand ; but when it is two or three times larger and heavier than its own body, it must

Ant-Lion's Pitfalls, in an experimenting-box.

have recourse to other means of removal. The larger stones it usually leaves till the last ; and when it has removed all the sand which it intends, it then proceeds to try what it can do with the less manageable obstacles. For this purpose it crawls backwards to the place where a stone may be, and thrusting its tail under it, is at great pains to get it properly balanced on its back, by an alternate motion of the rings composing its body. When it has succeeded in adjusting the stone, it crawls up the side of the pit with great care, and deposits its burthen on the outside of the circle. Should the stone happen to be round, the balance can be kept only with the greatest difficulty, as it has to travel with its load upon a slope of loose sand, which is ready to give way at every step; and often when the insect has carried it to the very brink, it rolls off its back and tumbles down to the bottom of the pit. This accident, so far from discouraging the ant-lion, only stimulates it to more persevering efforts. Bonnet observed it renew these attempts to dislodge a stone five or six times. It is only when it finds it utterly impossible to succeed, that it abandons the design and commences another pit in a fresh situation. When it succeeds in getting a stone beyond the line of its circle, it is not contented with letting it rest there; but, to prevent it from again rolling in, it goes on to push it to a considerable distance. We may be pardoned for pausing before we give full credence to these details.

The ant-lion feeds only on the blood or juice of insects; and as soon as it has extracted these, it tosses the dry carcase out of its den.

When it is about to change into a pupa, it proceeds in nearly the same manner as the caterpillar of the water-betony moth (*Cucullia scrophulariæ*). It first builds a case of sand, the particles of which are secured by threads of silk, and then tapestries the whole with a silken web. Within this it undergoes its transformation into a pupa, and in due time it emerges in form of a four-winged fly, closely resembling the dragon-flies (*Libellulæ*), vulgarly and erroneously called *horse-stingers*.

The instance of the ant-lion naturally leads us to consider the design of the Author of Nature in so nicely adjusting, in all animals, the means of destruction and of escape. As the larger quadrupeds of prey are provided with a most ingenious machinery for preying on the weaker, so are those furnished with the most admirable powers of evading their destroyers.

In the economy of insects, we constantly observe that the means of defence, not only of the individual creatures, but of their larvæ and pupæ, against the attacks of other insects, and of birds, is proportioned, in the ingenuity of their arrangements, to the weakness of the insect employing them. Those species which multiply the quickest have the greatest number of enemies. Bradley, an English naturalist, has calculated that two sparrows carry, in the course of a week, above three thousand caterpillars to the young in their nests. But though this is, probably, much beyond the truth, it is certain that there is a great and constant destruction of individuals going forward; and yet the species is never destroyed. In this way a balance is kept up, by which one portion of animated nature cannot usurp the means of life and enjoyment which the world offers to another portion. In all matters relating to reproduction, Nature is prodigal in her arrangements. Insects have more stages to pass through before they attain their perfect growth than other creatures. The continuation of the species is, therefore, in many cases, provided for by a much larger number of eggs being deposited than ever become fertile. How many larvæ are produced, in comparison with the number which pass into the pupa state; and how many pupæ perish before they become perfect insects! Every garden is covered with caterpillars; and yet how few moths and butterflies, comparatively, are seen, even in the most sunny season? Insects which lay few eggs are, commonly, most remarkable in their contrivances for their preservation. The dangers to which insect life is exposed are manifold; and therefore are the contrivances for its preservation of the most perfect kind, and invariably adapted to the peculiar habits of each tribe. The same wisdom determines the food of every species of insect; and thus some are found to delight in the rose-tree, and some in the oak. Had it been otherwise, the balance of vegetable life would not have been preserved. It is for this reason that the contrivances which an insect employs for obtaining its food are curious, in proportion to the natural difficulties of its structure. The ant-lion is carnivorous, but he has not the quickness of the spider, nor can he spread a net over a large surface, and issue from his citadel to seize a victim which he has caught in his out-works. He is therefore taught to dig a trap, where he sits like the unwieldy giants of fable, waiting for some feeble one to cross his path. How laborious and patient are his operations—how uncertain the chances of success! Yet he never shrinks from them, because his instinct tells him that by these contrivances alone can he preserve his own existence, and continue that of his species.

CHAPTER XII.

THERE are at least five different species of moths similar in manners and economy, the caterpillars of which feed upon animal substances, such as furs, woollen cloths, silk, leather, and, what to the naturalist is no less vexing, upon the specimens of insects and other animals preserved in his cabinet. The moths in question are of the family named *Tinea* by entomologists, such as the tapestry-moth (*Tinea tapetzella*), the fur-moth (*Tinea pellionella*), the wool-moth (*Tinea vestianella*), the cabinet-moth (*Tinea destructor*, STEPHENS), &c.

The moths themselves are, in the winged state, small and well fitted for making their way through the most minute hole or chink, so that it is scarcely possible to exclude them by the closeness of a wardrobe or a cabinet.* If they cannot effect an entrance when a drawer is out, or a door open, they will contrive to glide through the key-hole ; and if they once get in, it is no easy matter to dislodge or destroy them, for they are exceedingly agile, and escape out of sight in a moment. Moufet is of opinion that the ancients possessed an effectual method of preserving stuffs from the moth, because the robes of Servius Tullius were preserved up to the death of Sejanus, a period of more than five hundred years. On turning to Pliny to learn this secret, we find him relating that stuff laid upon a coffin will be ever after safe from moths; in the same way as a person once stung by a scorpion will never afterwards be stung by a bee, or a wasp, or a hornet ! Rhasis, again, says that cantharides suspended in a house drive away moths; and he adds that

* See Fig. *d*, p. 238.

they will not touch anything wrapped in a lion's skin!—the
poor little insects, says Réaumur, sarcastically, being probably
in bodily fear of so terrible an animal.* Such are the stories
which fill the imagination even of philosophers, till real
science entirely expels them.

The effluvium of camphor or turpentine, or fumigation by
sulphur or chlorine, may sometimes kill them, when in the
winged state, but this will have no effect upon their eggs,
and seldom upon the caterpillars ; for they wrap themselves
up too closely to be easily reached by any agent except heat.
This, when it can be conveniently applied, will be certain
either to dislodge or to kill them. When the effluvium of
turpentine, however, reaches the caterpillar, Bonnet says it
falls into convulsions, becomes covered with livid blotches,
and dies.†

The mother insect takes care to deposit her eggs on or
near such substances as she instinctively foreknows will be
best adapted for the food of the young, taking care to dis-
tribute them so that there may be a plentiful supply and
enough of room for each. We have found, for example,
some of those caterpillars feeding upon the shreds of cloth
used in training wall-fruit trees ; but we never saw more
than two caterpillars on one shred. This scattering of the
eggs in many places renders the effects of the caterpillars
more injurious, from their attacking many parts of a garment
or a piece of stuff at the same time. (J. R.)

When one of the caterpillars of this family issues from the
egg, its first care is to provide itself with a domicile, which
indeed seems no less indispensable to it than food ; for, like
all caterpillars that feed under cover, it will not eat while it
remains unprotected. Its mode of building is very similar
to that which is employed by other caterpillars that make use
of extraneous materials. The foundation or frame-work is
made of silk secreted by itself, and into this it interweaves
portions of the material upon which it feeds. It is said by
Bingley, that "after having spun a fine coating of silk

immediately around its body, it cuts the filaments of the wool or fur close by the thread of the cloth, or by the skin, with its teeth, which act in the manner of scissors, into convenient lengths, and applies the bits, one by one, with great dexterity, to the *outside* of its silken case."* This statement, however, is erroneous, and inconsistent with the proceedings not only of the clothes-moth, but of every caterpillar that constructs a covering. None of these build from within outwards, but uniformly commence with the exterior wall, and finish by lining the interior with the finest materials. Réaumur, however, found that the newly-hatched caterpillars lived at first in a case of silk.

We have repeatedly witnessed the proceedings of these insects from the very foundation of their structures; and, at the moment of writing this, we turned out one from the carcase of an "old lady moth" (*Mormo maura*, OCHSENHEIM) in our cabinet, and placed it on a desk covered with green cloth, where it might find materials for constructing another dwelling. It wandered about for half a day before it began its operations; but it did not, as is asserted by Bonnet, and Kirby and Spence, "in moving from place to place, seem to be as much incommoded by the long hairs which surround it, as we are by walking amongst high grass," nor, "accordingly, marching scythe in hand," did it, "with its teeth, cut out a smooth road."† On the contrary, it did not cut a single hair till it selected one for the foundation of its intended structure. This it cut very near the cloth, in order, we suppose, to have it as long as possible; and placed it on a line with its body. It then immediately cut another, and placing it parallel to the first, bound both together with a few threads of its own silk. The same process was repeated with other hairs, till the little creature had made a fabric of some thickness, and this it went on to extend till it was large enough to cover its body; which (as is usual with caterpillars) it employed as a model and measure for regulat-

* 'Animal Biography,' vol. iii. p. 330, Third Edition.
† Bonnet, xi. p. 204; Kirby and Spence, 'Introduction,' i. 464, Fifth Edition.

ing its operations. We remarked that it made choice of
longer hairs for the outside than for the parts of the interior,
which it thought necessary to strengthen by fresh additions;
but the chamber was ultimately finished by a fine and closely-
woven tapestry of silk. We could see the progress of its
work by looking into the opening at either of the ends; for at
this stage of the structure the walls are quite opaque, and the
insect concealed. It may be thus observed to turn round, by
doubling itself and bringing its head where the tail had just
been; of course, the interior is left wide enough for this
purpose, and the centre, indeed, where it turns, is always
wider than the extremities. (J. R.)

When the caterpillar increases in length, it takes care to
add to the length of its house, by working in fresh hairs at

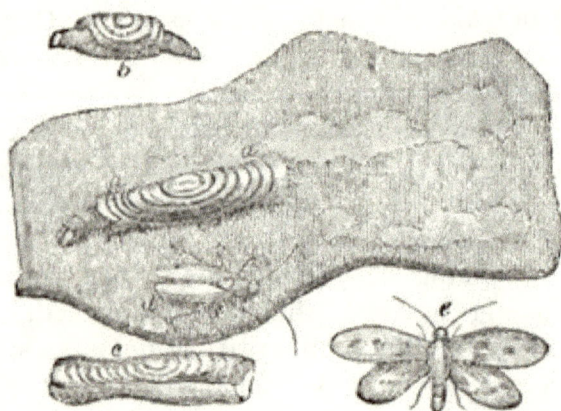

Cases, &c., of the Clothes-Moth (*Tinea pellionella*).—*a*, Caterpillar feeding in a case,
which has been lengthened by ovals of different colours; *b*, Case cut at the ends for
experiment; *c*, Case cut open by the insect for enlarging it; *d, e*, The clothes-moths in
their perfect state, when, as they cease to eat, they do no further injury.

either end; and if it be shifted to stuffs of different colours,
it may be made to construct a party-coloured tissue, like a
Scotch plaid. Réaumur cut off with scissors a portion at
each end, to compel the insect to make up the deficiency.
But the caterpillar increases in thickness as well as in length,
so that, its first house becoming too narrow, it must either
enlarge it, or build a new one. It prefers the former as less
troublesome, and accomplishes its purpose "as dexterously,"

says Bonnet, " as any tailor, and sets to work precisely as we
should do, slitting the case on the two opposite sides, and
then adroitly inserting between them two pieces of the
requisite size. It does not, however, cut open the case from
one end to the other at once; the sides would separate too
far asunder, and the insect be left naked. It therefore first
cuts each side about half-way down, beginning sometimes at
the centre and sometimes at the end (Fig. *c*), and then, after
having filled up the fissure, proceeds to cut the remaining
half; so that, in fact, four enlargements are made, and four
separate pieces inserted. The colour of the case is always
the same as that of the stuff from which it is taken. Thus,
if its original colour be blue, and the insect, previously to
enlarging it, be put upon red cloth, the circles at the end,
and two stripes down the middle, will be red."* Réaumur
found that they cut these enlargements in no precise order,
but sometimes continuously, and sometimes opposite each
other, indifferently.

The same naturalist says he never knew one leave its old
dwelling in order to build a new; though, when once ejected
by force from its house, it would never enter it again, as
some other species of caterpillars will do, but always
preferred building another. We, on the contrary, have
more than once seen them leave an old habitation. The
very caterpillar, indeed, whose history we have above given,
first took up its abode in a specimen of the ghost-moth
(*Hepialus humuli*), where, finding few suitable materials for
building, it had recourse to the cork of the drawer, with the
chips of which it made a structure almost as warm as it
would have done from wool. Whether it took offence at our
disturbing it one day, or whether it did not find sufficient
food in the body of the ghost-moth, we know not; but it left
its cork house, and travelled about eighteen inches, selected
".the old lady," one of the largest insects in the drawer, and
built a new apartment, composed partly of cork as before,
and partly of bits clipt out of the moth's wings. (J. R.)

We have seen these caterpillars form their habitations of

---

* Bonnet, vol. ix. p. 203.

every sort of insect, from a butterfly to a beetle; and the soft, feathery wings of moths answer their purpose very well: but when they fall in with such hard materials as the musk beetle (*Cerambyx moschatus*), or the large scolopendra of the West Indies, they find some difficulty in the building.

When the structure is finished, the insect deems itself secure to feed on the materials of the cloth or other animal matter within its reach, provided it is dry and free from fat

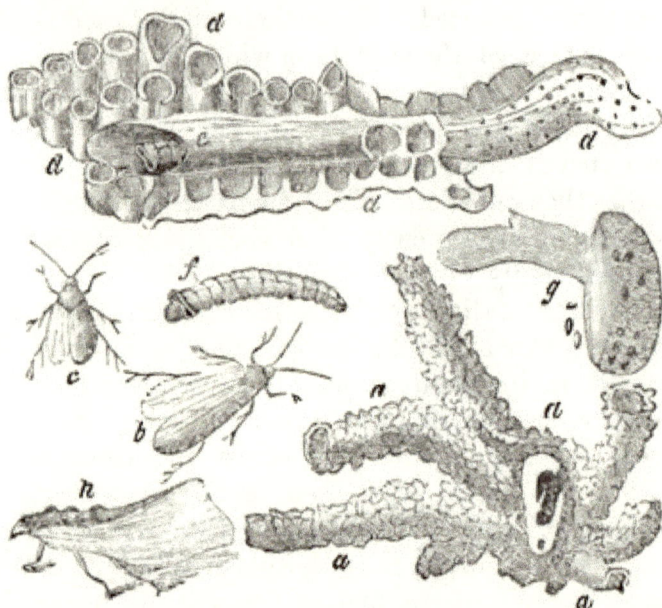

Transformations of the honeycomb-moths. *a a a*, Galleries of the cell-boring caterpillar; *b*, the female; *c*, the male moth (*Galleria alvearia*); *d d d d*, galleries of the wax-eating caterpillar, *e*, seen at the entrance; *f*, the same exposed; *g*, its cocoon; *h*, the moth (*Galleria cereana*).

or grease, which Réaumur found it would not touch. This may probably be the origin of the practice of putting a bit of candle with furs, &c., to preserve them from the moth. For building, it always selects the straightest and loosest pieces of wool, but for food it prefers the shortest and most compact; and to procure these it eats into the body of the stuff, rejecting the pile or nap, which it necessarily cuts across at the origin, and permits to fall, leaving it threadbare, as if it had been much worn. It must have been this circumstance which induced Bonnet to fancy (as we have already mentioned) that it cuts the hairs to make itself a smooth, comfortable path to walk upon. It would be equally correct to say that an ox or a sheep dislikes walking amongst long grass, and therefore eats it down in order to clear the way.

[There is a little insect closely allied to these moths, which does a vast amount of harm to the bee-combs. This is the honey-comb moth, of which there are in England two species, both belonging to the genus Galleria. This little creature is continually trying to make its way into the hives, and is as continually opposed by the bees, who instinctively know their enemy. If it once slips past the guards, the unfortunate bees are doomed to lose a considerable amount of their stored treasures, and have sometimes been so worried that they have been obliged to leave the hive altogether.

[As soon as it can hide itself in an empty cell—an easy matter enough for so tiny a moth, which harmonizes exactly in colour with the bee-combs—it proceeds to lay its eggs, and, having discharged its office, dies. The eggs soon hatch into little grubs and caterpillars with very hard horny heads and soft bodies. As soon as they come into the dark world of the hive, they begin to eat their way through the combs, spinning the while a tunnel of silk, which entirely protects them from the stings of the bees. They can traverse these tunnels with tolerable speed, so that the bees do not know where to find their enemies; and if perchance they should discover one of them at the mouth of its burrow, the hard, horny head is all that is visible, and against its polished surface the sting of the bee is useless. The rapidity with which they drive the silken tubes through the comb is really marvellous; and even if they get among a collection of empty bee-combs, they make as much havoc as if they were bred in the hive from which the combs were taken.

[In the accompanying illustration are seen figures of the two species of honey-moths, together with their tunnels. The species may be easily distinguished by the shape of the wings, *Galleria alvearia* having, as seen at Figs. *b*, *c*, the ends

R

of its wings rounded, and *Galleria cereana* having them squared.

[Some moths, also belonging to the vast Family Tineidæ, do much damage to grain, and have also the habit of spinning silken tissues as they eat their way through the grain. One of them is more plentiful on the Continent than in England, but is known in this country by the name of the mottled woollen moth (*Tinea granella*)].

The caterpillar, which is smooth and white, ties together with silk several grains of wheat, barley, rye, or oats, weaving a gallery between them, from which it projects its head while feeding; the grains, as Réaumur remarks, being prevented from rolling or slipping by the silk which unites them. He justly ridicules the absurd notion of its filing off the outer skin of the wheat by rubbing upon it with its body, the latter being the softer of the two, and he disproved, by experiment, Leeuwenhoeck's assertion that it will also feed on woollen cloth. It is from the end of May till the beginning of July that the moths, which are of a silvery grey, spotted with brown, appear and lay their eggs in granaries.

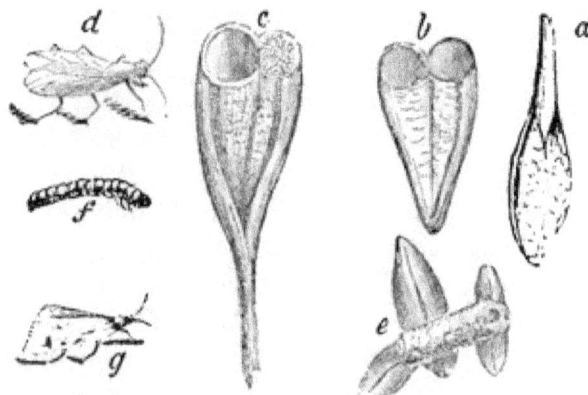

Transformations of the Grain-moths. *a*, Grain of barley, including a caterpillar, *b*, *c*, the grain cut across, seen to be hollowed out, and divided by a partition of silk ; *d*, the moth (*Tinea Hordei*) ; *e*, grains of wheat tied together by the caterpillar· *f*, *g*, the caterpillar and moth (*Euplocamus granella*).

The caterpillar of another still more singular grain-moth (*Tinea Hordei*, KIRBY and SPENCE) proves sometimes very destructive of granaries. The mother-moth, in May or June, lays about twenty or more eggs on a grain of barley or wheat; and when the caterpillars are hatched they disperse, each selecting a single grain. M. Réaumur imagines that sanguinary wars must sometimes arise, in cases of pre-occupancy, a single grain of barley being a rich heritage for one of these tiny insects; but he confesses he never saw such contests. When the caterpillar has eaten its way into the interior of the grain, it feeds on the farina, taking care not to gnaw the skin nor even to throw out its excrements, so that except the little hole, scarcely discernible, the grain appears quite sound. When it has eaten all the farina, it spins itself a case of silk within the now hollow grain, and changes to a pupa in November.

TENT-MAKING CATERPILLARS.

The caterpillars of a family of small moths (*Tineidæ*), which feed on the leaves of various trees, such as the hawthorn, the elm, the oak, and most fruit-trees, particularly the pear, form habitations which are exceedingly ingenious and elegant. They are so very minute that they require close inspection to discover them; and to the cursory observer, unacquainted with their habits, they will appear more like the withered leaf-scales of the tree, thrown off when the buds expand, than artificial structures made by insects. It is only, indeed, by seeing them move about upon the leaves, that we discover they are inhabited by a living tenant, who carries them as the snail does its shell.

These tents are from a quarter of an inch to an inch in length, and usually about the breadth of an oat-straw. That they are of the colour of a withered leaf is not surprising; for they are actually composed of a piece of leaf; not, however, cut out from the whole thickness, but artfully separated from the upper layer, as a person might separate one of the leaves of paper from a sheet of pasteboard.

The tents of this class of caterpillars, which are found on the elm, the alder, and other trees with serrated leaves, are

much in the shape of a minute goldfish. They are convex
on the back, where the indentations of the leaf out of which
they have been cut add to the resemblance, by appearing like
the dorsal fins of the fish. By depriving one of those caterpillars common on the hawthorn of its tents, for the sake of
experiment, we put it under the necessity of making another;
for, as Pliny remarks of the clothes-moth, they will rather
die of hunger than feed unprotected. When we placed it on

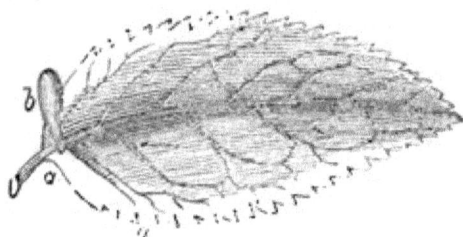

A Caterpillar's tent upon a leaf of the elm.—*a, a*, The part of the leaf from which the
tent has been cut out; *b*, the tent itself.

a fresh hawthorn leaf, it repeatedly examined every part of
it, as if seeking for its lost tent, though, when this was put
in its way, it would not again enter it; but, after some delay,
commenced a new one. (J. R.)

For this purpose, it began to eat through one of the two
outer membranes which compose the leaf and enclose the
pulp (*parenchyma*), some of which, also, it devoured, and
then thrust the hinder part of its body into the perforation.
The cavity, however, which it had formed, being yet too
small for its reception, it immediately resumed the task of
making it larger. By continuing to gnaw into the pulp
between the membranes of the leaf (for it took the greatest
care not to puncture or injure the membranes themselves), it
soon succeeded in mining out a gallery rather larger than
was sufficient to contain its body. We perceived that it did
not throw out as rubbish the pulp it dug into, but devoured
it as food—a circumstance not the least remarkable in its
proceedings.

As the two membranes of leaf thus deprived of the
enclosed pulp appeared white and transparent, every movement of the insect within could be distinctly seen; and it

was not a little interesting to watch its ingenious operations while it was making its tent from the membranes prepared as we have just described. These, as Réaumur has remarked, are in fact to the insect like a piece of cloth in the hands of a tailor; and no tailor could cut out a shape with more neatness and dexterity than this little workman does. As the caterpillar is furnished in its mandibles with an excellent pair of scissors, this may not appear to be a difficult task; yet, when we examine the matter more minutely, we find that the peculiar shape of the two extremities requires different curvatures, and this, of course, renders the operation no less complex, as Réaumur subjoins, than the shaping of the pieces of cloth for a coat.* The insect, in fact, shapes the membranes slightly convex on one side and concave on the other, and at one end twice as large as the other. In the instance which we observed, beginning at the larger end, it bent them gently on each side by pressing them with its body thrown into a curve. We have not said it *cuts*, but *shapes* its materials; for it must be obvious that if the insect had cut both the membranes at this stage of its operations, the pieces would have fallen and carried it along with them.

To obviate such an accident it proceeded to join the two edges, and secure them firmly with silk, before it made a single incision to detach them. When it had in this manner joined the two edges along one of the sides, it inserted its head on the outside of the joining, first at one end and then at the other, gnawing the fibres till that whole side was separated. It proceeded in the same manner with the other side, joining the edges before it cut them: and when it arrived at the last fibre, the only remaining support of its now finished tent, it took the precaution, before snipping it, to moor the whole to the uncut part of the leaf by a cable of its own silk. Consequently, when it does cut the last nervure, it is secure from falling, and can then travel along the leaf, carrying its tent on its back, as a snail does its shell. (J. R.)

* ' Mem. Hist. Insect.' iii. p. 106.

We have just discovered (Nov. 4th, 1829) upon the
nettle a tent of a very singular appearance, in consequence
of the materials of which it is made. The caterpillar
seems, indeed, to have proceeded exactly in the same
manner as those which we have described, mining first

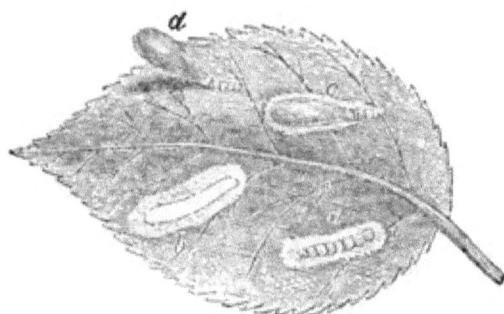

a, The Caterpillar occupying the space it has eaten between the cuticles of the leaf ;
b, a portion of the upper cuticle, cut out for the formation of the tent; *c*, the tent
nearly completed ; *d*, the perfect tent, with the caterpillar protruding its head.

between the two membranes of the leaf, and then uniting
these and cutting out his tent. But the tent itself looks
singular from being all over studded with the stinging
bristles of the nettle, and forming a no less formidable coat
of mail to the little inhabitant than the spiny hide of the
hedgehog. In feeding it does not seem to have mined into
the leaf, but to have eaten the whole of the lower membrane,
along with the entire pulp, leaving nothing but the upper
membrane untouched. (J. R.) During the summer of 1830
we discovered a very large tent which had been formed out
of a blade of grass; and another stuck all over with chips
of leaves upon the common maple.

TENTS OF STONE-MASON CATERPILLARS.

The caterpillar of a small moth (*Tinea*) which feeds upon
the lichens growing on walls, builds for itself a moveable
tent of a very singular kind. M. de la Voye was the first
who described these insects; but though they are frequently
overlooked, from being very small, they are by no means

uncommon on old walls. Réaumur observed them regularly
for twenty years together on the terrace-wall of the Tuileries
at Paris; and they may be found in abundance in similar
situations in this country. This accurate observer refuted
by experiment the notion of M. de la Voye that the caterpillars fed upon the stones of the wall; but he satisfied
himself that they detached particles of the stone for the
purpose of building their tents or sheaths (*fourreaux*), as he
calls their dwellings. In order to watch their mode of
building, Réaumur gently ejected half-a-dozen of them from
their homes, and observed them detach grain after grain
from a piece of stone, binding each into the wall of their

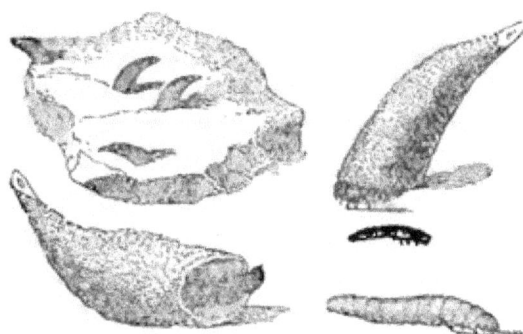

Lichen-Tents and Caterpillars, both of their natural size and magnified.

building with silk till the cell acquired the requisite magnitude, the whole operation taking about twenty-four hours
of continued labour. M. de la Voye mentions small granular
bodies of a greenish colour, placed irregularly on the exterior of the structure, which he calls eggs; but we agree
with Réaumur in thinking it more probable that they are
small fragments of moss or lichen intermixed with the stone:
in fact, we have ascertained that they are so. (J. R.)

When these little architects prepare for their change
into chrysalides before becoming moths, they attach their
tents securely to the stone over which they have hitherto
rambled, by spinning a strong mooring of silk, so as not
only to fill up every interstice between the main entrance
of the tent and the stone, but also weaving a close, thick

curtain of the same material, to shut up the entire aperture.

It is usual for insects which form similar structures to issue, when they assume the winged state, from the broader end of their habitation ; but our little stone-mason proceeds in a different manner. It leaves open the apex of the cone from the first, for the purpose of ejecting its excrements, and latterly it enlarges this opening a little, to allow of a free exit when it acquires wings ; taking care, however, to spin over it a canopy of silk, as a temporary protection, which it can afterwards burst through without difficulty. The moth itself is very much like the common clothes-moth in form, but is of a gilded-bronze colour, and considerably smaller.

In the same locality, M. de Maupertuis found a numerous brood of small caterpillars, which employed grains of stone, not, like the preceding, for building feeding-tents, but for their cocoons. This caterpillar was of a brownish-grey colour, with a white line along the back, on each side of which were tufts of hair. The cocoons which it built were oval, and less in size than a hazel-nut, the grains of the stone being skilfully woven into irregular meshes of silk.

In June, 1829, we found a numerous encampment of the tent-building caterpillars described by MM. de la Voye and Réaumur, on the brick wall of a garden at Blackheath, Kent. (J. R.) They were so very small, however, and so like the lichen on the wall, that had not our attention been previously directed to their habits, we should have considered them as portions of the wall ; for not one of them was in motion, and it was only by the neat, turbinated, conical form in which they had constructed their habitations that we detected them. We tried the experiment above mentioned, of ejecting one of the caterpillars from its tent, in order to watch its proceedings when constructing another ; but probably its haste to procure shelter, or the artificial circumstances into which it was thrown, influenced its operations, for it did not form so good a tent as the first, the texture of the walls being much slighter, while it was more rounded at the apex, and of course not so elegant. Réaumur found, in all his similar experiments, that the new structure equalled the old; but most of the trials of this kind which we have made correspond with the inferiority which we have here recorded. The process indeed is the same, but it seems to be done with more hurry and less care. It may be, indeed, in some cases, that the supply of silk necessary to unite the bits of stone, earth, or lichen employed, is too scanty for perfecting a second structure.

We remarked a very singular circumstance in the operations of our little architect, which seems to have escaped the minute and accurate attention of Réaumur. When it commenced its structure, it was indispensable to lay a foundation for the walls about to be reared; but as the tent was to be moveable like the shell of a snail, and not stationary, it would not have answered its end to cement the foundation to the wall. We had foreseen this difficulty, and felt not a little interested in discovering how it would be got over. Accordingly, upon watching its movements with some attention, we were soon gratified to perceive that it used its own body as the primary support of the building. It fixed a thread of silk upon one of its right feet, warped it over to the corresponding left foot, and upon the thread thus stretched between the two feet it glued grains of stone and chips of lichen, till the wall was of the required thickness. Upon this, as a foundation, it continued to work till it had formed a small portion in form of a parallelogram; and proceeding in a similar way, it was not long in making a ring a very little wider than sufficient to admit its body. It extended this ring in breadth, by working on the inside only, narrowing the diameter by degrees, till it began to take the form of a cone. The apex of this cone was not closed up, but left as an aperture through which to eject its excrements.

It is worthy of remark, that one of the caterpillars which we deprived of its tent attempted to save itself the trouble of building a new one, by endeavouring to unhouse one of its neighbours. For this purpose, it got upon the outside of the inhabited tent, and, sliding its head down to the entrance, tried to make its way into the interior. But the rightful owner did not choose to give up his premises so easily, and fixed his tent down so firmly upon the table where we had placed it, that the intruder was forced to abandon his attempt. The instant, however, that the other unmoored his tent and began to move about, the invader renewed his efforts to eject him, persevering in the struggle for several hours, but without a chance of success. At one time we imagined that he would have accomplished his felonious intentions; for he bound down the apex of the tent to the table with cables of silk. But he attempted his entrance at the wrong end. He ought to have tried the aperture in the apex, by enlarging which a little he would undoubtedly have made good his entrance; and as the inhabitant could not have turned upon him for want of room, the castle must have been surrendered. This experiment, however, was not tried, and there was no hope for him at the main entrance.

Muff-shaped Tents.

The ingenuity of man has pressed into his service not only the wool, the hair, and even the skins of animals, but has most extensively searched the vegetable kingdom for the materials of his clothing. In all this, however, he is rivalled by the tiny inhabitants of the insect world, as we have already seen; and we are about now to give an additional instance of the art of a species of caterpillars which select a warmer material for their tents than even the caterpillar of the clothes-moth. It may have been remarked by many who are not botanists, that the seed-catkins of the willow become, as they ripen, covered with a species of down or cotton, which, however, is too short in the fibre to be advantageously employed in our manufactures. But the caterpillars, to which we have alluded, find it well adapted for their habitations.

The muff-looking tent in which we find these insects does not require much trouble to construct; for the caterpillar does not, like the clothes-moth caterpillar, join the willow-cotton together, fibre by fibre—it is contented with the

a, Branch of the Willow, with seed-spikes covered with cotton ; *b,* Muff-tents, made of this cotton by *c,* the Caterpillar.

state in which it finds it on the seed. Into this it burrows, lines the interior with a tapestry of silk, and then detaches the whole from the branch where it was growing, and carries it about with it as a protection while it is feeding.*

An inquiring friend of Réaumur having found one of these insects floating about in its muff-tent upon water, concluded that they feed upon aquatic plants; but he was soon convinced that it had only been blown down by an accident, which must frequently happen, as willows so often hang over water. May it not be, that the buoyant materials of the tent were intended to furnish the little inhabitant

* Réaumur, iii. p. 130.

with a life-boat, in which, when it chanced to be blown into the water, it might sail safely ashore and regain its native tree ?

LEAF-MINING CATERPILLARS.

The process of mining between the two membranes of a leaf is carried on to a farther extent by minute caterpillars allied to the tent-makers above described. The tent-maker never deserts his house, except when compelled, and therefore can only mine to about half the length of his own body; but the miners now to be considered make the mine itself their dwelling-place, and as they eat their way, they lengthen and enlarge their galleries. A few of these mining caterpillars are the progeny of small weevils (*Curculionidæ*), some of two-winged flies (*Diptera*), but the greater number are produced from a genus of minute moth (*Œcophora*, LATR.), which, when magnified, appear to be amongst the most splendid and brilliant of Nature's productions, vying even with the humming-birds and diamond-beetles of the tropics in the rich metallic colours which bespangle their wings. Well may Bonnet call them "tiny miracles of Nature," and regret that they are not *en grand.**

There are few plants or trees whose leaves may not, at some season of the year, be found mined by these caterpillars, the track of whose progress appears on the upper surface in winding lines. Let us take one of the most common of these for an example,—that of the rose-leaf, produced by the caterpillar of Ray's golden-silver spot (*Argyromiges Rayella?* CURTIS), of which we have just gathered above a dozen specimens from one rose-tree. (J. R.)

It may be remarked that the winding line is black, closely resembling the tortuous course of a river on a map,—beginning like a small brook, and gradually increasing in breadth as it proceeds. This representation of a river exhibits, besides, a narrow white valley on each side of it, increasing as it goes, till it terminates in a broad delta. The valley is the

* Bonnet, 'Contempl. de la Nature,' part xii.

portion of the inner leaf from which the caterpillar has eaten the pulp (*parenchyma*), while the river itself has been formed by the liquid *ejectamenta* of the insect, the watery part becoming evaporated. In other species of miners,

Leaf of the Monthly Rose (*Rose Indica*), mined by Caterpillars of Argyromiges?

however, the dung is hard and dry, and consequently these only exhibit the valley without the river (see p. 255).

On looking at the back of the leaf, where the winding line begins, we uniformly find the shell of the very minute egg from which the caterpillar has been hatched, and hence perceive that it digs into the leaf the moment it escapes from the egg, without wandering a hair's-breadth from the spot; as if afraid lest the air should visit it too roughly. The egg is, for the most part, placed upon the mid-rib of the rose-leaf, but sometimes on one of the larger nervures. When once it has got within the leaf, it seems to pursue no certain direction, sometimes working to the centre, sometimes to the circumference, sometimes to the point, and sometimes to the base, and even, occasionally, crossing or keeping parallel to its own previous track.

The most marvellous circumstance, however, is the minuteness of its workmanship; for though a rose-leaf is thinner than this paper, the insect finds room to mine a tunnel to live in, and plenty of food, without touching

the two external membranes. Let any one try with the
nicest dissecting instruments to separate the two plates of
a rose-leaf, and he will find it impossible to proceed far
without tearing one or other. The caterpillar goes still
further in minute nicety; for it may be remarked, that
its track can only be seen on the upper, and not on the
under surface of the leaf, proving that it eats as it proceeds only half the thickness of the pulp, or that portion of
it which belongs to the upper membrane of the leaf.

We have found this little miner on almost every sort of
rose-tree, both wild and cultivated, including the sweet-briar,
in which, the leaf being very small, it requires nearly the
whole parenchyma to feed one caterpillar. They seem, however, to prefer the foreign monthly rose to any of our native
species, and there are few trees of this where they may not
be discovered.

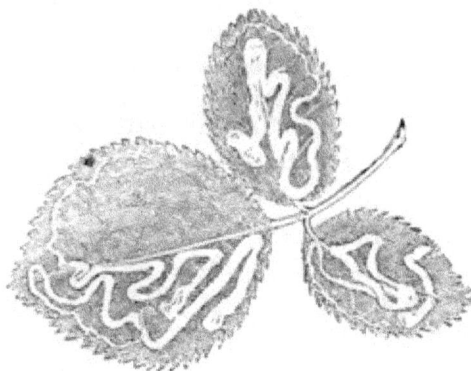

Leaf of the Dew-berry Bramble (*Rubus cœsius*), mined by Caterpillars.

Tunnels very analogous to the preceding may be found
upon the common bramble (*Rubus fruticosus*); and on the
holly, early in spring, one which is in form of an irregular
whitish blotch. But in the former case, the little miner
seems to proceed more regularly, always, when newly
hatched, making directly for the circumference, upon or near
which also the mother-moth deposits her egg, and winding
along for half the extent of the leaf close upon the edge,
following, in some cases, the very indentations formed by
the terminating nervures.

The bramble-leaf miner seems also to differ from that of the rose-leaf, by eating the pulp both from the upper and under surface, at least the track is equally distinct above and below; yet this may arise from the different consistence of the leaf pulp, that in the rose being firm, while that of the bramble is soft and puffy.

On the leaves of the common primrose (*Primula veris*), as well as on the garden variety of it, the polyanthus, one of these mining caterpillars may very frequently be found. It is, however, considerably different from the preceding, for there is no black trace—no river to the valley which it excavates: its ejectamenta, being small and solid, are seen,

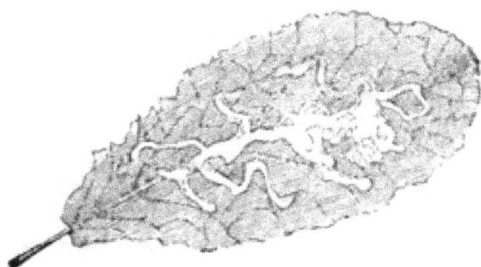

Leaf of the Primrose (*Primula veris*), mined by a Caterpillar.

when the leaf is dried, in little black points like grains of sand. This miner also seems more partial than the preceding to the mid-rib and its vicinity, in consequence of which its path is seldom so tortuous, and often appears at its extremity to terminate in an area comparatively extensive, arising from its recrossing its previous tracks. (J. R.)

Swammerdam describes a mining caterpillar which he found on the leaves of the alder, though it did not, like those we have just described, excavate a winding gallery; it kept upon the same spot, and formed only an irregular area. A moth was produced from this, whose upper wings, he says, "shone and glittered most gloriously with crescents of gold, silver, and brown, surrounded by borders of delicate black." Another area miner which he found on the leaves of willows, as many as seventeen on one leaf, producing what appeared to be rusty spots, was metamorphosed into a very minute weevil (*Curculio Rhionoc.*). He says he has been informed that, in warm climates, worms an inch long are found in leaves, and adds, with great simplicity, "on these many fine experiments might have been made, if the inhabitants had not laboured under the cursed thirst of gold."*

The vine-leaf miner, when about to construct its cocoon, cuts, from the termination of its gallery, two pieces of the membrane of the leaf, deprived of their pulp, in a similar manner to the tent-makers described above, uniting them and lining them with silk. This she carries to some distance before she lays herself up to undergo her change. Her mode of walking under her burthen is peculiar, for, not contented with the security of a single thread of silk, she forms, as Bonnet says, "little mountains (*monticules*) of silk, from distance to distance, and seizing one of these with her teeth, drags herself forward, and makes it a scaffolding from which she can build another."† Some of the miners, however, do not leave their galleries, but undergo their transformations there, taking the precaution to mine a cell, not in the upper, but in the under surface; others only shift to another portion of the leaf.

Social Leaf-Miners.

The preceding descriptions apply to caterpillars who construct their mines in solitude, there being seldom more than one on a leaf or leaflet, unless when two mother-flies happen to lay their eggs on the same leaf; but there are others, such as the miners of the leaves of the henbane (*Hyoscyamus niger*), which excavate a common area in concert—from four to eight forming a colony. These are very like flesh-maggots, being larger than the common miners; the leaves of this plant, from being thick and juicy, giving them space to work and plenty to eat.

Most of the solitary leaf-miners either cannot or will not construct a new mine, if ejected by an experimenter from the old, as we have frequently proved; but this is not the

* Swammerd., ' Book of Nature,' vol. ii. p. 84.
† ' Contempl. de la Nature,' part xii. p. 197.

case with the social miners of the henbane-leaf. Bonnet ejected one of these, and watched it with his glass till it commenced a new tunnel, which it also enlarged with great expedition; and in order to verify the assertion of Réaumur, that they neither endeavour nor fear to meet one another, he introduced a second. Neither of them manifested any knowledge of the other's contiguity, but both worked hard at the gallery, as did a third and a fourth which he afterwards introduced; for though they seemed uneasy, they never attacked one another, as the solitary ones often do when they meet.*

BARK-MINING CATERPILLARS.

A very different order of mining caterpillars are the progeny of various beetles, which excavate their galleries in the soft inner bark of trees, or between it and the young wood (*alburnum*). Some of these, though small, commit extensive ravages, as may readily be conceived when we are told that as many as eighty thousand are occasionally found on one tree. In 1783 the trees thus destroyed by the printer-beetle (*Tomicus typographus*, LATR.), so called from its tracks resembling letters, amounted to above a million and a half in the Hartz forest. It appears there periodically, and confines its ravages to the fir. This insect is said to have been found in the neighbourhood of London.

On taking off the bark of decaying poplars and willows, we have frequently met with the tracks of a miner of this order, extending in tortuous pathways, about a quarter of an inch broad, for several feet and even yards in length. The excavation is not circular, but a compressed oval, and crammed throughout with a dark-coloured substance like sawdust—the excrement no doubt of the little miner, who is thereby protected from the attacks of *Staphylinidæ*, and other predaceous insects from behind. But though we have found a great number of these subcortical tracks, we have never discovered one of the miners, though they are very probably

* Bonnet, 'Observ. sur les Insectes,' vol. ii. p. 425.

s

the grubs of the pretty musk-beetle (*Cerambyx moschatus*), which are so abundant in the neighbourhood of the trees in question, that the very air in summer is perfumed with their odour. (J. R.)

[Mr. Rennie is undoubtedly right in his suggestion. I have found similar holes in old willow trees, and have traced them throughout their varied ramifications. They contain the larvæ and pupæ of the musk-beetle, some of which may be seen in the Museum at Oxford. On these trees, which mostly grow along the banks of the Cherwell, the perfect beetle was so abundant that it might be taken in any number, and, as described by Mr. Rennie, the air was perfumed with its powerful and agreeable odour. So strong is the scent of this beetle, that I have known it adhere to gloves after the lapse of many weeks, and I have often caught the scent when passing along the road, and merely by the aid of the nostrils discovered the insect.

[On account of the vast number of carpenter-beetles, it is impossible to notice more than a few of them, and we will therefore select some of the most conspicuous. One of them, belonging to the genus *Ptilinus*, is very familiar to us as boring into wooden furniture, and producing the effect which is popularly called " worm-eaten." Fortunately, the little creatures can be easily ejected, and the wood rendered free of them ever afterwards. All that is needed is to take a syringe with a very fine aperture—an injecting syringe is the best—and by its aid to force into the holes a solution of corrosive sublimate in spirits of wine—say a large teaspoonful of the powdered salt to a pint of spirits. The rapidity with which the poisoned spirit permeates the wood is wonderful, and in a short time it may be seen oozing out of twenty or thirty holes at once. This solution is peculiarly effective, as it kills all the insects, destroys every egg that it touches, and renders the wood poisonous to the grubs that happen to escape. I used to be greatly plagued with the Ptilinus among my ethnological collection, until I tried the corrosive sublimate, and ever since my spears, bows and arrows, and clubs have remained intact.

[Another troublesome insect is the *Scolytus destructor*,

which makes its radiating tunnels between the bark and the tree. Whole forests have been destroyed by this voracious little beetle, the bark having been completely detached, and the tree necessarily killed. The habits of this beetle are well described in the following passage.]

We have frequently observed a very remarkable instinct in the grubs of a species of beetle (*Scolytus destructor.*

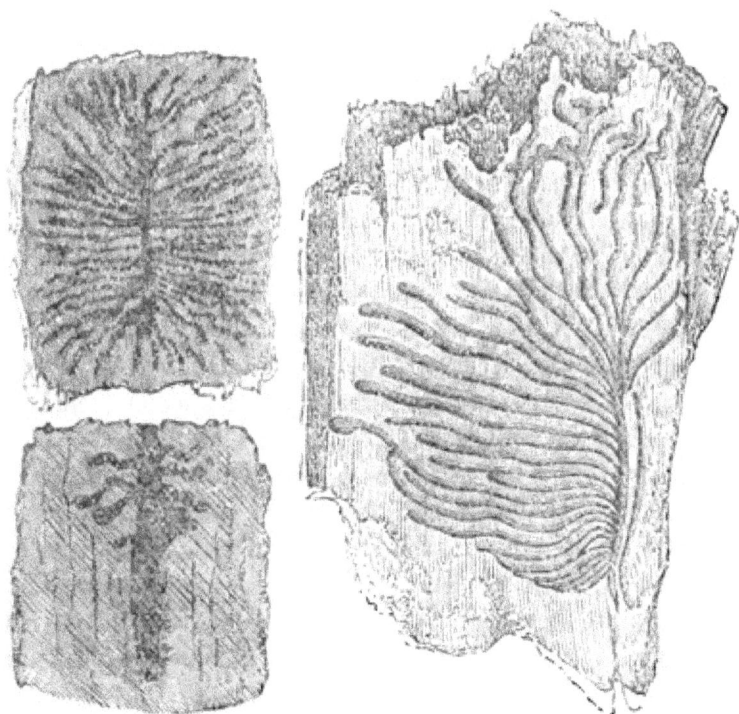

Bark mined in rays by beetle-grubs.

Geoffroy), which lives under the dead bark of trees. The mother insect, as is usual with beetles, deposits her eggs in a patch or cluster in a chink or hole in the bark; and when the brood is hatched, they begin feeding on the bark which had formed their cradle. There is, of course, nothing wonderful in their eating the food selected by their mother; but it appears that, like the caterpillars of the clothes-moth, and the tent insects, they cannot feed except under cover. They dig, therefore, long tubular galleries between the bark and the wood; and, in order not to interfere with the *runs* of their brethren, they branch off from the place of hatching like rays from the centre of a circle : though these are not always in a right line, yet, however near they may approach to the contiguous ones, none of them ever break into each other's premises. We cannot but admire the remarkable instinct implanted in these grubs by their Creator; which guides them thus in lines diverging farther and farther as they increase in size, so that they are prevented from interfering with the comforts of one another.

[We now come to one or two of the beetles which bore deeply into the very wood of the trees. As a rule, the musk-beetle keeps rather towards the exterior of the tree, but there are many that are not so cautious, and which besides damage the tree additionally by nibbling a quantity of chips, wherewith they strengthen their cocoons. We will first take the two insects which are shown in the accompanying illustration. That on the right hand is a species of weevil, or Curculio, and is an undescribed species belonging to the genus Rhyncophorus. It is a native of Australia. The insect and its cocoon are drawn one third less than their real size. The colour of the beetle is warm chestnut brown, and the bold marks on the thorax are jet black. In its larval state it burrows into the palm-trees, and when about to assume the pupal condition it makes the remarkable cocoon which is figured. Generally, these wooden cocoons are made of little chips which are bitten from the wood, and woven together with silk. This cocoon, however, is made of long fibres, which are torn rather than bitten, and are so long that one of them will sometimes encircle the cell three times, making an average length of nine inches. It is tolerably compact in structure, and the colour is pale brown.

[On the left hand is an opened cocoon of an English beetle belonging to the genus Rhagium. Like the last-mentioned insect, the Rhagium prefers long fibres to short chips, though it does not use them of such a length as the Rhyncophorus. The cocoon is generally made between the bark and the wood, from the latter of which the fibres are torn. In consequence of the mode of structure, the cocoon is pale straw colour, while the hollow in which it rests is quite dark.

[All entomologists are familiar with the pretty little wasp-beetle (*Clytus arietis*), which derives its popular name from the wasp-like colours of its body. In the larval state it is one of the carpenter-grubs, and may be found in posts, fir-trees, and similar localities. In this country, although

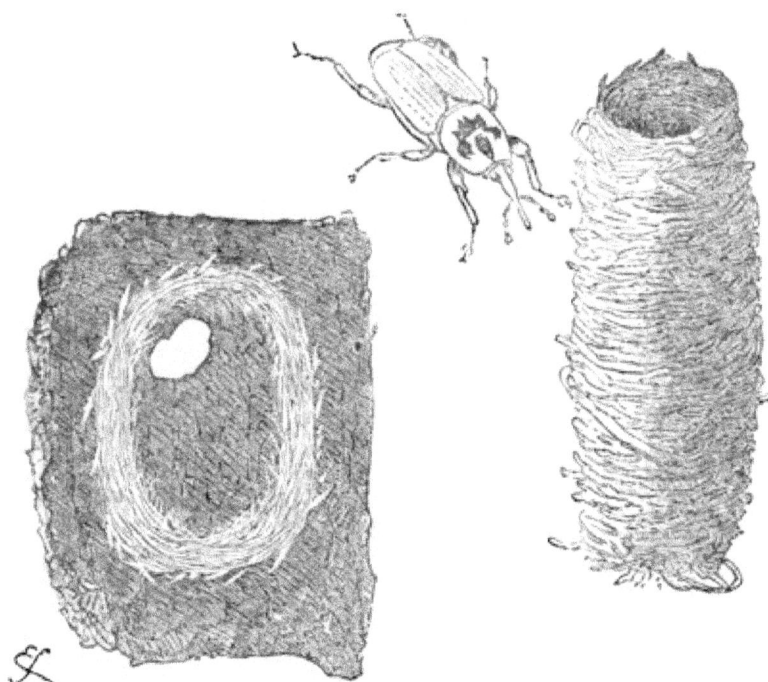

Cocoon of Rhagium.   Rhyncophorus and cocoon.

plentiful, it is not numerous enough to do much harm; but in Ceylon, a closely-allied species is one of the pests of the island. It is popularly known by the title of coffee-borer, from its habit of boring into the stems of the coffee-plant. The landowner looks with absolute horror on this pretty but destructive insect, and would pay a heavy sum annually to any one who would undertake to extirpate the tiny foe. Whole plantations have been swept off by it, and up to the present time no remedy has had more than a temporary and partial success.]

Another capricorn beetle of this family is no less destructive to bark in its perfect state than the above are when grubs, as from its habit of eating round a tree, it cuts the course of the returning sap, and destroys it.

[The late Mr. Waterton once showed me a stout branch which had fallen on his head while he was standing under a tree, the branch having been cut completely through by the jaws of some large longicorn beetle. The mode in which the insect had severed the branch was exactly like that which is practised by the beaver when it cuts down a tree.

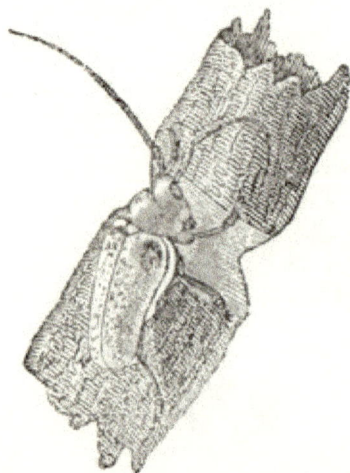

Capricorn Beetle (*Cerambyx Lamia amputator*) rounding off the bark of a tree.

[The burrows and cocoons of two other species of Cerambyx are shown in the accompanying illustration, and in both cases may be seen the provision which is made for the exit of the beetle after it has attained the perfect condition. The double burrow of the left is that of *Cerambyx carcharias*, and those on the right-hand figure are the habitations of *Cerambyx populneus*. The reader will see how these insects cut up the wood of the branch, and can well understand the infinite mischief which can be done to a coffee plantation by the Clytus.

The last wood-boring beetle which will be mentioned is

the stag-beetle of our own country. In the larval state this insect resides in tree trunks, mostly towards the roots, and therefore escapes observation more successfully than would be the case if it inhabited a higher portion of the tree. When full-grown, the larva is of enormous size, and the hole which it bores is necessarily of corresponding dimensions. In

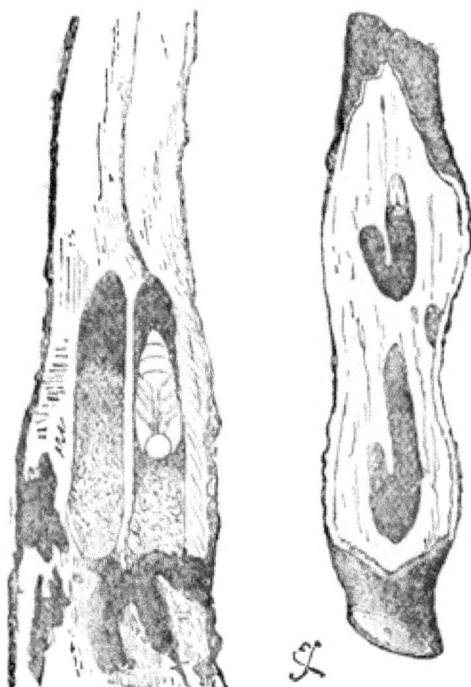

Cerambyx carcharias.   *Cerambyx populneus.*

some parts of England, the larva are popularly called "Joe Bassetts," and are said to turn into "Pincher Bobs." The latter title is a very appropriate one, as any one can testify who has allowed a fine male stag-beetle to grip his finger between its jaws. As to the Joe Bassett, it is simply a local name.]

CHAPTER XIII.

STRUCTURES OF GRASSHOPPERS, CRICKETS, AND BEETLES.

GRASSHOPPERS, locusts, crickets, and beetles are, in many respects, no less interesting than the insects whose architectural proceedings we have already detailed. They do not, indeed, build any edifice for the accommodation of themselves or their progeny ; but most, if not all of them, excavate retreats in walls or in the ground.

The house-cricket (*Acheta domestica*) is well known for its habit of picking out the mortar of ovens and kitchen fire-places, where it not only enjoys warmth, but can procure abundance of food. It is usually supposed that it feeds on bread. M. Latreille says it only eats insects, and it certainly thrives well in houses infested by the cockroach ; but we have also known it eat and destroy lamb's-wool stockings, and other wollen stuffs, hung near a fire to dry. It is evidently not fond of hard labour, but prefers those places where the mortar is already loosened, or at least is new, soft, and easily scooped out; and in this way it will dig covert ways from room to room. In summer, crickets often make excursions from the house to the neighbouring fields, and dwell in the crevices of rubbish, or the cracks made in the ground by dry weather, where they chirp as merrily as in the snuggest chimney-corner. Whether they ever dig retreats in such circumstances we have not ascertained : though it is not improbable they may do so for the purpose of making nests. M. Bory St. Vincent tells us that the Spaniards are so fond of crickets that they keep them in cages like singing birds.*

* Dict. Classique d'Hist. Nat. Art. Grillon.

THE MOLE-CRICKET.

The insect, called, from its similarity of habits to the mole, the mole-cricket (*Gryllotalpa vulgaris*, LATR.), is but too well known in gardens, corn-fields, and the moist banks of rivers and ponds, in some parts of England, such as Wilt-shire and Hampshire, though it is comparatively rare or unknown in others. It burrows in the ground, and forms extensive galleries similar to those of the mole, though

The Mole-Cricket, with a separate outline of one of its hands.

smaller; and these may always be recognized by a slightly elevated ridge of mould · for the insect does not throw up the earth in hillocks like the mole, but gradually, as it digs along, in the manner of the field-mouse. In this way it commits great ravages, in hotbeds and in gardens, upon peas, young cabbages, and other vegetables, the roots of which it is said to devour. It is not improbable, we think, that, like its congener, the house-cricket, it may also prey upon under-ground insects, and undermine the plants to get at them, as the mole has been proved to do. Mr Gould, indeed, fed a mole-cricket for several months upon ants.

The structure of the mole-cricket's arms and hands (if we

may call them so) is admirably adapted for these operations, being both very strong, and moved by a peculiar apparatus of muscles. The breast is formed of a thick, hard, horny substance, which is further strengthened within by a double framework of strong gristle, in front of the extremities of which the shoulder-blades of the arms are firmly jointed: a structure evidently intended to prevent the breast from being injured by the powerful action of the muscles of the arms in digging. The arms themselves are strong and broad, and the hand is furnished with four large sharp claws, pointed somewhat obliquely outwards, this being the direction in which it digs, throwing the earth on each side of its course. So strongly indeed does it throw out its arms, that we find it can thus easily support its own weight when

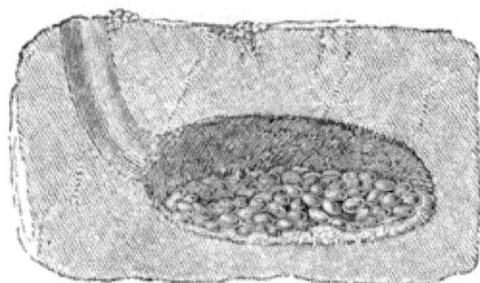

Nest of the Mole-Cricket.

held between the finger and thumb, as we have tried upon half-a-dozen of the living insects now in our possession.

The nest which the female constructs for her eggs, in the beginning of May, is well worthy of attention. The Rev. Mr. White, of Selborne, tells us that a gardener, at a house where he was on a visit, while mowing grass by the side of a canal, chanced to strike his scythe too deep, and pared off a large piece of turf, laying open to view an interesting scene of domestic economy. There was a pretty chamber dug in the clay, of the form and about the dimensions it would have had if moulded by an egg, the walls being neatly smoothed and polished. In this little cell were deposited about a hundred eggs, of the size and form of caraway comfits, and of a dull tarnished white colour. The eggs were not very deep, but just under a little heap of fresh mould, and within the influence of the sun's heat.* The dull tarnished white colour, however, scarcely agrees with a parcel of these eggs now before us, which are translucent, gelatinous, and greenish.

Like the eggs and young of other insects, however, those of the mole-cricket are exposed to depredation, and particularly to the ravages of a black beetle which burrows in similar localities. The mother insect, accordingly, does not think her nest secure till she has defended it, like a fortified town, with labyrinths, intrenchments, ramparts, and covert ways. In some part of these outworks she stations herself as an advanced guard, and when the beetle ventures within her circumvallations, she pounces upon him and kills him.

THE FIELD-CRICKET.

Another insect of this family, the field-cricket (*Acheta campestris*), also forms burrows in the ground, in which it lodges all day, and comes out chiefly about sunset to pipe its evening song. It is so very shy and cautious, however, that it is by no means easy to discover either the insect or its burrow. "The children in France amuse themselves with hunting after the field-cricket; they put into its hole an ant fastened by a long hair, and as they draw it out the cricket does not fail to pursue it, and issue from its retreat. Pliny informs us it might be captured in a much more expeditious and easy manner. If, for instance, a small and slender piece of stick were to be thrust into the burrow, the insect, he says, would immediately get upon it for the purpose of demanding the occasion of the intrusion: whence arose the proverb, *stultior grillo* (more foolish than a cricket), applied to one who, upon light grounds, provokes his enemy, and falls into the snares which might have been laid to entrap him."†

The Rev. Mr. White, who attentively studied their habits

* Natural History of Selborne, ii. 82.
† Entomologie, par R. A. E. 18mo., Paris, 1826, p. 168.

and manners, at first made an attempt to dig them out with
a spade, but without any great success ; for either the bottom
of the hole was inaccessible, from its terminating under a
large stone, or else in breaking up the ground the poor
creature was inadvertently squeezed to death. Out of one
thus bruised a great number of eggs were taken, which were
long and narrow, of a yellow colour, and covered with a very
tough skin. More gentle means were then used, and these
proved successful. A pliant stalk of grass, gently insinuated
into the caverns, will probe their windings to the bottom, and
bring out the inhabitant ; and thus the humane inquirer may
gratify his curiosity without injuring the object of it.

When the males meet, they sometimes fight very fiercely,
as Mr. White found by some that he put into the crevices of
a dry stone wall, where he wished to have them settle. For
though they seemed distressed by being taken out of their
knowledge, yet the first that got possession of the chinks
seized on all the others that were obtruded upon him with
his large row of serrated fangs. With their strong jaws,
toothed like the shears of a lobster's claws, they perforate
and round their curious regular cells, having no fore-claws
to dig with, like the mole-cricket. When taken into the
hand, they never attempt to defend themselves, though
armed with such formidable weapons. Of such herbs as
grow about the mouths of their burrows they eat indiscriminately, and never in the day-time seem to stir more than
two or three inches from home. Sitting in the entrance
of their caverns, they chirp all night as well as day, from
the middle of the month of May to the middle of July. In
hot weather, when they are most vigorous, they make the
hills echo ; and, in the more still hours of darkness, may be
heard to a very considerable distance. "Not many summers ago," says Mr. White, " I endeavoured to transplant a
colony of these insects to the terrace in my garden, by boring
deep holes in the sloping turf. The new inhabitants stayed
some time, and fed and sang ; but they wandered away by
degrees, and were heard at a greater distance every morning ;
so it appears that on this emergency they made use of their

wings in attempting to return to the spot from which they were taken."* The manner in which these insects lay their eggs is represented in the following figure, which is that of an insect nearly allied to the crickets, though of a different genus.

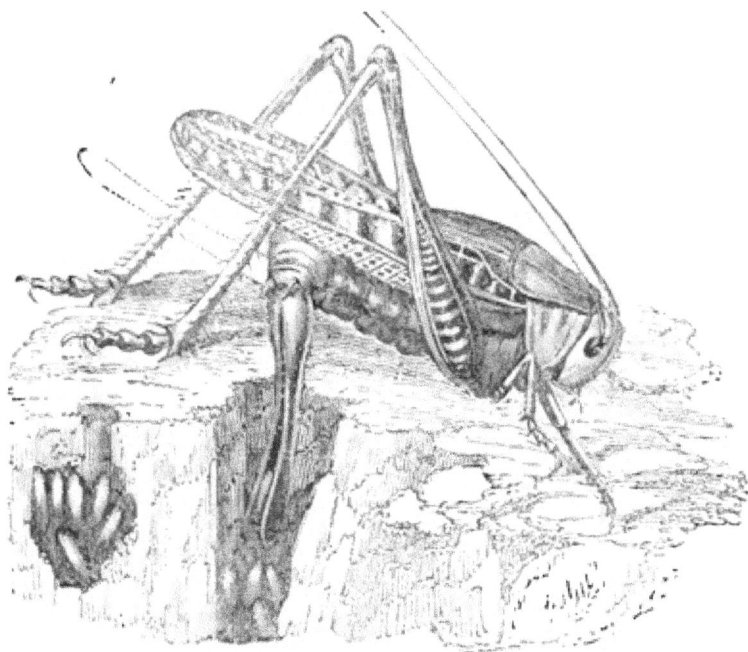

Acrida verrucivora depositing her eggs.
The usual position of the ovipositor is represented by dots.

A more laborious task is performed by an insect by no means uncommon in Britain, the burying beetle (*Necrophorus vespillo*), which may be easily recognized by its longish body, of a black colour, with two broad and irregularly indented bands of yellowish brown. A foreign naturalist, M. Gleditsch, gives a very interesting account of its industry. He had "often remarked that dead moles, when laid upon the ground, especially if upon loose earth, were almost sure to disappear in the course of two or three days, often of twelve hours. To ascertain the cause, he placed a mole upon one of the beds in his garden. It had vanished

* Natural History of Selborne.

by the third morning; and on digging where it had been laid, he found it buried to the depth of three inches, and under it four beetles, which seemed to have been the agents in this singular inhumation. Not perceiving anything particular in the mole, he buried it again; and on examining it at the end of six days, he found it swarming with maggots, apparently the issue of the beetles, which M. Gleditsch now naturally concluded had buried the carcase for the food of their future young. To determine these points more clearly, he put four of these insects into a glass vessel, half filled with earth and properly secured, and upon the surface of the earth two frogs. In less than twelve hours one of the frogs was interred by two of the beetles; the other two ran about the whole day, as if busied in measuring the dimensions of the remaining corpse, which on the third day was also found buried. He then introduced a dead linnet. A pair of the beetles were soon engaged upon the bird. They began their operations by pushing out the earth from under the body, so as to form a cavity for its reception; and it was curious to see the efforts which the beetles made, by dragging at the feathers of the bird from below, to pull it into its grave. The male, having driven the female away, continued the work alone for five hours. He lifted up the bird, changed its place, turned it, and arranged it in the grave, and from time to time came out of the hole, mounted upon it, and trod it under foot, and then retired below, and pulled it down. At length, apparently wearied with this uninterrupted labour, it came forth, and leaned its head upon the earth beside the bird, without the smallest motion, as if to rest itself, for a full hour, when it again crept under the earth. The next day, in the morning, the bird was an inch and a half under ground, and the trench remained open the whole day, the corpse seeming as if laid out upon a bier, surrounded with a rampart of mould. In the evening it had sunk half an inch lower, and in another day the work was completed, and the bird covered. M. Gleditsch continued to add other small dead animals, which were all sooner or later buried; and the result of his experiment was, that in fifty days four beetles had interred, in the very small space of earth allotted to them, twelve carcases, viz., four frogs, three small birds, two fishes, one mole, and two grasshoppers, besides the entrails of a fish, and two morsels of the lungs of an ox. In another experiment, a single beetle buried a mole, forty times its own bulk and weight, in two days."*

In the summer of 1826, we found on Putney Heath, in Surrey, four of these beetles, hard at work in burying a dead crow, precisely in the manner described by M. Gleditsch. (J. R.)

DUNG-BEETLE.

A still more common British insect, the dorr, clock, or dung-beetle (*Geotrupes stercorarius*), uses different materials for burying along with its eggs. "It digs," to use the words of Kirby and Spence, "a deep cylindrical hole, and carrying down a mass of the dung to the bottom, in it deposits its eggs. And many of the species of the genus *Ateuchus* roll together wet dung into round pellets, deposit an egg in the midst of each, and when dry push them backwards, by their hind feet, to holes of the surprising depth of three feet, which they have previously dug for their reception, and which are often several yards distant. The attention of these insects to their eggs is so remarkable, that it was observed in the earliest ages, and is mentioned by ancient writers, but with the addition of many fables; as that they were all of the male sex ; that they became young again every year ; and that they rolled the pellets containing their eggs from sunrise to sunset every day, for twenty-eight days, without intermission."†

"We frequently notice in our evening walks," says Mr. Knapp, "the murmuring passage, and are often stricken by the heedless flight of the great dorr-beetle (*Geotrupes stercorarius*), clocks, as the boys call them. But this evening my

* Act. Acad. Berolin. 1752, et Gleditsch, Phys. Botan., quoted by Kirby and Spence, ii. 353.
† Moufet, 153. Kirby and Spence, ii. 350.

attention was called to them in particular, by the constant
passing of such a number as to constitute something like a
little stream ; and I was led to search into the object of their
direct flight, as in general it is irregular and seemingly
inquisitive. I soon found that they dropped on some recent
nuisance : but what powers of perception must these creatures possess, drawn from all distances and directions by the
very little fetor which, in such a calm evening, could be
diffused around, and by what inconceivable means could
odours reach this beetle in such a manner as to rouse so
inert an insect into action ! But it is appointed one of the
great scavengers of the earth, and marvellously endowed
with powers of sensation, and means of effecting this purpose of its being. Exquisitely fabricated as it is to receive
impressions, yet probably it is not more highly gifted than
any of the other innumerable creatures that wing their way
around us, or creep about our · paths, though by this one
perceptible faculty, thus ' dimly seen,' it excites our wonder
and surprise. How wondrous then the whole !

"The perfect cleanliness of these creatures is a very
notable circumstance, when we consider that nearly their
whole lives are passed in burrowing in the earth, and removing nuisances ; yet such is the admirable polish of their
coating and limbs, that we very seldom find any soil adhering to them. The meloe, and some of the scarabæi, upon
first emerging from their winter's retreat, are commonly
found with earth clinging to them ; but the removal of this
is one of the first operations of the creature ; and all the
beetle race, the chief occupation of which is crawling about
the soil, and such dirty employs, are, notwithstanding,
remarkable for the glossiness of their covering, and freedom
from defilements of any kind. But purity of vesture seems
to be a principal precept of Nature, and observable throughout creation. Fishes, from the nature of the element in
which they reside, can contract but little impurity. Birds
are unceasingly attentive to neatness and lustration of their
plumage. All the slug race, though covered with slimy
matter calculated to collect extraneous things, and reptiles,

are perfectly free from soil. The fur and hair of beasts, in a state of liberty and health, is never filthy or sullied with dirt. Some birds roll themselves in dust, and, occasionally, particular beasts cover themselves with mire; but this is not from any liking or inclination for such things, but to free themselves from annoyances, or to prevent the bites of insects. Whether birds in preening, and beasts in dressing themselves, be directed by any instinctive faculty, we know not; but they evidently derive pleasure from the operation, and thus this feeling of enjoyment, even if the sole motive, becomes to them an essential source of comfort and of health."[*]

The rose or green chafer (*Cetonia aurata*), which is one of our prettiest native insects, is one of the burrowers, and, for the purpose of depositing her eggs, digs, about the middle of June, into soft light ground. When she is seen at this operation, with her broad and delicate wings folded up in their shining green cases, speckled with white, it could hardly be imagined that she had but just descended from the air, or dropped down from some neighbouring rose.

The proceedings of the Tumble-Dung Beetle of America (*Scarabæus pilularius*, LINN.) are described in a very interesting manner by Catesby, in his 'Carolina.' "I have," says he, "attentively admired their industry, and mutual assisting of each other in rolling their globular balls from the place where they made them to that of their interment, which is usually the distance of some yards, more or less. This they perform breech foremost, by raising their hind parts, and forcing along the ball with their hind feet. Two or three of them are sometimes engaged in trundling one ball, which, from meeting with impediments on account of the unevenness of the ground, is sometimes deserted by them. It is, however, attempted by others with success, unless it happens to roll into some deep hollow or chink, where they are constrained to leave it; but they continue their work by rolling off the next ball that comes in their way. None of them

---

* Journal of a Naturalist, p. 311.

seem to know their own balls, but an equal care for the whole appears to affect all the community. They form these pellets while the dung remains moist, and leave them to harden in the sun before they attempt to roll them. In their moving of them from place to place, both they and the balls may frequently be seen tumbling about the little eminences that are in their way. They are not, however, easily discouraged ; and, by repeating their attempts, usually surmount the difficulties."

He further informs us that they " find out their subsistence by the excellency of their noses, which direct them in their flight to newly-fallen dung, on which they immediately go to work, tempering it with a proper mixture of earth. So intent are they always upon their employment, that, though handled or otherwise interrupted, they are not to be deterred, but immediately, on being freed, persist in their work without any apprehension of danger. They are said to be so exceedingly strong and active as to move about, with the greatest ease, things that are many times their own weight. Dr. Brichell was supping one evening in a planter's house of North Carolina, when two of them were conveyed, without his knowledge, under the candlesticks. A few blows were struck on the table, and, to his great surprise, the candlesticks began to move about, apparently without any agency ; and his surprise was not much lessened when, on taking one of them up, he discovered that it was only a chafer that moved it."

We have often found the necklace-beetle (*Carabus monilis*) inhabiting a chamber dug out in the earth of a garden, just sufficient to contain its body, and carefully smoothed and polished. From the form of this little nest, it would seem as if it were constructed, not by digging out the earth and removing it, but chiefly by the insect pushing its body forcibly against the walls. The beetles which we have found nestling in this manner have been all males ; and therefore it cannot be intended for a breeding-cell ; for male insects are never, we believe, sufficiently generous to their mates to assist them in such labours. The beetle in question appears to be partial to celery trenches (J. R.); probably from the loose earth of which they are composed yielding, without much difficulty, to the pressure of its body.

[Many of the subterranean larvæ which are turned up by the spade or the plough are the imperfect conditions of earth-burrowing beetles, and many of them are among the most insidious pests of the farmer, their ravages being all the more dangerous because they are unseen.]

The most destructive, perhaps, of the creatures usually called grubs are the larvæ of the may-bug or cockchafer (*Melolontha vulgaris*), but too well known, particularly in the southern and midland districts of England, as well as in Ireland, where the grub is called the Connaught worm;[*] but fortunately not abundant in the north. We only once met with the cockchafer in Scotland, at Sorn, in Ayrshire. (J. R.) Even in the perfect state, this insect is not a little destructive to the leaves of both forest and fruit trees. In 1823, we remember to have observed almost all the trees about Dulwich and Camberwell defoliated by them; and Salisbury says, the leaves of the oaks in Richmond Park were so eaten by them, that scarcely an entire leaf was left. But it is in their previous larva state that they are most destructive, as we shall see by tracing their history.

The mother cockchafer, when about to lay her eggs, digs into the earth of a meadow or corn-field to the depth of a span, and deposits them in a cluster at the bottom of the excavation. Rösel, in order to watch the proceedings, put some females into glasses half-filled with earth, covered with a tuft of grass and a piece of thin muslin. In a fortnight, he found some hundreds of eggs deposited, of an oval shape and a pale-yellow colour. Placing the glass in a cellar, the eggs were hatched towards autumn, and the grubs increased remarkably in size. In the following May they fed so voraciously that they required a fresh turf every second day; and even this proving too scanty provender, he sowed in several garden pots a crop of peas, lentils, and salad, and when the plants came up he put a pair of grubs in each pot;

---

[*] Bingley, Anim. Biog., vol. iii. p. 230.

and in this manner he fed them through the second and third years. During this period, they cast their skins three or four times, going for this purpose deeper into the earth, and burrowing out a hole where they might effect their change undisturbed; and they do the same in winter, during which they become torpid and do not eat.

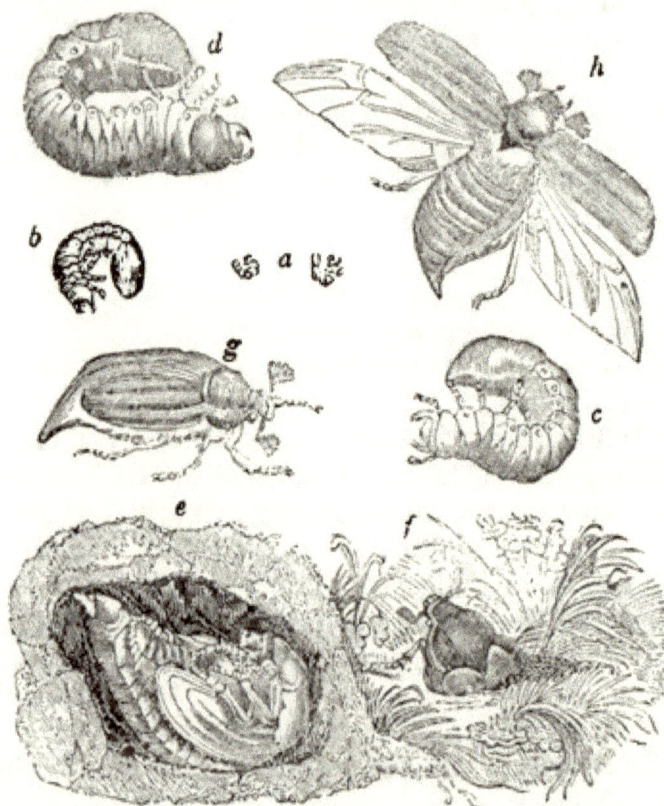

Transformations of the Cockchafer (*Melolontha vulgaris*). *a*, Newly-hatched larvæ. *b*, larva, one year old. *c*, the same larva at the second year of its growth. *d*, the same three years old. *e*, section of a bank of earth, containing the chrysalis of the fourth year. *f*, the chafer first emerging from the earth. *g*, the perfect chafer in a sitting posture. *h*, the same flying

When the grub changes into a pupa, in the third autumn after it is hatched, it digs a similar burrow about a yard deep; and when kept in a pot, and prevented from going deep enough, it shows great uneasiness and often dies. The perfect beetle comes forth from the pupa in January or

February ; but it is then as soft as it was whilst still a grub, and does not acquire its hardness and colour for ten or twelve days, nor does it venture above ground before May, in the fourth year from the time of its hatching. At this time, the beetles may be observed issuing from their holes in the evening, and dashing themselves about in the air as if blind.

During the three summers then of their existence in the grub state, these insects do immense injury, burrowing between the turf and the soil, and devouring the roots of grass and other plants ; so that the turf may easily be rolled off, as if cut by a turfing spade, while the soil underneath for an inch or more is turned into soft mould like the bed of a garden. Mr. Anderson, of Norwich, mentions having seen a whole field of fine flourishing grass so undermined by these grubs, that in a few weeks it became as dry, brittle, and withered as hay.* Bingley also tells us that "about sixty years ago, a farm near Norwich was so infested with cock-chafers, that the farmer and his servants affirmed they gathered eighty bushels of them ; and the grubs had done so much injury, that the court of the city, in compassion to the poor fellow's misfortune, allowed him twenty-five pounds."† In the year 1785, a farmer, near Blois, in France, employed a number of children and poor persons to destroy the cock-chafers at the rate of two liards a hundred, and in a few days they collected fourteen thousand.‡

"I remember," says Salisbury, "seeing in a nursery near Bagshot, several acres of young forest trees, particularly larch, the roots of which were completely destroyed by it, so much so that not a single tree was left alive."§ We are doubtful, however, whether this was the grub of the cock-chafer, and think it more likely to have been that of the green rose-beetle (*Cetonia aurata*), which feeds on the roots of trees.

* Phil. Trans., vol. xliv. p. 579.   † Anim. Biog., vol. iii. p. 233.
‡ Anderson's Recr. in Agricult., vol. iii. p. 420.   § Hints, p. 74.

CHAPTER XIV.

ARCHITECTURE OF ANTS.—MASON-ANTS.

ALL the species of ants are social. There are none solitary, as is the case with bees and wasps. They are all more or less skilful in architecture, some employing masonry, and others being carpenters, wood-carvers, and miners. They consequently afford much that is interesting to naturalists who observe their operations. The genuine history of ants has only been recently investigated, first by Gould in 1747, and subsequently by Linnæus, De Geer, Huber, and Latreille. Previous to that time their real industry and their imagined foresight were held up as moral lessons, without any great accuracy of observation; and it is probable that, even now, the mixture of truth and error in Addison's delightful papers in the Guardian (Nos. 156, 157) may be more generally attractive than the minute relation of careful naturalists. Gould disproved, most satisfactorily, the ancient fable of ants storing up corn for winter provision, no species of ants ever eating grain, or feeding in the winter upon anything. It is to Huber the younger, however, that we are chiefly indebted for our knowledge of the habits and economy of ants; and to Latreille for a closer distinction of the species. Some of the more interesting species, whose singular economy is described by the younger Huber, have not been hitherto found in this country. We shall, however, discover matter of very considerable interest in those which are indigenous; and as our principal object is to excite inquiry and observation with regard to those insects which may be easily watched in our own gardens and fields, we shall chiefly confine ourselves to the ants of these islands. We shall begin with the labours of those native ants which may be called earth-masons, from their digging in the ground, and forming structures with pellets of moistened loam, clay, or sand.

MASON-ANTS.

We have used in the preceding pages the terms *mason-bees* and *mason-wasps*, for insects which build their nests of earthy materials. On the same principle, we have followed the ingenious M. Huber the younger, in employing the term mason-ants for those whose nests on the exterior appear to be hillocks of earth, without the admixture of other materials, whilst in the interior they present a series of labyrinths, lodges, vaults, and galleries constructed with considerable skill. Of these mason-ants, as of the mason-wasps and bees already described, there are several species, differing from one another in their skill in the art of architecture.

One of the most common of the ant-masons is the turf-ant (*Formica cæspitum*, LATR.), which is very small and of a blackish-brown colour. Its architecture is not upon quite so extensive a scale as some of the others; but, though slight, it is very ingenious. Sometimes they make choice of the shelter of a flat stone or other covering, beneath which they hollow out chambers and communicating galleries; at other times they are contented with the open ground; but most commonly they select a tuft of grass or other herbage, the stems of which serve for columns to their earthen walls.

We had a small colony of these ants accidentally established in a flower-pot, in which we were rearing some young plants of the tiger-lily (*Lilium tigrinum*), the stems of which being stronger than the grass where they usually build, enabled them to rear their edifice higher, and also to make it more secure, than they otherwise might. It was wholly formed of small grains of moist earth, piled up between the stems of the lily without any apparent cement; indeed it has been ascertained by Huber, as we shall afterwards see, that they use no cement beside water. This is not always to be procured, as they depend altogether on rains and dew; but they possess the art of joining grains of dry sand so as to support one another, on some similar principle, no doubt, to that of the arch.

The nest which our turf-ants constructed in the flower-pot was externally of an imperfect square form, in consequence of its situation; for they usually prefer a circular plan. The principal chambers were placed under the arches, and, when inspected, contained a pile of cocoons and pupæ. Beneath those upper chambers there were others dug out deeper down, in which were also a numerous collection of eggs and cocoons in various stages of advancement. (J. R.)

Mr. Knapp describes a still more curious structure of another species of ant common in this country:—"One year," says he, "on the third of March, my labourer being employed in cutting up ant-hills, or tumps as we call them, exposed to view multitudes of the yellow species (*Formica flava*) in their winter's retirement. They were collected in numbers in little cells and compartments, communicating with others by means of narrow passages. In many of the cells they had deposited their larvæ, which they were surrounding and attending, but not brooding over or covering. Being disturbed by our rude operations, they removed them from our sight to more hidden compartments. The larvæ were small. Some of these ant-hills contained multitudes of the young of the wood-louse (*Oniscus armadillo*), inhabiting with perfect familiarity the same compartments as the ants, crawling about with great activity with them, and perfectly domesticated with each other. They were small and white; but the constant vibration of their antennæ, and the alacrity of their motions, manifested a healthy vigour. The ants were in a torpid state; but on being removed into a temperate room, they assumed much of their summer's animation. How these creatures are supported during the winter season it is difficult to comprehend; as in no one instance could we perceive any store or provision made for the supply of their wants. The minute size of the larvæ manifested that they had been recently deposited; and consequently that their parents had not remained during winter in a dormant state, and thus free from the calls of hunger. The preceding month of February, and part of January, had been remarkably severe; the frost had penetrated deep into the earth, and long held it frozen; the ants were in many cases not more than four inches beneath the surface, and must have been enclosed in a mass of frozen soil for a long period; yet they, their young, and the ouisci, were perfectly uninjured by it: affording another proof of the fallacy of the commonly received opinion, that cold is *universally* destructive to insect life."*

The earth employed by mason-ants is usually moist clay, either dug from the interior parts of their city or moistened by rain. The mining-ants and the ash-coloured (*Formica fusca*) employ earth which is probably not selected with so much care, for it forms a much coarser mortar than what we see used in the structure of the yellow ants (*F. flava*) and the brown ants (*F. brunnea*). We have never observed them bringing their building materials of this kind from a distance, like the mason-bees and like the wood or hill ant (*F. rufa*); but they take care, before they fix upon a locality, that it shall produce them all that they require. We are indebted to Huber the younger for the most complete account which has hitherto been given of these operations, of which details we shall make free use.

" To form," says this shrewd observer, " a correct judgment of the interior arrangement or distribution of an ant-hill, it is necessary to select such as have not been accidentally spoiled, or whose form has not been too much altered by local circumstances; a slight attention will then suffice to show that the habitations of the different species are not all constructed after the same system. Thus, the hillock raised by the ash-coloured ants will always present thick walls, fabricated with coarse earth, well-marked stories, and large chambers, with vaulted ceilings, resting upon a solid base. We never observe roads, or galleries, properly so called, but large passages, of an oval form, and all around considerable cavities and extensive embankments of earth. We further notice, that the little architects observe a certain proportion between the large arched ceilings and the pillars that are to support them.

* Journal of a Naturalist, p. 304.

"The brown ant (*Formica brunnea*), one of the smallest of the ants, is particularly remarkable for the extreme finish of its work. Its body is of a reddish shining brown, its head a little deeper, and the antennæ and feet a little lighter in colour. The abdomen is of an obscure brown, the scale narrow, of a square form, and slightly scolloped. The body is one line and two-fifths in length.*

"This ant, one of the most industrious of its tribe, forms its nest of stories four or five lines in height. The partitions are not more than half a line in thickness; and the substance of which they are composed is so finely grained, that the inner walls present one smooth unbroken surface. These stories are not horizontal; they follow the slope of the ant-hill, and lie one upon another to the ground-floor, which communicates with the subterranean lodges. They are not always, however, arranged with the same regularity, for these ants do not follow an invariable plan; it appears, on the contrary, that nature has allowed them a certain latitude in this respect, and that they can, according to circumstances, modify them to their wish; but however fantastical their habitations may appear, we always observe they have been formed by concentrical stories. On examining each story separately, we observe a number of cavities or halls, lodges of narrower dimensions, and long galleries, which serve for general communication. The arched ceilings covering the most spacious places are supported either by little columns, slender walls, or by regular buttresses. We also notice chambers, that have but one entrance, communicating with the lower story, and large open spaces, serving as a kind of cross-road (*carrefour*), in which all the streets terminate.

"Such is the manner in which the habitations of these ants are constructed. Upon opening them, we commonly find the apartments, as well as the large open spaces, filled with adult ants; and always observed their pupæ collected in the apartments more or less near the surface. This, however, seems regulated by the hour of the day, and the tempera-

* A line is the twelfth part of the old French inch. *See* Companion to the Almanac for 1830, p. 114.

ture : for in this respect these ants are endowed with great
sensibility, and know the degree of heat best adapted for
their young. The ant-hill contains, sometimes, more than
twenty stories in its upper portion, and at least as many
under the surface of the ground. By this arrangement the
ants are enabled, with the greatest facility, to regulate the
heat. When a too-burning sun overheats their upper apartments, they withdraw their little ones to the bottom of the
ant-hill. The ground-floor becoming, in its turn, uninhabitable during the rainy season, the ants of this species transport what most interests them to the higher stories ; and it
is there we find them more usually assembled, with their
eggs and pupæ, when the subterranean apartments are
submerged."*

Ants have a great dislike to water, when it exceeds that
of a light shower to moisten their building materials. One
species, mentioned by Azara as indigenous to South America,
instinctively builds a nest from three to six feet high,† to
provide against the inundations during the rainy season.
Even this, however, does not always save them from sub-
mersion ; and, when that occurs, they are compelled, in order
to prevent themselves from being swept away, to form a
group somewhat similar to the curtain of the wax-workers of
hive-bees (see p. 133). The ants constituting the basis of
this group lay hold of some shrub for security, while their
companions hold on by them ; and thus the whole colony,
forming an animated raft, floats on the surface of the water
till the inundation (which seldom continues longer than a
day or two) subsides. We confess, however, that we are
somewhat sceptical respecting this story, notwithstanding
the very high character of the Spanish naturalist.

It is usual with architectural insects to employ some
animal secretion, by way of mortar or size, to temper the
materials with which they work ; but the whole economy of
ants is so different, that it would be wrong to infer from
analogy a similarity in this respect, though the exquisite

* M. P. Huber on Ants, p. 20.
† Stedman's Surinam, vol. i. p. 160.

polish and extreme delicacy of finish in their structures lead,
naturally, to such a conclusion. M. P. Huber, in order to
resolve this question, at first thought of subjecting the
materials of the walls to chemical analysis, but wisely (as we
think) abandoned it for the surer method of observation.
The details which he has given, as the result of his researches, are exceedingly curious and instructive. He began
by observing an ant-hill till he could perceive some change
in its form.

"The inhabitants," says he, "of that which I selected, kept
within during the day, or only went out by subterranean
galleries which opened at some feet distance in the meadow.
There were, however, two or three small openings on the
surface of the nest; but I saw none of the labourers pass out
this way, on account of their being too much exposed to the
sun, which these insects greatly dread. This ant-hill,
which had a round form, rose in the grass, at the border of
a path, and had sustained no injury. I soon perceived that
the freshness of the air and the dew invited the ants to walk
over the surface of their nest; they began making new
apertures; several ants might be seen arriving at the same
time, thrusting their heads from the entrances, moving about
their antennæ, and at length adventuring forth to visit the
environs.

"This brought to my recollection a singular opinion of
the ancients. They believed that ants were occupied in their
architectural labours during the night, when the moon was
at its full."*

M. Latreille discovered a species of ants which were, so
far as he could ascertain, completely blind,† and of course it
would be immaterial to them whether they worked by night
or during the day. All observers indeed agree that ants
labour in the night, and a French naturalist is therefore of
opinion that they never sleep—a circumstance which is well
ascertained with respect to other animals, such as the shark,

* M. P. Huber on Ants, p. 23.
† Latreille, Hist. Nat. des Fourmis.

which will track a ship in full sail for weeks together.* The ingenious historian of English ants, Gould, says they never intermit their labours by night or by day, except when compelled by excessive rains. It is probable the ancients were mistaken in asserting that they only work when the moon shines ;† for, like bees, they seem to find no difficulty in building in the dark, their subterranean apartments being as well finished as the upper stories of their buildings. But to proceed with the narrative of M. P. Huber.

"Having thus noticed the movements of these insects during the night, I found they were almost always abroad and engaged about the dome of their habitation after sunset. This was directly the reverse of what I had observed in the conduct of the wood-ants (*F. rufa*), who only go out during the day, and close their doors in the evening. The contrast was still more remarkable than I had previously supposed ; for, upon visiting the brown ants some days after, during a gentle rain, I saw all their architectural talents in full play.

"As soon as the rain commenced, they left in great numbers their subterranean residence, re-entered it almost immediately, and then returned, bearing between their teeth pellets of earth, which they deposited on the roof of their nest. I could not at first conceive what this was meant for, but at length I saw little walls start up on all sides with spaces left between them. In several places, columns, ranged at regular distances, announced halls, lodges, and passages which the ants proposed establishing ; in a word, it was the rough beginning of a new story.

"I watched with a considerable degree of interest the most trifling movements of my masons, and found they did not work after the manner of wasps and humble-bees, when occupied in constructing a covering to their nest. The latter sit, as it were, astride on the border or margin of the

---

* Dr. Cleghorn, Thesis de Somno.

† Aristotle Hist. Animal. ix. 38. Pliny says, " Operantur et noctu plenâ lunâ ; eadem interlunio cessant," *i.e.*, They work in the night at full moon. but they leave off between moon and moon. It is the latter that we think doubtful.

covering, and take it between their teeth to model and at-
tenuate it according to their wish. The wax of which it is
composed, and the paper which the wasp employs, moistened
by some kind of glue, are admirably adapted for this purpose,
but the earth of which the ants make use, from its often
possessing little tenacity, must be worked up after some
other manner.

" Each ant, then, carried between its teeth the pellet of
earth it had formed by scraping with the end of its mandibles
the bottom of its abode, a circumstance which I have frequently witnessed in open day. This little mass of earth,
being composed of particles but just united, could be readily
kneaded and moulded as the ants wished; thus when they
had applied it to the spot where they had to rest, they divided and pressed against it with their teeth, so as to fill up
the little inequalities of their wall. The antennæ followed
all their movements, passing over each particle of earth as
soon as it was placed in its proper position. The whole was
then rendered more compact by pressing it lightly with the
fore-feet. This work went on remarkably fast. After
having traced out the plan of their masonry, in laying here
and there foundations for the pillars and partitions they
were about to erect, they raised them gradually higher, by
adding fresh materials. It often happened that two little
walls, which were to form a gallery, were raised opposite,
and at a slight distance from each other. When they had
attained the height of four or five lines, the ants busied
themselves in covering in the space left between them by a
vaulted ceiling.

" As if they judged all their partitions of sufficient elevation, they then quitted their labours in the upper part of
the building ; they affixed to the interior and upper part of
each wall fragments of moistened earth, in an almost horizontal direction, and in such a way as to form a ledge,
which, by extension, would be made to join that coming
from the opposite wall. These ledges were about half a
line in thickness; and the breadth of the galleries was, for
the most part, about a quarter of an inch. On one side

several vertical partitions were seen to form the scaffolding of a lodge, which communicated with several corridors by apertures formed in the masonry; on another, a regularly-formed hall was constructed, the vaulted ceiling of which was sustained by numerous pillars; further off, again, might be recognised the rudiments of one of those cross roads of which I have before spoken, and in which several avenues terminate. These parts of the ant-hill were the most spacious; the ants, however, did not appear embarrassed in constructing the ceiling to cover them in, although they were often more than two inches in breadth.

" In the upper part of the angles formed by the different walls, they laid the first foundations of this ceiling, and from the top of each pillar, as from so many centres, a layer of earth, horizontal and slightly convex, was carried forward to meet the several portions coming from different points of the large public thoroughfare.

" I sometimes, however, laboured under an apprehension that the building could not possibly resist its own weight, and that such extensive ceilings, sustained only by a few pillars, would fall into ruin from the rain which continually dropped upon them; but I was quickly convinced of their stability, from observing that the earth brought by these insects adhered at all points, on the slightest contact; and that the rain, so far from lessening the cohesion of its particles, appeared even to increase it. Thus, instead of injuring the building, it even contributed to render it still more secure.

" These particles of moistened earth, which are only held together by juxtaposition, require a fall of rain to cement them more closely, and thus varnish over, as it were, those places where the walls and galleries remain uncovered. All inequalities in the masonry then disappear. The upper part of these stories, formed of several pieces brought together, presents but one single layer of compact earth. They require for their complete consolidation nothing but the heat of the sun. It sometimes, however, happens that a violent rain will destroy the apartments, especially should they be but slightly arched; but under these circumstances the ants reconstruct them with wonderful patience.

"These different labours were carried on at the same time, and were so closely followed up in the different quarters, that the ant-hill received an additional story in the course of seven or eight hours. All the vaulted ceilings being formed upon a regular plan, and at equal distances from one wall to the other, constituted, when finished, but one single roof. Scarcely had the ants finished one story than they began to construct another; but they had not time to finish it—the rain ceasing before the ceiling was fully completed. They still, however, continued their work for a few hours, taking advantage of the humidity of the earth; but a keen north wind soon sprung up, and hastily dried the collected fragments, which, no longer possessing the same adherence, readily fell into powder. The ants, finding their efforts ineffectual, were at length discouraged, and abandoned their employment; but what was my astonishment when I saw them destroy all the apartments that were yet uncovered, scattering here and there over the last story the materials of which they had been composed! These facts incontestably prove that they employ neither gum, nor any kind of cement, to bind together the several substances of their nest; but in place of this avail themselves of the rain, to work or knead the earth, leaving the sun and wind to dry and consolidate it."*

Dr. Johnson of Bristol observed very similar proceedings in the case of a colony of red ants (*Myrmica rubra?*), the roof of whose nest was formed by a flat stone. During dry weather, a portion of the side walls fell in; but the rubbish was quickly removed, though no repairs were attempted till a shower of rain enabled them to work. As soon as this occurred, they worked with extraordinary rapidity, and in a short time the whole of the fallen parts were rebuilt, and rendered as smooth as if polished with a trowel.

When a gardener wishes to water a plot of ground where he has sown seeds that require nice management, he dips a

* M. P. Huber on Ants, p. 31.

strong brush into water, and passes his hand backwards and forwards over the hairs for the purpose of producing a fine artificial shower. Huber successfully adopted the same method to excite his ants to recommence their labours, which had been interrupted for want of moisture. But sometimes, when they deem it unadvisable to wait for rain, they dig down (as we remarked to be the practice of the mason-bees) till they arrive at earth sufficiently moist for their purpose. They do not, however, like these bees, merely dig for materials ; for they use the excavations for apartments, as well as what they construct with the materials thence derived. They appear, in short, to be no less skilful in mining than in building.

Such is the general outline of the operations of this singular species ; but we are still more interested with the history which M. P. Huber has given of the labours of an individual ant. "One rainy day," he says, "I observed a labourer of the dark ash-coloured species (*Formica fusca*) digging the ground near the aperture which gave entrance to the ant-hill. It placed in a heap the several fragments it had scraped up, and formed them into small pellets, which it deposited here and there upon the nest. It returned constantly to the same place, and appeared to have a particular design, for it laboured with ardour and perseverance. I remarked a slight furrow, excavated in the ground in a straight line, representing the plan of a path or gallery. The labourer (the whole of whose movements fell under my immediate observation) gave it greater depth and breadth, and cleared out its borders ; and I saw, at length—in which I could not be deceived—that it had the intention of establishing an avenue which was to lead from one of the stories to the underground chambers. This path, which was about two or three inches in length, and formed by a single ant, was opened above, and bordered on each side by a buttress of earth. Its concavity, in the form of a pipe (*gouttière*), was of the most perfect regularity : for the architect had not left an atom too much. The work of this ant was so well followed and understood, that I could almost to a certainty

U

guess its next proceeding, and the very fragment it was about to remove. At the side of the opening where this path terminated was a second opening, to which it was necessary to arrive by some road. The same ant began and finished this undertaking without assistance. It furrowed out and opened another path, parallel to the first, leaving between each a little wall of three or four lines in height."

Like the hive-bees, ants do not seem to work in concert, but each individual separately. There is, consequently, an occasional want of coincidence in the walls and arches; but this does not much embarrass them, for a worker, on discovering an error of this kind, seems to know how to rectify it, as appears from the following observations :—

" A wall," says M. Huber, " had been erected, with the view of sustaining a vaulted ceiling, still incomplete, that had been projected towards the wall of the opposite chamber. The workman who began constructing it had given it too little elevation to meet the opposite partition, upon which it was to rest. Had it been continued on the original plan, it must infallibly have met the wall at about one-half of its height; and this it was necessary to avoid. This state of things very forcibly claimed my attention; when one of the ants arriving at the place, and visiting the works, appeared to be struck by the difficulty which presented itself; but this it as soon obviated, by taking down the ceiling, and raising the wall upon which it reposed. It then, in my presence, constructed a new ceiling with the fragments of the former one.

" When the ants commence any undertaking, one would suppose that they worked after some preconceived idea, which, indeed, would seem verified by the execution. Thus, should any ant discover upon the nest two stalks of plants which lie crossways, a disposition favourable to the construction of a lodge, or some little beams that may be useful in forming its angles and sides, it examines the several parts with attention; then distributes, with much sagacity and address, parcels of earth in the spaces, and along the stems, taking from every quarter materials adapted to its object, sometimes not caring to destroy the work that others had commenced; so much are its motions regulated by the idea it has conceived, and upon which it acts, with little attention to all else around it. It goes and returns, until the plan is sufficiently understood by its companions.

"In another part of the same ant-hill," continues M. Huber, "several fragments of straw seemed expressly placed to form the roof of a large house: a workman took advantage of this disposition. These fragments lying horizontally, at half-an-inch distance from the ground, formed, in crossing each other, an oblong parallelogram. The industrious insect commenced by placing earth in the several angles of this framework, and all along the little beams of which it was composed. The same workman afterwards placed several rows of the same materials against each other, when the roof became very distinct. On perceiving the possibility of profiting by another plant to support a vertical wall, it began laying the foundations of it; other ants having by this time arrived, finished in common what this had commenced." *

M. Huber made most of his observations upon the processes followed by ants in glazed artificial hives or formicaries. The preceding figure represents a view of one of his formicaries of mason-ants.

We have ourselves followed up his observations, both on natural ant-hills and in artificial formicaries. On digging

* Huber on Ants, p. 43.

cautiously into a natural ant-hill, established upon the edge
of a garden-walk, we were enabled to obtain a pretty complete view of the interior structure.   There were two stories,
composed of large chambers, irregularly oval, communicating with each other by arched galleries, the walls of all
which were as smooth and well-polished as if they had been
passed over by a plasterer's trowel.   The floors of the
chambers, we remarked, were by no means either horizontal
or level, but all more or less sloped, and exhibiting in each
chamber at least two slight depressions of an irregular
shape.   We left the under story of this nest untouched, with
the notion that the ants might repair the upper galleries, of
which we had made a vertical section; but instead of doing
so they migrated during the day to a large crack formed by

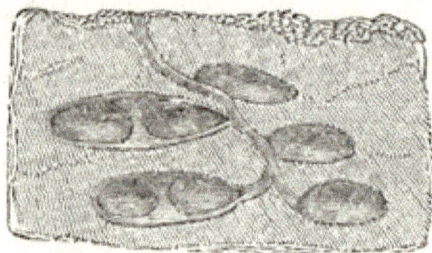

the dryness of the weather, about a yard from their old nest.
(J. R.)

We put a number of yellow ants (*Formica flava*), with
their eggs and cocoons, into a small glass frame, more than
half full of moist sand taken from their native hill, and
placed in a sloping position, in order to see whether they
would bring the nearly vertical, and therefore insecure, portion to a level by masonry.   We were delighted to perceive
that they immediately resolved upon performing the task
which had been assigned them, though they did not proceed
very methodically in their manner of building; for instead
of beginning at the bottom and building upwards, many of
them went on to add to the top of the outer surface, which
increased rather than diminished the insecurity of the whole.
Withal, however, they seemed to know how far to go, for no
portion of the newly-built wall fell, and in two days they

had not only reared a pyramidal mound to prop the rest, but had constructed several galleries and chambers for lodging the cocoons, which we had scattered at random amongst the sand. The new portion of this building is represented in the figure as supporting the upper and insecure parts of the nest.

We are sorry to record that our ingenious little masons were found upon the third day strewed about the outside of the building dead or dying, either from over-fatigue or perhaps from surfeit, as we had supplied them with as much honey as they could devour. A small colony of turf-ants have at this moment (July 28th, 1829) taken possession of the premises of their own accord. (J. R.)

CHAPTER XV.

THE largest of our British ants is that called the Hill-ant by Gould, the Fallow-ant by the English translator of Huber, and popularly the Pismire; but which we think may be more appropriately named the Wood-ant (*Formica rufa*, LATR.), from its invariable habit of living in or near woods and forests. This insect may be readily distinguished from other ants by the dusky black colour of its head and hinder parts, and the rusty brown of its middle. The structures reared by this species are often of considerable magnitude, and bear no small resemblance to a rook's nest thrown upon the ground bottom upwards. They occur in abundance in the woods near London, and in many other parts of the country: in Oak of Honour Wood alone, we are acquainted with the localities of at least two dozen,—some in the interior, and others on the hedge-banks on the outskirts of the wood. (J. R.)

The exterior of the nest is composed of almost every trans-portable material which the colonists can find in their vicinity; but the greater portion consists of the stems of withered grass and short twigs of trees, piled up in apparent confusion, but with sufficient regularity to render the whole smooth, conical, and sloping towards the base, for the pur-pose, we may infer, of carrying off rain-water. When within reach of a corn-field, they often also pick up grains of wheat, barley, or oats, and carry them to the nest as building materials, and not for food, as was believed by the ancients. There are wonders enough observable in the economy of ants, without having recourse to fancy—wonders which made Aristotle extol the sagacity of bloodless animals, and Cicero ascribe to them not only sensation, but mind, reason, and memory.* Ælian, however, describes, as if he had actually witnessed it, the ants ascending a stalk of growing corn, and throwing down "the ears which they bit off to their companions below." Aldrovand assures us that he had seen their granaries; and others pretend that they shrewdly bite off the ends of the grain to prevent it from germinating.† These are fables which accurate observation has satisfactorily contradicted.

But these errors, as it frequently happens, have contributed to a more perfect knowledge of the insects than we might otherwise have obtained; for it was the wish to prove or dis-

Nest of Wood-Ant.

prove the circumstance of their storing up and feeding upon grain which led Gould to make his observations on English ants; as the notion of insects being produced *from* putrid carcases had before led Redi to his ingenious experiments on their generation. Yet, although it is more than eighty years

* In formicâ non modo sensus, sed etiam mens, ratio, memoria.

† Aldrovandus de Formicis, and Johnston, Thaumaturg. Nat. p. 356.

since Gould's book was published, we find the error still repeated in very respectable publications.*

The coping which we above described as forming the
exterior of the wood-ant's nest, is only a small portion of the
structure, which consists of a great number of interior chambers and galleries, with funnel-shaped avenues leading to
them. The coping, indeed, is one of the most essential
parts, and we cannot follow a more delightful guide than the
younger Huber in detailing its formation.

"The labourers," he says, " of which the colony is composed, not only work continually on the outside of their nest,
but, differing very essentially from other species, who willingly remain in the interior, sheltered from the sun, they
prefer living in the open air, and do not hesitate to carry on,
even in our presence, the greater part of their operations.

" To have an idea how the straw or stubble-roof is formed,
let us take a view of the ant-hill at its origin, when it is
simply a cavity in the earth. Some of its future inhabitants
are seen wandering about in search of materials fit for the
exterior work, with which, though rather irregularly, they
cover up the entrance; whilst others are employed in mixing
the earth, thrown up in hollowing the interior, with fragments of wood and leaves, which are every moment brought
in by their fellow-assistants ; and this gives a certain consistence to the edifice, which increases in size daily. Our little
architects leave here and there cavities, where they intend
constructing the galleries which are to lead to the exterior,
and as they remove in the morning the barriers placed at the
entrance of their nest the preceding evening, the passages are
kept open during the whole time of its construction. We
soon observed the roof to become convex ; but we should be
greatly deceived did we consider it solid. This roof is destined to include many apartments or stories. Having observed
the motions of these little builders through a pane of glass,
adjusted against one of their habitations, I am thence enabled
to speak with some degree of certainty upon the manner in
which they are constructed. I ascertained that it is by exca-

* See Professor Paxton's Illustrations of Scripture, i. 307.

vating or mining the under portion of their edifice that they form their spacious halls—low, indeed, and of heavy construction, yet sufficiently convenient for the use to which they are appropriated, that of receiving, at certain hours of the day, the larvæ and pupæ.

" These halls have a free communication by galleries, made in the same manner. If the materials of which the ant-hill is composed were only interlaced, they would fall into a confused heap every time the ants attempted to bring them into regular order. This, however, is obviated by their tempering the earth with rain-water, which, afterwards hardened in the sun, so completely and effectually binds together the several substances, as to permit the removal of certain fragments from the ant-hill without any injury to the rest ; it, moreover, strongly opposes the introduction of the rain. I never found, even after long and violent rains, the interior of the nest wetted to more than a quarter of an inch from the surface, provided it had not been previously out of repair, or deserted by its inhabitants.

" The ants are extremely well sheltered in their chambers, the largest of which is placed nearly in the centre of the building ; it is much loftier than the rest, and traversed only by the beams that support the ceiling ; it is in this spot that all the galleries terminate, and this forms, for the most part, their usual residence.

" As to the underground portion, it can only be seen when the ant-hill is placed against a declivity ; all the interior may be then readily brought in view, by simply raising up the straw roof. The subterranean residence consists of a range of apartments, excavated in the earth, taking an horizontal direction."*

[It seems rather surprising that the wood-ants should be able, with such materials as they employ, to make a dome-shaped structure, which shall be furnished with cells and galleries, and yet shall endure rain and wind, without being penetrated by the one or blown away by the other. If the hill be closely examined, the little sticks of which it is

* Huber on Ants, p. 15.

composed will be seen to have a definite, though not very
regular arrangement; and it is. a noteworthy circumstance
that the longest are preserved for the galleries, being laid
across each other in a very ingenious manner, so as to prevent
the material from falling and filling up the galleries. This
structure was shown very clearly in a huge ant-hill in Bagshot
Park. We introduced a sheet of plate glass into the nest, so
as to divide it perpendicularly into two halves, and having
given the insects six weeks to repair damages, we removed
one half of the hill, so that the whole interior of the other half
could be seen through the glass. The whole economy of the
nest was thus made clear, and the artificial arrangement of
the materials showed itself very plainly on the roofs of the
cells and galleries.]

M. P. Huber, in order to observe the operations of the
wood-ant with more attention, transferred colonies of them
to his artificial formicaries, plunging the feet of the stand into
water to prevent their escape till they were reconciled to their
abode, and had made some progress in repairing it.

[Under the glass shade on the top of the formicary may

be seen the mound which the wood-ants have raised, according to their custom, and below, through the glass front, the reader may see the various passages and cells which communicate with the hill above. As the ants require that the lower part of their dwelling should be in darkness, a stout wooden door can be shut over the glass to exclude the light.]

There is this remarkable difference in the nest of the wood-ants, that they do not construct a long covert way as if for concealment, as the yellow and the brown ants do. The wood-ants are not, like them, afraid of being surprised by enemies, at least during the day, when the whole colony is either foraging in the vicinity or employed on the exterior. But the proceedings of the wood-ants at night are well worthy of notice ; and when M. Huber began to study their economy, he directed his entire attention to their night proceedings. " I remarked," says he, " that their habitations changed in appearance hourly, and that the diameter of those spacious avenues, where so many ants could freely pass each other during the day, was, as night approached, gradually lessened. The aperture, at length, totally disappeared, the dome was closed on all sides, and the ants retired to the bottom of their nest.

" In further noticing the apertures of these ant-hills, I fully ascertained the nature of the labour of its inhabitants, of which I could not before even guess the purport ; for the surface of the nest presented such a constant scene of agitation, and so many insects were occupied in carrying materials in every direction, that the movement offered no other image than that of confusion.

" I saw then clearly that they were engaged in stopping up passages ; and for this purpose they at first brought forward little pieces of wood, which they deposited near the entrance of those avenues they wished to close ; they placed them in the stubble ; they then went to seek other twigs and fragments of wood, which they disposed above the first, but in a different direction, and appeared to choose pieces of less size in proportion as the work advanced. They, at length,

brought in a number of dried leaves, and other materials of
an enlarged form, with which they covered the roof: an exact
miniature of the art of our builders, when they form the
covering of any building. Nature, indeed, seems everywhere
to have anticipated the inventions of which we boast, and
this is doubtless one of the most simple.

"Our little insects, now in safety in their nest, retire
gradually to the interior before the last passages are closed;
one or two only remain without, or concealed behind the
doors on guard, while the rest either take their repose, or
engage in different occupations in the most perfect security.
I was impatient to know what took place in the morning
upon these ant-hills, and therefore visited them at an early
hour. I found them in the same state in which I had left
them the preceding evening. A few ants were wandering
about on the surface of the nest, some others issued from
time to time from under the margin of their little roofs
formed at the entrance of the galleries; others afterwards
came forth, who began removing the wooden bars that block-
aded the entrance, in which they readily succeeded. This
labour occupied them several hours. The passages were at
length free, and the materials with which they had been
closed scattered here and there over the ant-hill. Every day,
morning and evening, during the fine weather, I was a witness
to similar proceedings. On days of rain the doors of all the
ant-hills remained closed. When the sky was cloudy in
the morning, or rain was indicated, the ants, who seemed
to be aware of it, opened but in part their several avenues,
and immediately closed them when the rain commenced."*

The galleries and chambers which are roofed in as thus
described are very similar to those of the mason-ants,
being partly excavated in the earth, and partly built with
the clay thence procured. It is in these they pass the
night, and also the colder months of the winter, when they
become torpid, or nearly so, and of course require not the
winter granaries of corn with which the ancients fabulously
furnished them.

* Huber on Ants, p. 11.

CARPENTER-ANTS.

The ants that work in wood perform much more extensive operations than any of the other carpenter insects which we have mentioned. Their only tools, like those of bees and wasps, are their jaws or mandibles; but though these may not appear so curiously constructed as the ovipositor file of the tree-hopper (*Cicada*), or the rasp and saw of the saw-flies (*Tenthredinidæ*), they are no less efficient in the performance of what is required. Among the carpenter-ants, the emmet or jet-ant (*F. fuliginosa*) holds the first rank, and is easily known by being rather less in size than the wood-ant, and by its fine shining black colour. It is less common in Britain than some of the preceding, though its colonies may occasionally be met with in the trunks of decaying oak or willow trees in hedges.

"The labourers," says Huber, "of this species work always in the interior of trees, and are desirous of being screened from observation : thus every hope on our part is precluded of following them in their several occupations. I tried every expedient I could devise to surmount this difficulty ; I endeavoured to accustom these ants to live and work under my inspection, but all my efforts were unsuccessful ; they even abandoned the most considerable portion of their nest to seek some new asylum, and spurned the honey and sugar which I offered them for nourishment. I was now, by necessity, limited to the inspection only of their edifices : but, by decomposing some of the fragments with care, I hoped to acquire some knowledge of their organization.

"On one side I found horizontal galleries, hidden in great part by their walls, which follow the circular direction of the layers of the wood ; and on another, parallel galleries, separated by extremely thin partitions, having no communication except by a few oval apertures. Such is the nature of these works, remarkable for their delicacy and lightness.

"In other fragments I found avenues which opened laterally, including portions of walls and transverse partitions, erected here and there within the galleries, so as

to form separate chambers.   When the work is further
advanced, round holes are always observed, encased, as it
were, between two pillars cut out in the same wall.   These
holes in course of time become square, and the pillars,
originally arched at both ends, are worked into regular
columns by the chisel of our sculptors.   This, then, is the
second specimen of their art.   This portion of the edifice will
probably remain in this state.

"But in another quarter are fragments differently wrought,
in which these same partitions, pierced now in every part,
and hewn skilfully, are transformed into colonnades, which
sustain the upper stories, and leave a free communication
throughout the whole extent.   It can readily be perceived
how parallel galleries, hollowed out upon the same plan, and
the sides taken down, leaving only from space to space what
is necessary to sustain their ceilings, may form an entire
story; but as each has been pierced separately, the flooring
cannot be very level: this, however, the ants turn to their
advantage, since these furrows are better adapted to retain
the larvæ that may be placed there.

"The stories constructed in the great roots offer greater
irregularity than those in the very body of the tree, arising

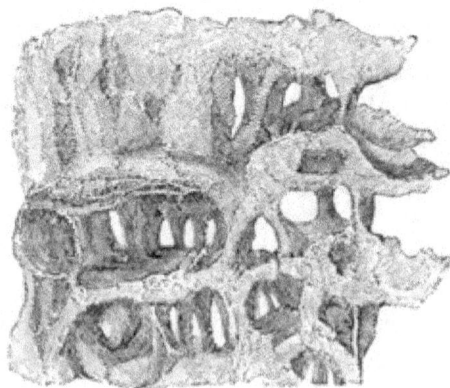

Portion of a Tree, with Chambers and Galleries chiseled out by Jet-Ants.

either from the hardness and interlacing of the fibres, which
renders the labour more difficult, and obliges the labourers

to depart from their accustomed manner, or from their not observing in tho extremities of their edifice the same arrangement as in tho centro: whatever it be, horizontal stories and numerous partitions are still found. If the work be less regular, it becomes more delicate; for the ants, profiting by the hardness and solidity of the materials, give to their building an extreme degree of lightness. I have seen fragments of from eight to ten inches in length, and of equal height, formed of wood as thin as paper, containing a number of apartments, and presenting a most singular appearance. At the entrance of these apartments, worked out with so much care, are very considerable openings; but in place of chambers and extensive galleries, the layers of the wood are hewn in arcades, allowing the ants a free passage in every direction. These may be regarded as tho gates or vestibules conducting to the several lodges."*

It is a singular circumstance in the structures of these ants, that all the wood which they carve is tinged of a black colour, as if it were smoked; and M. Huber was not a little solicitous to discover whence this arose. It certainly does not add to the beauty of their streets, which look as sombre as the most smoke-dyed walls in the older lanes of the metropolis. M. Huber could not satisfy himself whether it was caused by the exposure of the wood to the atmosphere, by some emanation from the ants, or by the thin layers of wood being acted upon or decomposed by the formic acid.† But if any or all of these causes operated in blackening the wood, we should be ready to anticipate a similar effect in the case of other species of ants which inhabit trees; yet the black tint is only found in the excavations of the jet-ant.

We are acquainted with several colonies of the jet-ants (*Formica fuliginosa*)—one of which, in the roots and trunk of an oak on the road from Lewisham to Sydenham, near Brockley, in Kent, is so extremely populous, that the numbers of its inhabitants appeared to us beyond any reasonable estimate. None of the other colonies of this species which we have seen appear to contain many hundreds.

* Huber, p. 56.   † The acid of ants.

On cutting into the root of the before-mentioned tree, we found the vertical excavations of much larger dimensions, both in width and depth, than those represented by Huber in the preceding cut (page 302). What surprised us the most was to see the tree growing vigorously and fresh, though its roots were chiseled in all directions by legions of workers, while every leaf, and every inch of the bark, was also crowded by parties of foragers. On one of the low branches we found a deserted nest of the white-throat (*Sylvia cinerea*, TEMMINCK), in the cavity of which they were piled upon one another as close as the unhappy negroes in the hold of a slave-ship; but we could not discover what had attracted them hither. Another dense group, collected on one of the branches, led us to the discovery of a very singular oak gall, formed on the bark in the shape of a pointed cone, and

F fuliginosa.

crowded together. It is probable that the juice which they extracted from these galls was much to their taste. (J.R.)

Beside the jet-ant, several other species exercise the art of carpentry,—nay, what is more wonderful still, they have the ingenuity to knead up, with spider's-web for a cement, the chips which they chisel out into a material with which they construct entire chambers. The species which exercise this singular art are the Ethiopian (*Formica nigra*) and the yellow ant (*F. flava*).*

We once observed the dusky ants (*F. fusca*) at Blackheath, in Kent, busily employed in carrying out chips from the interior of a decaying black poplar, at the root of which a colony was established; but, though it thence appears that this species can chisel wood if they choose, yet they usually burrow in the earth, and by preference, as we have remarked, at the root of a tree, the leaves of which supply them with food.

———

Among the foreign ants we may mention a small yellow ant of South America, described by Dampier, which seems, from his account, to construct a nest of green leaves. "Their sting," he says, "is like a spark of fire; and they are so thick among the boughs in some places, that one shall be covered with them before he is aware. These creatures have nests on great trees, placed on the body between the limbs: some of their nests are as big as a hogshead. This is their winter habitation; for in the wet season they all repair to these their cities, where they preserve their eggs. In the dry season, when they leave their nests, they swarm all over the woodlands, for they never trouble the savannahs. Great paths, three or four inches broad, made by them, may be seen in the woods. They go out light, but bring home heavy loads on their backs, all of the same substance, and equal in size. I never observed anything besides pieces of green leaves, so big that I could scarcely see the insect for his burthen; yet they would march stoutly, and so many were pressing forward that it was a very pretty sight, for the path looked perfectly green with them."

Ants observed in New South Wales, by the gentlemen in the expedition under Captain Cook, are still more interesting.

* Huber.

"Some," we are told, "are as green as a leaf, and live upon trees, where they build their nests of various sizes, between that of a man's head and his fist. These nests are of a very curious structure: they are formed by bending down several of the leaves, each of which is as broad as a man's hand, and gluing the points of them together, so as to form a purse. The viscous matter used for this purpose is an animal juice which nature has enabled them to elaborate. Their method of first bending down the leaves we had no opportunity to observe; but we saw thousands uniting all their strength to hold them in this position, while other busy multitudes were employed within in applying this gluten that was to prevent their returning back. To satisfy ourselves that the leaves were bent and held down by the efforts of these diminutive artificers, we disturbed them in their work; and as soon as they were driven from their stations, the leaves on which they were employed sprang up with a force much greater than we could have thought them able to conquer by any combination of their strength. But, though we gratified our curiosity at their expense, the injury did not go unrevenged; for thousands immediately threw themselves upon us, and gave us intolerable pain with their stings, especially those which took possession of our necks and hair, from whence they were not easily driven. Their sting was scarcely less painful than that of a bee; but, except it was repeated, the pain did not last more than a minute.

"Another sort are quite black, and their operation and manner of life are not less extraordinary. Their habitations are the inside of the branches of a tree, which they contrive to excavate by working out the pith almost to the extremity of the slenderest twig, the tree at the same time flourishing as if it had no such inmate. When we first found the tree we gathered some of the branches, and were scarcely less astonished than we should have been to find that we had profaned a consecrated grove, where every tree, upon being wounded, gave signs of life; for we were instantly covered with legions of these animals, swarming from every broken bough, and inflicting their stings with incessant violence.

"A third kind we found nested in the root of a plant, which grows on the bark of trees in the manner of mistletoe, and which they had perforated for that use. This root is commonly as big as a large turnip, and sometimes much bigger. When we cut it we found it intersected by innumerable winding passages, all filled with these animals, by which, however, the vegetation of the plant did not appear to have suffered any injury. We never cut one of these roots that was not inhabited, though some were not bigger than a hazel-nut. The animals themselves are very small, not more than half as big as the common red ant in England. They had stings, but scarcely force enough to make them felt: they had, however, a power of tormenting us in an equal, if not in a greater degree; for the moment we handled the root, they swarmed from innumerable holes, and running about those parts of the body that were uncovered, produced a titillation more intolerable than pain, except it is increased to great violence."[*]

The species called sugar-ants in the West Indies are particularly destructive to the sugar-cane, as well as to lime, lemon, and orange-trees, by excavating their nests at the roots, and so loosening the earth that they are frequently uprooted and blown down by the winds. If this does not happen, the roots are deprived of due nourishment, and the plants become sickly and die.[†]

[One or two examples of foreign ants are well worthy of notice. The first of them is an insect whose habits bear strongly upon the familiar passage in Proverbs, ch. vi. v. 6 :—

"Go to the ant, thou sluggard ; consider her ways, and be wise :
" Which having no guide, overseer, or ruler,
" Provideth her meat in the summer, and gathereth her food in the harvest."

This passage is one that has been often mentioned as a proof that the Bible is not to be implicitly trusted. Judging from all the species of ants known to entomologists, some writers argue that the author of the proverb in question

---

[*] Hawkesworth's Account of Cook's First Voyage.
[†] Phil. Trans., xxx. p. 346.

was ignorant of the real history of the ant, and was taking up a popular fallacy.

[Still, although the ants of the old world are chiefly carnivorous, or feed on soft substances, and in consequence have not the least idea of hoarding food for the winter, there is one species of Brazilian ant which absolutely builds houses, prepares ground, sows seed, reaps the grain, and stores it away for future consumption. It is the Agricultural Ant, *Atta malefaciens*, first described by Dr. Lincecum, who watched the insect for twelve years before publishing an account that he knew would at first be received with incredulity. The following abstract of his paper appeared in the Journal of the Linnæan Society.

["The species which I have named 'Agricultural' is a large brownish ant. It dwells in what may be termed paved cities, and like a thrifty, diligent, provident farmer, makes suitable and timely arrangements for the changing seasons. It is, in short, endowed with skill, ingenuity, and untiring patience, sufficient to enable it successfully to contend with the varying exigencies which it may have to encounter in the life conflict.

["When it has selected a situation for its habitation, if on ordinary dry ground, it bores a hole, around which it raises the surface three and sometimes six inches, forming a low circular mound, having a very gentle inclination from the centre to the outer border, which on an average is three or four feet from the entrance. But if the location is chosen on low, flat, wet land, liable to inundation, though the ground may be perfectly dry at the time the ant sets to work, it nevertheless elevates the mound, in the form of a pretty sharp cone, to the height of fifteen or twenty inches or more, and makes the entrance near the summit. Around the mound, in either case, the ant clears the ground of all obstructions, and levels and marks the surface to the distance of three or four feet from the gate of the city, giving the space the appearance of a handsome pavement, as it really is.

["Within this paved area not a blade of any green thing is allowed to grow, except a single species of grain-bearing grass. Having planted this crop in a circle around, and two or three feet from the centre of the mound, the insect tends and cultivates it with constant care, cutting away all other grasses and weeds that may spring up amongst it, and all around outside the farm circle to the extent of one or two feet more. The cultivated grass grows luxuriantly, and produces a heavy crop of small, white, flinty seeds, which under the microscope very closely resemble ordinary rice. When ripe, it is carefully harvested and carried by the workers, chaff and all, into the granary cells, where it is divested of the chaff and packed away. The chaff is taken out and thrown beyond the limits of the paved area.

["During protracted wet weather it sometimes happens that the provision-stores become damp, and are liable to sprout and spoil. In this case, on the first fine day, the ants bring out the damp and damaged grain, and expose it to the sun till it is dry, when they carry it back and pack away all the sound seeds, leaving those that had sprouted to waste.

[" In a peach orchard not far from my house is a considerable elevation, on which is an extensive bed of rock. In the sand-beds overlying portions of this rock are five cities of the agricultural ants, evidently very ancient. My observations on their manners and customs have been limited to the last twelve years, during which time the inclosure surrounding the orchard has prevented the approach of cattle to the ant-farms. The cities which are outside the inclosure, as well as those protected in it, are at the proper season invariably planted with the ant-rice. The crop may accordingly always be seen springing up within the circle about the 1st of November every year. Of late years, however, since the number of farms and cattle has greatly increased, and the latter are eating off the grass much closer than formerly, thus preventing the ripening of the seeds, I notice that the agricultural ant is placing its cities along the turn-rows in the fields, walks in gardens, inside about the gates, &c., where they can cultivate their farms without molestation from the cattle.

[" There can be no doubt that the particular species of

grain-bearing grass mentioned above is intentionally planted. In farmer-like manner the ground upon which it stands is carefully divested of all other grasses and weeds during the time it is growing. When it is ripe, the grain is taken care of, the dry stubble cut away and carried off, the paved area being left unencumbered until the ensuing autumn, when the same ant-rice reappears within the same circle, and receives the same agricultural attention as was bestowed upon the previous crop—and so on, year after year, as I *know* to be the case, in all situations where the ants' settlements are protected from granivorous animals."

[This interesting account is simply the result of twelve years' patient investigation on the part of Dr. Lincecum, who took special care not to invent a theory and to twist facts in accordance with it, but watched the entire proceedings of the insects for a series of years.

Crematogaster

[The preceding illustration represents the rather remarkable nest of an Australian ant, belonging to the genus Crematogaster. This word signifies "hanging-belly," and the name has been applied to the ant in consequence of the manner in which its abdomen is held up in the air, so that it overhangs the back.

[As may be seen, the nest is of considerable size, and might from its external appearance be mistaken for that of a wasp. The interior of it, however, is even more elaborate, being full of little covered passages interlacing with each other in a most intricate manner, but all leading to the internal galleries.

[The two nests which are shown in the next illustration are, if possible, still more remarkable.

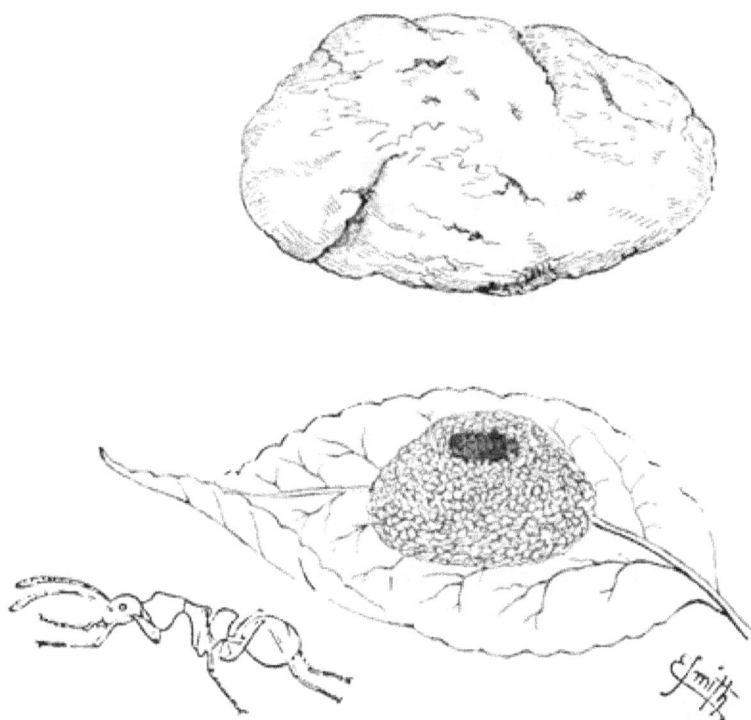

[The upper one is found in Cayenne, and is made by an insect called the fungus ant (*Polyrachis bispinosa*), because the nest looks as if it were made of fungus. It is not, however, composed of that material, but of the fibre of the cotton-tree (*Bombax ceiba*).

[The fibre is in itself very short, barely exceeding an inch in length, but it is cut very much shorter by the ant, who contrives to felt it together in a most curious manner, so that it is hardly possible to trace the course of any one fibre. The size of the nest is, on an average, about eight or nine inches in diameter. The insect itself is given in the preceding illustration, but very much enlarged. If the reader will look at the centre of the body, he will see the projections which have given it the name of *bispinosa*, or two-spined.

[The lower figure represents the nest of another species of ant belonging to the same genus, and called scientifically, *Polyrachis textor*. The nest is most ingeniously made of little pieces of wood and tendrils, put together so as to form a kind of open network, through which the interior of the nest is plainly visible. This insect inhabits Malacca.]

CHAPTER XVI.

STRUCTURES OF WHITE ANTS, OR TERMITES.

WHEN we look back upon the details which we have given of the industry and ingenuity of numerous tribes of insects, both solitary and social, we are induced to think it almost impossible that they could be surpassed. The structures of wasps and bees, and still more those of the wood-ant (*Formica rufa*), when placed in comparison with the size of the insects, equal our largest cities compared with the stature of man. But when we look at the buildings erected by the white ants of tropical climates, all that we have been surveying dwindles into insignificance. Their industry appears greatly to surpass that of our ants and bees, and they are certainly more skilful in architectural contrivances. The elevation, also, of their edifices is more than five hundred times the height of the builders. Were our houses built according to the same proportions, they would be twelve or fifteen times higher than the London Monument, and four or five times higher than the pyramids of Egypt, with corresponding dimensions in the basements of the edifices. These statements are, perhaps, necessary to impress the extraordinary labours of ants upon the mind; for we are all more or less sensible to the force of comparisons. The analogies between the works of insects and of men are not perfect; for insects are all provided with instruments peculiarly adapted to the end which they instinctively seek, while man has to form a plan by progressive thought, and upon the experience of others, and to complete it with tools which he also invents.

The termites do not stand above a quarter of an inch high, while their nests are frequently twelve feet, and Jobson mentions some which he had seen as high as twenty feet; "of compass," he adds, "to contain a dozen men, with the heat of the sun baked into that hardness, that we used to hide ourselves in the ragged tops of them when we took up stands to shoot at deer or wild beasts."* Bishop Heber saw a number of these high ant-hills in India, near the principal entrance of the Sooty or Moorshedabad river. "Many of them," he says, "were five or six feet high, and probably seven or eight feet in circumference at the base, partially overgrown with grass and ivy, and looking at a distance like the stumps of decayed trees. I think it is Ctesias, among the Greek writers, who gives an account alluded to by Lucian in his 'Cock,' of monstrous ants in India, as large as foxes. The falsehood probably originated in the stupendous fabrics which they rear here, and which certainly might be supposed to be the work of a much larger animal than their real architect."† Herodotus has a similar fable of the enormous size and brilliant appearance of the ants of India.

Nor is it only in constructing dwellings for themselves that the termites of Africa and of other hot climates employ their masonic skill. Though, like our ants and wasps, they are almost omnivorous, yet wood, particularly when felled and dry, seems their favourite article of food; but they have an utter aversion to feeding in the light, and always eat their way with all expedition to the interior. It thence would seem necessary for them either to leave the bark of a tree, or the outer portion of the beam or door of a house, undevoured, or to eat in open day. They do neither; but are at the trouble of constructing galleries of clay, in which they can conceal themselves, and feed in security. In all their foraging excursions, indeed, they build covert ways, by which they can go out and return to their encampment.‡

Others of the species (for there are several), instead of building galleries, exercise the art of miners, and make their approaches under ground, penetrating beneath the foundation of houses or areas, and rising again either through the floors, or by entering the bottom of the posts that support the building, when they follow the course of the fibres, and

---

* Jobson's Gambia, in Purchas's Pilgrim, ii. p. 1570.

† Heber's Journal, vol. i. p. 248.

‡ Smeathman, in Phil. Trans., vol. lxxi.

make their way to the top, boring holes and cavities in different places as they proceed. Multitudes enter the roof, and intersect it with pipes or galleries, formed of wet clay, which serve for passages in all directions, and enable them more readily to fix their habitations in it. They prefer the softer woods, such as pine and fir, which they hollow out with such nicety, that they leave the surface whole, after having eaten away the inside. A shelf or plank attacked in this manner looks solid to the eye, when, if weighed, it will not out-balance two sheets of pasteboard of the same dimensions. It sometimes happens that they carry this operation so far on stakes in the open air, as to render the bark too flexible for their purpose; when they remedy the defect by plastering the whole stick with a sort of mortar which they make with clay, so that, on being struck, the form vanishes, and the artificial covering falls in fragments on the ground. In the woods, when a large tree falls from age or accident, they enter it on the side next the ground, and devour it at leisure, till little more than the bark is left. But in this case they take no precaution of strengthening the outward defence, but leave it in such a state as to deceive an eye unaccustomed to see trees thus gutted of their insides: and "you may as well," says Mr. Smeathman, "step upon a cloud." It is an extraordinary fact, that when these creatures have formed pipes in the roof of a house, instinct directs them to prevent its fall, which would ensue from their having sapped the posts on which it rests; but as they gnaw away the wood, they fill up the interstices with clay, tempered to a surprising degree of hardness, so that, when the house is pulled down, these posts are transformed from wood to stone. They make the walls of their galleries of the same composition as their nests, varying the materials according to their kind; one species using the red clay, another black clay, and the third a woody substance, cemented with gums, as a security from the attacks of their enemies, particularly the common ant, which, being defended by a strong, horny shell, is more than a match for them, and when it can get at them, rapaciously seizes them, and drags them to its nest for food for its young brood. If any accident breaks down part of their walls, they repair the breach with all speed. Instinct guides them to perform their office in the creation, by mostly confining their attacks to trees that are beginning to decay, or such timber as has been severed from its roots for use, and would decay in time. Vigorous, healthy trees do not require to be destroyed, and accordingly, these consumers have no taste for them.*

M. Adanson describes the termites of Senegal as constructing covert ways along the surface of wood which they intend to attack; but though we have no reason to distrust so excellent a naturalist, in describing what he saw, it is certain that they more commonly eat their way into the interior of the wood, and afterwards form the galleries, when they find that they have destroyed the wood till it will no longer afford them protection.

But it is time that we should come to their principal building, which may, with some propriety, be called a city; and, according to the method we have followed in other instances, we shall trace their labours from the commencement. We shall begin with the operations of the species which may be appropriately termed the Warrior (*Termes fatalis*, LINN.; *T. bellicosus*, SMEATH.).

We must premise, that though they have been termed white *ants*, they do not belong to the same order of insects with our ants; yet they have a slight resemblance to ants in their form, but more in their economy. Smeathman, to whom we owe our chief knowledge of the genus, describes them as consisting of kings, queens, soldiers, and workers, and is of opinion that the workers are larvæ, the soldiers nymphæ, and the kings and queens the perfect insects. In this opinion he coincides with Sparrmann † and others; but Latreille is inclined to think, from what he observed in a European species (*Termes lucifugus*) found near Bordeaux, that the soldiers form a distinct race, like the neuter workers among bees and ants, while the working termites are larvæ,‡

---

* Smeathman.　　　　　† Quoted by De Geer, vol. vii.
　　‡ Hist. Nat. Générale, vol. xiii. p. 66.

which are furnished with strong mandibles for gnawing; when they become nymphs, the rudiments of four wings appear, which are fully developed in the perfect insects.

Termes bellicosus in the winged state.

[It is now known that the differences of form among the termites are accounted for as follows. The winged specimens are the fully developed males and females, popularly called kings and queens. These crawl to the aperture of their house and take flight, retiring to earth after a short time. When a male and female meet each other, they cast off their wings exactly as do the ants of our own country, and become the founders of a new colony. Their soldiers are undeveloped males, and the workers are undeveloped females.]

In the winged state, they migrate to form new colonies, but the greater number of them perish in a few hours, or become the prey of birds, and even of the natives, who fry them as delicacies. " I have discoursed with several gentlemen," says Smeathman, " upon the taste of the white ants, and on comparing notes, we have always agreed that they are most delicious and delicate eating. One gentleman compared them to sugared marrow, another to sugared cream and a paste of sweet almonds."*

Mr. Smeathman's very interesting paper affords us the most authentic materials for the further description of these wonderful insects ; and we therefore continue partly to extract from, and partly to abridge, his account.

The few pairs that are so fortunate as to survive the various casualties that assail them, are usually found by workers (larvæ), which, at this season, are running con-

* Smeathman, in Phil. Trans., vol. lxxi. p. 169, note.

tinually on the surface of the ground, on the watch for them.
As soon as they discover the objects of their search, they
begin to protect them from their surrounding enemies, by
inclosing them in a small chamber of clay, where they
become the parents of a new community, and are distinguished from the other inhabitants of the nest by the title of
king and queen. Instinct directs the attention of these
labouring insects to the preservation of their race, in the
protection of this pair and their offspring. The chamber
that forms the rudiment of a new nest is contrived for their
safety, but the entrances to it are too small to admit of their
ever leaving it; consequently, the charge of the eggs devolves upon the labourers, who construct nurseries for their
reception. These are small, irregularly-shaped chambers,
placed at first round the apartment of the king and queen,
and not exceeding the size of a hazel-nut; but in nests of
long standing they are of great comparative magnitude, and
distributed at a greater distance. The receptacles for hatching the young are all composed of wooden materials, apparently joined together with gum, and, by way of defence,
cased with clay. The chamber that contains the king and
queen is nearly on a level with the surface of the ground;
and as the other apartments are formed about it, it is generally situated at an equal distance from the sides of the nest,
and directly beneath its conical point. Those apartments
which consist of nurseries and magazines of provisions, form
an intricate labyrinth, being separated by small, empty
chambers and galleries, which surround them, or afford a
communication from one to another. This labyrinth extends
on all sides to the outward shells, and reaches up within it
to two-thirds or more of its height, leaving an open area
above, in the middle, under the dome, which reminds the
spectator of the nave of an old cathedral. Around this are
raised three or four large arches, which are sometimes two
or three feet high, next the front of the area, but diminish as
they recede further back, and are lost amidst the innumerable
chambers and nurseries behind them.

Every one of these buildings consists of two distinct parts, the exterior and the interior. The exterior is one large shell, in the manner of a dome, large and strong enough to inclose and shelter the interior from the vicissitudes of the weather, and the inhabitants from the attacks of natural or accidental enemies. It is always, therefore, much stronger than the interior building, which is the habitable part, divided, with a wonderful kind of regularity and contrivance, into an amazing number of apartments for the residence of the king and queen, and the nursing of the numerous progeny; or for magazines, which are always found well filled with stores and provisions. The hills make their first appearance above ground by a little turret or two, in the shape of sugar-loaves, which are run a foot high or more. Soon after, at some little distance, while the former are increasing in height and size, they raise others, and so go on increasing their number, and widening them at the base, till their works below are covered with these turrets, of which they always raise the highest and largest in the middle, and by filling up the intervals between each turret, collect them into one dome. They are not very curious or exact in the workmanship, except in making them very solid and strong; and when, by their joining them, the dome is completed, for which purpose the turrets answer as scaffolds, they take away the middle ones entirely, except the tops, which, joined together, make the crown of the cupola, and apply the clay to the building of the works within, or to erecting fresh turrets for the purpose of raising the hillock still higher; so that some part of the clay is probably used several times, like the boards and posts of a mason's scaffold.

When these hills are little more than half their height, it is a common practice of the wild bulls to stand as sentinels on them, while the rest of the herd are ruminating below. They are sufficiently strong for that purpose, and at their full height answer excellently well as places of look-out; and Mr. Smeathman has been, with four more, on the top of one of these hillocks, to watch for a vessel in sight. The outward shell, or dome, is not only of use to protect and support the interior buildings from external violence and

the heavy rains, but to collect and preserve a regular degree
of the warmth and moisture necessary for hatching the eggs
and cherishing the young. The royal chamber occupied by
the king and queen appears to be, in the opinion of this
little people, of the most consequence, being always situated
as near the centre of the interior building as possible. It is
always nearly in the shape of half an egg, or an obtuse oval,
within, and may be supposed to represent a long oven. In
the infant state of the colony it is but about an inch in
length ; but in time will be increased to six or eight inches,
or more, in the clear, being always in proportion to the size
of the queen, who, increasing in bulk as in age, at length
requires a chamber of such dimensions.

Queen distended with Eggs.

Its floor is perfectly horizontal, and in large hillocks,
sometimes more than an inch thick of solid clay. The
roof, also, which is one solid and well-turned oval arch, is
generally of about the same solidity ; but in some places it
is not a quarter of an inch thick on the sides where it joins
the floor, and where the doors or entrances are made level
with it, at nearly equal distances from each other. These
entrances will not admit any animal larger than the soldiers
or labourers ; so that the king and the queen (who is, at full
size, a thousand times the weight of a king) can never
possibly go out, but remain close prisoners.

[There is a good series of the queen cells of the Termite in
the British Museum, and the reader is strongly recommended
to go and examine them. Some of them are as large as
cocoa-nuts. Around the cell are a number of small holes,
looking as if they had been bored with a bradawl. Now, if

the cell be carefully opened, a most curious arrangement will be seen. Each of the little holes serves as an opening into a passage which communicates with the interior of the cell. The apartment, if we may so call it, which contains the queen, is only just large enough to hold her, and there is no door or opening for her egress. This, however, is not required, as her enormous size prevents her from moving. Through these passages runs incessantly a stream of worker termites, some of them carrying eggs which the queen has just laid, and others returning to the royal chamber for a fresh supply.]

The royal chamber, if in a large hillock, is surrounded by a countless number of others, of different sizes, shapes, and dimensions; but all of them arched in one way or another— sometimes elliptical or oval. These either open into each other, or communicate by passages as wide as, and are evidently made for, the soldiers and attendants, of whom great numbers are necessary, and always in waiting. These apartments are joined by the magazines and nurseries. The former are chambers of clay, and are always well filled with provisions, which, to the naked eye, seem to consist of the raspings of wood, and plants which the termites destroy, but are found by the microscope to be principally the gums or inspissated juices of plants. These are thrown together in little masses, some of which are finer than others, and resemble the sugar about preserved fruits; others are like tears of gum, one quite transparent, another like amber, a third brown, and a fourth quite opaque, as we see often in parcels of ordinary gums. These magazines are intermixed with the nurseries, which are buildings totally different from the rest of the apartments; for these are composed entirely of wooden materials, seemingly joined together with gums. Mr. Smeathman calls them the nurseries because they are invariably occupied by the eggs and young ones, which appear at first in the shape of labourers, but white as snow These buildings are exceedingly compact, and divided into many very small irregular-shaped chambers, not one of which is to be found of half an inch in width. They are placed all round, and as near as possible to the royal apartments.

Y

When the nest is in the infant state, the nurseries are close to the royal chambers; but as, in process of time, the queen enlarges, it is necessary to enlarge the chamber for her accommodation; and as she then lays a greater number of eggs, and requires a greater number of attendants, so it is necessary to enlarge and increase the number of the adjacent apartments; for which purpose the small nurseries which are first built are taken to pieces, rebuilt a little further of a size larger, and the number of them increased at the same time. Thus they continually enlarge their apartments, pull down, repair, or rebuild, according to their wants, with a degree of sagacity, regularity, and foresight, not even imitated by any other kind of animals or insects.

All these chambers, and the passages leading to and from them, being arched, they help to support each other; and while the interior large arches prevent them from falling into the centre, and keep the area open, the exterior building supports them on the outside.   There are, comparatively speaking, few openings into the great area, and they, for the most part, seem intended only to admit into the nurseries that genial warmth which the dome collects.   The interior building, or assemblage of nurseries, chambers, &c., has a flattish top or roof, without any perforation, which would keep the apartments below dry, in case through accident the dome should receive any injury, and let in water; and it is never exactly flat and uniform, because the insects are always adding to it by building more chambers and nurseries; so that the division or columns between the future arched apartment resemble the pinnacles on the fronts of some old buildings, and demand particular notice, as affording one proof that for the most part the insects project their arches, and do not make them by excavation.   The area has also a flattish floor, which lies over the royal chamber, but sometimes a good height above it, having nurseries and magazines between.   It is likewise waterproof, and contrived to let the water off if it should get in, and run over by some short way into the subterraneous passages, which run under the lowest apartments in the hill in various directions, and are of an astonishing size, being wider than the bore of a great cannon. One that Mr. Smeathman measured was perfectly cylindrical, and thirteen inches in diameter. These subterraneous passages, or galleries, are lined very thick with the same kind of clay of which the hill is composed, and ascend the inside of the outward shell in a spiral manner; and winding round the whole building up to the top, intersect each other at different heights, opening either immediately in the dome in various places, and into the interior building, the new turrets, &c., or communicating with them by other galleries of different diameters, either circular or oval.

From every part of these large galleries are various small covert ways, or galleries leading to different parts of the building. Under ground there are a great many that lead downward by sloping descents, three and four feet perpendicular among the gravel, whence the workers cull the finer parts, which, being kneaded up in their mouths to the consistence of mortar, become that solid clay or stone of which their hills and all their buildings, except their nurseries, are composed. Other galleries again ascend, and lead out horizontally on every side, and are carried under ground near to the surface a vast distance : for if all the nests are destroyed within a hundred yards of a house, the inhabitants of those which are left unmolested farther off will still carry on their subterraneous galleries, and, invading it by sap and mine, do great mischief to the goods and merchandise contained in it.

It seems there is a degree of necessity for the galleries under the hills being thus large, since they are the great thoroughfares for all the labourers and soldiers going forth or returning, whether fetching clay, wood, water, or provisions ; and they are certainly well calculated for the purposes to which they are applied by the spiral slope which is given them ; for if they were perpendicular the labourers would not be able to carry on their building with so much facility, as they ascend a perpendicular with great difficulty, and the soldiers can scarcely do it at all. It is on this account that sometimes a road like a ledge is made on the perpendicular side of any part of the building within their hill, which is flat on the upper surface and half an inch wide, and ascends gradually like a staircase, or like those winding roads which are cut on the sides of hills and mountains, that would otherwise be inaccessible ; by which and similar con-

a, A covered way and nest, on the branch of a tree, of the *Termites arborum*.   *b*, Section of the Hill-nest of the *Termites bellicosi*, to show the interior.   *c*, Hill-nest of the *Termites bellicosi*, entire.

trivances they travel with great facility to every interior part.

This, too, is probably the cause of their building a kind of bridge of one great arch, which answers the purpose of a flight of stairs from the floor of the area to some opening on the side of one of the columns that support the great arches. This contrivance must shorten the distance exceedingly to those labourers who have the eggs to carry from the royal chamber to some of the upper nurseries, which in some hills would be four or five feet in the straightest line, and much more if carried through all the winding passages leading through the inner chambers and apartments. Mr. Smeathman found one of these bridges, half an inch broad, a quarter of an inch thick, and ten inches long, making the side of an elliptic arch of proportionable size; so that it is wonderful it did not fall over or break by its own weight before they got it joined to the side of the column above.

It was strengthened by a small arch at the bottom, and had a hollow or groove all the length of the upper surface, either made purposely for the inhabitants to travel over with more safety, or else, which is not improbable, worn by frequent treading.

TURRET-BUILDING WHITE ANTS.

Apparently more than one species smaller than the preceding, such as the *Termes mordax* and *T. atrox* of Smeathman, construct nests of a very different form, the figures of which resemble a pillar, with a large mushroom for a capital. These turrets are composed of well-tempered black earth, and stand nearly three feet high. The conical mushroom-shaped roof is composed of the same material, and the brims hang over the column, being three or four inches wider than its perpendicular sides. Most of them, says Smeathman, resemble in shape the body of a round windmill, but some of the roofs have little elevation in the middle. When one of these turrets is completed, the insects do not afterwards enlarge or alter it; but if it be found too small for them, they lay the foundation of another at a few inches' distance. They sometimes, but not often, begin the second before the first is finished, and a third before they have completed the second. Five or six of these singular turrets in a group may be seen in the thick woods at the foot of a tree. They are so very strongly built, that in case of violence, they will sooner tear up the gravel and solid heart of their foundation than break in the middle. When any of them happen to be thus thrown down, the insects do not abandon them; but, using their overturned column as a basis, they run up another perpendicularly from it to the usual height, fastening the under part at the same time to the ground, to render it the more secure.

The interior of a turret is pretty equally divided into innumerable cells, irregular in shape, but usually more or

Turret Nests of White Ants. One nest is represented cut through, with the upper part lying on the ground.

less angular, generally quadrangular or pentagonal, though the angles are not well defined. Each shell has at least two entrances; but there are no galleries, arches, nor wooden nurseries, as in the nests of the warrior (*T. bellicosus*). The two species which build turret nests are very different in size, and the dimensions of the nests differ in proportion.

THE WHITE ANTS OF TREES.

Latreille's species of white ant (*Termes lucifugus*, Rossi), formerly mentioned as found in the south of Europe, appear to have more the habits of the jet-ant, described page 301, than their congeners of the tropics. They live in the interior of the trunks of trees, the wood of which they eat, and form their habitations of the galleries which they thus excavate. M. Latreille says they appear to be furnished with an acid for the purpose of softening the wood, the odour of which is exceedingly pungent. They prefer the part of the wood nearest to the bark, which they are careful not to injure, as it affords them protection. All the walls of their galleries are moistened with small globules of a gelatinous substance, similar to gum Arabic. They are chiefly to be found in the trunks of oak and pine trees, and are very numerous.[*]

Another of the species (*Termes arborum*), described by Smeathman, builds a nest on the exterior of trees, altogether different from any of the preceding. These are of a spherical or oval shape, occupying the arm or branch of a tree sometimes from seventy to eighty feet from the ground, and as large, in a few instances, as a sugar-cask. The composition used for a building material is apparently similar to that used by the warriors for constructing their nurseries, being the gnawings of wood in very small particles, kneaded into a paste with some species of cement or glue, procured, as Smeathman supposes, partly from gummiferous trees, and partly from themselves: but it is more probable, we think, that it is wholly secreted, like the wax of bees, by the insects themselves. With this cement, whatever may be its composition, they construct their cells, in which there is nothing very wonderful except their great numbers. They are very firmly built, and so strongly attached to the trees, that they will resist the most violent tornado. It is impossible, indeed, to detach them, except by cutting them in pieces, or sawing off the branch, which is frequently done to procure the insects for young turkeys. (See engraving, p. 324, for a figure of this nest.)

This species very often, instead of selecting the bough of a tree, builds in the roof or wall of a house, and unless

---

[*] Latreille, Hist. Nat. Générale, tom. xiii. p. 64.

observed in time, and expelled, occasions considerable damage. It is easier, in fact, to shut one's door against a fox or a thief, than to exclude such insidious enemies, whose aversion to light renders it difficult to trace them even when they are numerous.

[There are also termites in Europe, and the city of La Rochelle has suffered terribly from them. They eat the trees in the gardens, and not a stake can be driven into the ground, or even a plank left for twenty-four hours, without being attacked. They also enter the houses and utterly ruin them by eating every bit of timber that is used in them. In one instance, where a room had been repaired, the stalactitic galleries of the termites showed themselves the very day after the workmen had left the room.

[They invaded the prefecture, and did exceeding damage, one of their feats of voracity being so extraordinary as to deserve mention. The archives of the department were left in boxes, and privately inspected. One day, when a paper was needed, the whole of the documents fell to pieces, and were metamorphosed as if by magic into a heap of clay. The termites had got into the boxes by boring through the wainscot of the room, and had then penetrated among the papers. They consumed every particle of them except the uppermost sheet and the edges, supplying their place with clay. The consequence was, that although the heap of documents seemed to be correct, there was nothing but a mass of clay galleries and a single sheet of paper at the top.

[So voracious are they, that even a piece of paper wrapped round a bottle was eaten, the termites building a gallery of clay in order to reach it under cover.]

If we reflect on the prodigious numbers of those insects, and their power and rapidity of destroying, we cannot but admire the wisdom of Providence in creating so indefatigable and useful an agent in countries where the decay of vegetable substances is rapid in proportion to the heat of the climate. We have already remarked that they always prefer decaying or dead timber; and it is indeed a very general law among insects which feed on wood to prefer what is unsound ; the same principle holds with respect to fungi, lichens, and other parasitical plants.

All the species of Termites are not social ; but the solitary ones do not, like their congeners, distinguish themselves in architecture. In other respects, their habits are more similar; for they destroy almost every substance, animal and vegetable. The most common of the solitary species must be familiar to all our readers by the name of wood-louse (*Termes pulsatorium*, LINN.; *Atropos lignarius*, LEACH)—one of the insects which produces the ticking superstitiously termed the *death-watch*. It is not so large as the common wood-louse, but whiter and more slender, having a red mouth and yellow eyes. It lives in old books, the paper on walls, collections of insects and dried plants, and is extremely agile in its movements, darting, by jerks, into dark corners for the purpose of concealment. It does not like to run straight forward without resting every half-second, as if to listen or look about for its pursuer, and at such resting times it is easily taken. The ticking noise is made by the insect beating against the wood with its head, and it is supposed by some to be peculiar to the female, and to be connected with the laying of her eggs. M. Latreille, however, thinks that the wood-louse is only the grub of the *Psocus abdominalis*, in which case it could not lay eggs ; but this opinion is somewhat questionable. Another death-watch is a small beetle (*Anobium tesselatum*).

CHAPTER XVII.

STRUCTURES OF SILK SPUN BY CATERPILLARS, INCLUDING THE
SILK-WORM.

———

" Millions of spinning-worms,
That in their green shops weave the smooth-hair'd silk."
MILTON's *Comus.*

———

ALL the caterpillars of butterflies, moths, and, in general,
of insects with four wings, are capable of spinning silk;
of various degrees of fineness and strength, and differing in
colour, but usually white, yellow, brown, black, or grey.
This is not only of advantage in constructing nests for themselves, and particularly for their pupæ, as we have so
frequently exemplified in the preceding pages, but it enables
them, the instant they are excluded from the egg, to protect
themselves from innumerable accidents, as well as from
enemies. If a caterpillar, for instance, be exposed to a gust
of wind, and blown off from its native tree, it lets itself
gently down, and breaks its fall, by immediately spinning a
cable of silk, along which, also, it can reascend to its former
station when the danger is over. In the same way it frequently disappoints a bird that has marked it out for prey,
by dropping hurriedly down from a branch, suspended to its
never-failing delicate cord. The leaf-rollers, formerly described, have the advantage of other caterpillars in such cases,
by being able to move as quickly backwards as forwards ; so
that when a bird puts in its bill at one end of the roll, the
insect makes a ready exit at the other, and drops along its
thread as low as it judges convenient. We have seen caterpillars drop in this way from one to six feet or more ; and
by means of their cable, which they are careful not to break,

they climb back with great expedition to their former place.

The structure of their legs is well adapted for climbing up their singular rope—the six fore-legs being furnished with a curved claw; while the pro-legs (as they have been termed) are no less fitted for holding them firm to the branch when they have regained it, being constructed on the principle of forming a vacuum, like the leather sucker with which boys lift and drag stones. The foot of the common fly has a similar sucker, by which it is enabled to walk on glass, and otherwise support itself against gravity. The different forms of the leg and pro-leg of a spinning caterpillar are represented in the figure.

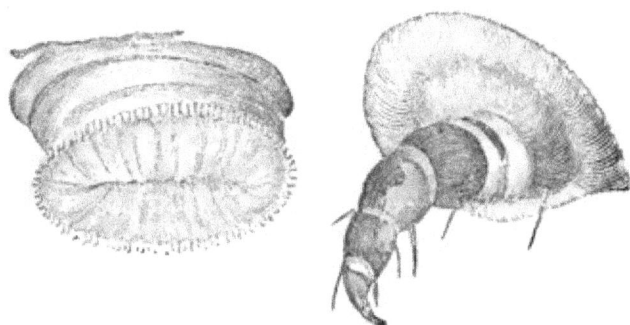

Leg and Pro-leg of a Caterpillar, greatly magnified.

In order to understand the nature of the apparatus by which a caterpillar spins its silk, it is to be recollected that its whole interior structure differs from that of warm-blooded animals. It has, properly speaking, no heart, though a long tubular *dorsal vessel*, which runs along the back, and pulsates from twenty to one hundred times per minute, has been called so by Malpighi and others, but neither Lyonnet nor Cuvier could detect any vessel issuing from it, and consequently the fluid which is analogous to blood has no circulation. It differs also from the higher orders of animals in having no brain, the nerves running along the body being only united by little knobs, called ganglions. Another circumstance is, that it has no lungs, and does not breathe by the mouth, but by air-holes, or spiracles, eighteen in number, situated along the sides, in the middle of the rings, as may be seen in the following figure from Lyonnet.

These spiracles communicate on each side with tubes, that have been called the wind-pipes (*tracheæ*). The spinning apparatus is placed near the mouth, and is connected with the silk-bags, which are long, slender, floating vessels, containing a liquid gum. The bags are closed at their lower extremity, become wider towards the middle, and more slender towards the head, where they unite to form the spinning-tube, or spinneret. The bags being in most cases longer than the body of the caterpillar, necessarily lie in a convoluted state, like the intestines of quadrupeds. The capacity, or rather the length, of the silk-bags is in proportion to the quantity of silk required for spinning; the *Cossus ligniperda*, for example, from living in the wood of trees. spins little,

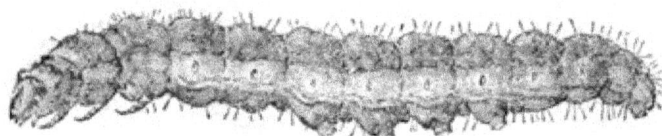

Caterpillar of the Goat Moth (*Cossus ligniperda*).

having a bag only one-fourth the length of that of the silk-worm, though the caterpillar is at least twice the dimensions of the latter. The following figure, taken from the admirable treatise of Lyonnet on the anatomy of the *Cossus*, will render these several organs more easily understood than any description.

The spinneret itself was supposed by Réaumur to have two outlets for the silk; but Lyonnet, upon minute dissection, found that the two tubes united into one before their termination; and he also assured himself that it was composed of alternate slips of horny and membranaceous substance,—the one for pressing the thread into a small diameter, and the other for enlarging it at the insect's pleasure. It is cut at the end somewhat like a writing-pen, though with less of a slope, and is admirably fitted for being applied to objects to

which it may be required to attach silk. The following are
magnified figures of the spinneret of the *Cossus*, from Lyonnet.

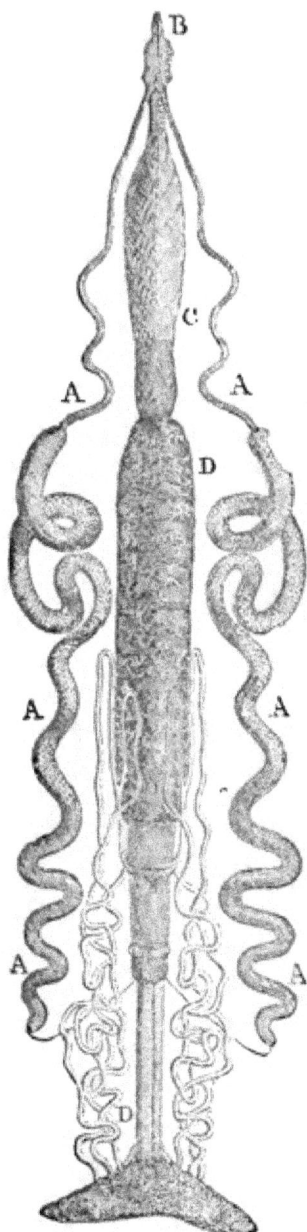

Interior Structure of the Cossus.—A, silk bags; B, silk tube, through which the viscid matter, of which the silk threads are formed, is forced by a peristaltic motion; C, stomach; D D, intestines, with the coil of bile vessels.

"You may sometimes have seen," says the Abbé de la
Pluche, "in the work-rooms of goldsmiths or gold-wire-

drawers, certain iron plates, pierced with holes of different
calibres, through which they draw gold and silver wire, in
order to render it finer.   The silk-worm has under her

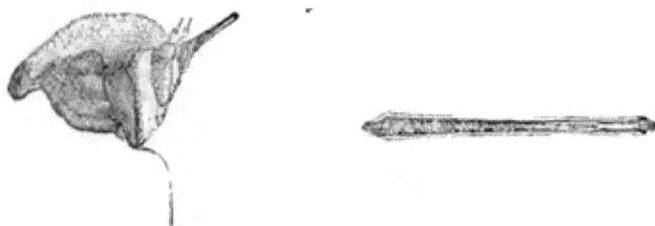

Side-view of the Silk-tube.   Section of the Silk-tube, magnified 22,000 times.

mouth such a kind of instrument, perforated with a pair of
holes [united into one on the outside*,] through which she
draws two drops of the gum that fills her two bags.   These
instruments are like a pair of distaffs for spinning the gum
into a silken thread.   She fixes the first drop of gum that
issues where she pleases, and then draws back her head, or
lets herself fall, while the gum, continuing to flow, is drawn

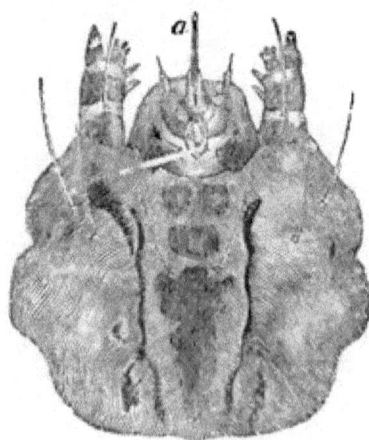

Labium, or lower lip of Cossus.—*a*, Silk-tube.

out and lengthened into a double stream.   Upon being
exposed to the air, it immediately loses its fluidity, becomes
dry, and acquires consistence and strength.   She is never

* Lyonnet.

deceived in adjusting the dimensions of the [united] apertures, or in calculating the proper thickness of the thread, but invariably makes the strength of it proportionable to the weight of her body.

It would be a very curious thing to know how the gum which composes the silk is separated and drawn off from the other juices that nourish the animal. It must be accomplished like the secretions formed by glands in the human body. I am therefore persuaded that the gum-bags of the silk-worm are furnished with a set of minute glands, which being impregnated with gum, afford a free passage to all the juices of the mulberry-leaf corresponding with this glutinous matter, while they exclude every fluid of a different quality."* When confined in an open glass vessel, the goat-moth caterpillar will effect its escape by constructing a curious silken ladder, as represented by Roesel.

Caterpillars, as they increase in size, cast their skins as lobsters do their shells, and emerge into renewed activity under an enlarged covering. Previous to this change, when the skin begins to gird and pinch them, they may be observed to become languid, and indifferent to their food, and at length they cease to eat, and await the sloughing of their skin. It is now that the faculty of spinning silk seems to be of great advantage to them; for, being rendered inactive and helpless by the tightening of the old skin around their expanding body, they might be swept away by the first puff of wind, and made prey of by ground beetles or other carnivorous prowlers. To guard against such accidents, as soon as they feel that they can swallow no more food, from being half choked by the old skin, they take care to secure themselves from danger by moorings of silk spun upon the leaf or the branch where they may be reposing. The caterpillar of the white satin-moth (*Leucoma salicis*, STEPHENS) in this way draws together with silk one or two leaves, similar to the leaf-rollers (*Tortricidæ*), though it always feeds openly without any covering. The caterpillar of the puss-moth again, which, in its third skin, is large and heavy, spins a

Spectacle de la Nature, vol. i.

thick web on the upper surface of a leaf, to which it adheres till the change is effected.

The most important operation, however, of silk-spinning is performed before the caterpillar is transformed into a chrysalis, and is most remarkable in the caterpillars of moths and other four-winged flies, with the exception of those of butterflies; for though these exhibit, perhaps, greater ingenuity, they seldom spin more than a few threads to secure the chrysalis from falling, whereas the others spin for it a complete envelope or shroud. We have already seen, in the preceding pages, several striking instances of this operation, when, probably for the purpose of husbanding a scanty supply of silk, extraneous substances are worked into the texture. In the case of other caterpillars, silk is the only material employed.

Of this the cocoon of the silk-worm is the most prominent example, in consequence of its importance in our manufactures and commerce, and on that account will demand from us somewhat minute details, though it would require volumes to incorporate all the information which has been published on the subject.

SILK-WORM.

The silk-worm, like most other caterpillars, changes its skin four times during its growth. The intervals at which the four moultings follow each other depend much on climate or temperature, as well as on the quality and quantity of food. It is thence found, that if they are exposed to a high temperature, say from 81° to 100° Fahrenheit, the moultings will be hastened; and only five days will be consumed in moulting the third or fourth time, whilst those worms that have not been hastened take seven or eight days.*

The period of the moultings is also influenced by the temperature in which the eggs have been kept during the winter. When the heat of the apartment has been regulated, the first moulting takes place on the fourth or fifth day after hatching, the second begins on the eighth day, the third

* Cours d'Agriculture, par M. Rozier. Paris, 1801.

takes up the thirteenth and fourteenth days, and the last occurs on the twenty-second and twenty-third days. The fifth age, in such cases, lasts ten days, at the end of which, or thirty-two days after hatching, the caterpillars attain their full growth, and ought to be three inches in length; but if they have not been properly fed, they will not be so long.

With the age of the caterpillar, its appetite increases, and is at its maximum after the fourth moulting, when it also attains its greatest size. The silk gum is then elaborated in the reservoirs, while the caterpillar ceases to eat, and soon diminishes again in size and weight. This usually requires a period of nine or ten days, commencing from the fourth moulting, after which it begins to spin its shroud of silk. In this operation it proceeds with the greatest caution, looking carefully for a spot in which it may be most secure from interruption.

" We usually," says the Abbé de la Pluche, " give it some little stalks of broom, heath, or a piece of paper rolled up, into which it retires, and begins to move its head to different places, in order to fasten its thread on every side. All this work, though it looks to a bystander like confusion, is not without design. The caterpillar neither arranges its threads nor disposes one over another, but contents itself with distending a sort of cotton or floss to keep off the rain; for Nature having ordained silk-worms to work under trees, they never change their method even when they are reared in our houses.

" When my curiosity led me to know how they spun and placed their beautiful silk, I took one of them, and frequently removed the floss with which it first attempted to make itself a covering; and as by this means I weakened it exceedingly, when it at last became tired of beginning anew, it fastened its threads on the first thing it encountered, and began to spin very regularly in my presence, bending its head up and down, and crossing to every side. It soon confined its movements to a very contracted space, and, by degrees, entirely surrounded itself with silk; and the remainder of its operations became invisible, though these may

z

be understood from examining the work after it is finished. In order to complete the structure, it must draw out of the gum-bag a more delicate silk, and then with a stronger gum bind all the inner threads over one another.

" Here, then, are three coverings entirely different, which afford a succession of shelter. The outer loose silk, or floss, is for keeping off the rain ; the fine silk in the middle prevents the wind from causing injury ; and the glued silk, which composes the tapestry of the chamber where the insect lodges, repels both air and water, and prevents the intrusion of cold.

" After building her cocoon, she divests herself of her fourth skin, and is transformed into a chrysalis, and subsequently into a moth (*Bombyx mori*), when, without saw or centre-bit, she makes her way through the shell, the silk, and the floss ; for the Being who teaches her how to build herself a place of rest, where the delicate limbs of the moth may be formed without interruption, instructs her likewise how to open a passage for escape.

" The cocoon is like a pigeon's egg, and more pointed at one end than the other ; and it is remarkable that the caterpillar does not interweave its silk towards the pointed end, nor apply its glue there as it does in every other part,* by bending itself all around with great pliantness and agility : what is more, she never fails, when her labour is finished, to fix her head opposite to the pointed extremity. The reason of her taking this position is, that she has purposely left this part less strongly cemented, and less exactly closed. She is instinctively conscious that this is to be the passage for the perfect insect which she carries in her bowels, and has therefore the additional precaution never to place this pointed extremity against any substance that might obstruct the moth at the period of its egress.

" When the caterpillar has exhausted herself to furnish the labour and materials of the three coverings, she loses the form of a worm ; her spoils drop all around the chrysalis ; first throwing off the skin, with the head and jaws attached to

* This is denied by recent observers.

it, and the new skin hardening into a sort of leathery consistence. Its nourishment is already in its stomach, and consists of a yellowish mucus, but gradually the rudiments of the moth unfold themselves,—the wings, the antennæ, and the legs becoming solid. In about a fortnight or three weeks. a slight swelling in the chrysalis may be remarked, which at length produces a rupture in the membrane that covers it, and by repeated efforts the moth bursts through the leathery envelope into the chamber of the cocoon.

"The moth then extends her antennæ, together with her head and feet, towards the point of the cone, which not being thickly closed up in that part gradually yields to her efforts ; she enlarges the opening, and at last comes forth, leaving at the bottom of the cone the ruins of its former state—namely, the head and entire skin of the caterpillar, which bear some resemblance to a heap of foul linen."*

Réaumur was of opinion that the moth makes use of its eyes as a file, in order to effect its passage through the silk ; while Malpighi, Peck, and others, believe that it is assisted by an acid which it discharges in order to dissolve the gum that holds the fibres of the silk together (see p. 338). Mr. Swayne denies that the threads are broken at all, either by filing or solution ; for he succeeded in unwinding a whole cocoon from which the moth had escaped. The soiling of the cocoon by a fluid, however, we may remark, is no proof of the acid ; for all moths and butterflies discharge a fluid when they assume wings, whether they be inclosed in a cocoon or not ; but it gives no little plausibility to the opinion, that "the end of the cocoon is observed to be wetted for an hour, and sometimes several hours, before the moth makes its way out."† Other insects employ different contrivances for escape, as we have already seen, and shall still further exemplify.

It is the middle portion of the cocoon, after removing the floss or loose silk on the exterior, which is used in our manufactures ; and the first preparation is to throw the cocoons

* Spectacle de la Nature, vol. i.
† Count Dandolo's Art of Rearing Silk-Worms, Eng. Transl., p. 215.

into warm water, and to stir them about with twigs, to dissolve any slight gummy adhesions which may have occurred when the caterpillar was spinning. The threads of several cones, according to the strength of the silk wanted, are then taken and wound off upon a reel. The refuse, consisting of what we may call the tops and bottoms of the cones, are not wound, but carded, like wool or cotton, in order to form coarser fabrics. We learn from the fact of the cocoons being generally unwound without breaking the thread, that the insect spins the whole without interruption. It is popularly supposed, however, that if it be disturbed during the operation by any sort of noise, it will take alarm, and break its thread; but Latreille says this is a vulgar error.*

The length of the unbroken thread in a cocoon varies from six hundred to a thousand feet; and as it is all spun double by the insect, it will amount to nearly two thousand feet of silk, the whole of which does not weigh above three grains and a half; five pounds of silk from ten thousand cocoons is considerably above the usual average. When we consider, therefore, the enormous quantity of silk which is used at present, the number of worms employed in producing it will almost exceed our comprehension. The manufacture of the silk, indeed, gives employment, and furnishes subsistence, to several millions of human beings; and we may venture to say, that there is scarcely an individual in the civilized world who has not some article made of silk in his possession.

In ancient times, the manufacture of silk was confined to the East Indies and China, where the insects that produce it are indigenous. It was thence brought to Europe in small quantities, and in early times sold at so extravagant a price, that it was deemed too expensive even for royalty. The Emperor Aurelian assigned the expense as a reason for refusing his empress a robe of silk; and our own James I., before his accession to the crown of England, had to borrow of the Earl of Mar a pair of silk stockings to appear in before

* On a tort de croire que le bruit nuise à ces insectes, Hist. Nat. Générale, vol. xiii. p. 170.

the English ambassador, a circumstance which probably led him to promote the cultivation of silk in England.* The Roman authors were altogether ignorant of its origin,—some supposing it to be grown on trees as hair grows on animals,—others that it was produced by a shell-fish similar to the mussel, which is known to throw out threads for the purpose of attaching itself to rocks,—others that it was the entrails of a sort of spider, which was fed for four years with paste, and then with the leaves of the green willow, till it burst with fat,—and others that it was the produce of a worm which built nests of clay and collected wax. The insect was at length spread into Persia ; and eggs were afterwards, at the instance of the Emperor Justinian, concealed in hollow canes by two monks, and conveyed to the Isle of Cos. This emperor, in the sixth century, caused them to be introduced into Constantinople, and made an object of public utility. They were thence successively cultivated in Greece, in Arabia, in Spain, in Italy, in France, and in all places where any hope could be indulged of their succeeding. In America the culture of the silk-worm was introduced into Virginia in the time of James I., who himself composed a book of instructions on the subject, and caused mulberry-trees and silk-worms' eggs to be sent to the colony. In Georgia, also, lands were granted on condition of planting one hundred white mulberry-trees on every ten acres of cleared land.†

The growth of the silk-worm has also been tried, but with no great success, in this country. Evelyn computed that one mulberry-tree would feed as many silk-worms annually as would produce seven pounds of silk. "According to that estimate," says Barham,‡ "the two thousand trees already planted in Chelsea Park (which take up one-third of it) will make 14.000 lbs. weight of silk ; to be commonly worth but twenty shillings a pound, those trees must make 14,000*l.* per annum." During the last century, some French refugees in the south of Ireland made considerable plantations of the

* Shaw's Gen. Zoology, vol. vi.
† North American Review, Oct. 1828, p. 449.
‡ Essay on the Silk-Worm, p. 95. London, 1719.

mulberry, and had begun the cultivation of silk with every appearance of success; but since their removal the trees have been cut down.* In the vicinity of London, also, a considerable plantation of mulberry-trees was purchased by the British, Irish, and Colonial Silk Company in 1825; but we have not learned whether this Company have any active measures now in operation.

The manufacture of silk was introduced into this country in 1718, at Derby, by Mr. John Lombe, who travelled into Italy to obtain the requisite information; but so jealous were the Italians of this, that according to some statements which have obtained belief, he fell a victim to their revenge, having been poisoned at the early age of twenty-nine.†

There are not only several varieties of the common silk-worm (*Bombyx mori*), but other species of caterpillars, which spin silk capable of being manufactured, though not of so good qualities as the common silk. None of our European insects, however, seem to be well fitted for the purpose, though it has been proposed by Fabricius and others to try the crimson under-wing (*Catocala sponsa*, Schrank), &c. M. Latreille quotes from the 'Recreations of Natural History,' by Wilhelm, the statement that the cocoons of the emperor-moth (*Saturnia pavonia*) had been successfully tried in Germany, by M. Wentzel Hegeer de Berchtoldsdorf, under an imperial patent.

Emperor-Moth.

The emperor-moth, indeed, is no less worthy of our attention with respect to the ingenuity of its architecture than the beauty of its colours, and has consequently attracted the attention of every Entomologist. The caterpillar feeds on fruit-trees and on the willow, and spins a cocoon, in the form of a Florence flask, of strong silk, so thickly woven that it appears almost like damask or leather. It differs from most other cocoons in not being closed at the upper or smaller end, which terminates in a narrow circular aperture, formed

---

* Preface to Dandolo on the Silk-Worm, Eng. Transl., p. xiii.
† Glover's Directory of the County of Derby, Introd., p. xvi.

by the convergence of little bundles of silk, gummed together, and almost as elastic as whalebone. In consequence of all these terminating in needle-shaped points, the entrance of depredators is guarded against, upon the principle which prevents the escape of a mouse from a wire trap. The insect, however, not contented with this protection, constructs another in form of a canopy or dome, within the external aperture, so as effectually to shield the chrysalis from danger. We have formerly remarked (page 210) that the caterpillar of the *Ægeria asiliformis* of Stephens in a similar way did not appear to be contented with a covering of thin wood, without an additional bonnet of brown wax. The cocoon of

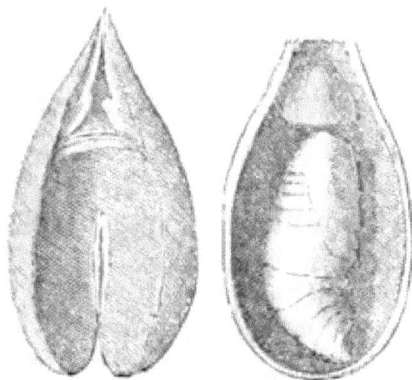

Cocoons of the Emperor-moth, cut open to show their structure.

the emperor-moth, though thus in some measure impenetrable from without, is readily opened from within; and when the moth issues from its pupa case, it easily makes its way out without either the acid or eye-files ascribed to the silk-worm. The elastic silk gives way upon being pushed from within, and when the insect is fairly out, it shuts again of its own accord, like a door with spring hinges,—a circumstance which at first puzzled Roesel not a little when he saw a fine large moth in his box, and the cocoon apparently in the same state as when he had put it there. Another naturalist conjectures that the converging threads are intended to compress the body of the moth as it emerges, in order to force the

fluids into the nervures of the wings; for when he took the
chrysalis previously out of the cocoon, the wings of the moth
never expanded properly.* Had he been much conversant
with breeding insects, he would rather, we think, have imputed
this to some injury which the chrysalis had received. We
have witnessed the shrivelling of the wings which he alludes
to, in many instances, and not unfrequently in butterflies
which spin no cocoon. The shrivelling, indeed, frequently
arises from the want of a sufficient supply of food to the
caterpillar in its last stage, occasioning a deficiency in the
fluids.

The elasticity of the cocoon is not peculiar to the emperor-
moth. A much smaller insect, the green cream-border-moth
(*Tortrix chlorana*) before mentioned (page 190), for its
ingenuity in bundling up the expanding leaves of the willow,
also spins an elastic shroud for its chrysalis, of the singular
shape of a boat with the keel uppermost. Like the caterpillar of *Pyralis strigulalis* (page 217), whose building,
though of different materials, is exactly of the same form,
its first spins two approximating walls of whitish silk, of the
form required, and when these are completed, it draws them
forcibly together with elastic threads, so placed as to retain
them closely shut. The passage of the moth out of this
cocoon might have struck Roesel as still more marvellous
than that of his emperor, in which there was at least a small
opening; while in the boat cocoon there is none. We have
now before us two of these, which we watched the caterpillars
through the process of building, in the summer of 1828, and
from one only a moth issued—the other, as often happens,
having died in the chrysalis. But what is most remarkable,
it is impossible by the naked eye to tell which of these two
has been opened by the moth, so neatly has the joining been
finished. (J. R.)

Some species of moths spin a very slight silken tissue for
their cocoons, being apparently intended more to retain
them from falling than to afford protection from other accidents. The gipsy-moth (*Hypogymna dispar*), rare in most

* Meinecken, quoted by Kirby and Spence, iii. 280.

parts of Britain, is one of these. It selects for its retreat a
crack in the bark of the tree upon which it feeds, and over
this spins only a few straggling threads. We found last
summer (1829), in the hole of an elm-tree in the Park at
Brussels, a group of half a dozen of these, that did not seem
to have spun any covering at all, but trusted to a curtain of
moss (*Hypna*) which margined the entrance. (J. R.) In a
species nearly allied to this, the yellow-tussock (*Dasychira
pudibunda*, STEPHENS), the cocoon, one of which we have
now before us, is of a pretty close texture, and interwoven
with the long hairs of the caterpillar itself (see figure *b*, page

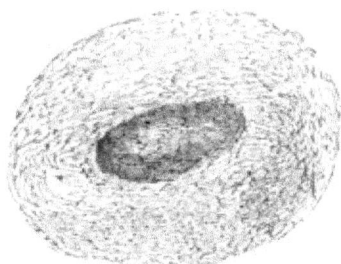

Cocoon of Arctia villica.

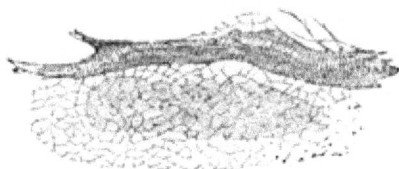

Net-work cocoon.

17), which it plucks out piecemeal during the process of
building,—as is also done by the vapourer (*Orgyia antiqua*,
HÜBNER), and many others.

These are additional instances of the remarks we formerly
made, that caterpillars which spin a slight web are transformed into perfect insects in a much shorter period than
those which spin more substantial ones. Thus the cream-
spot tiger (*Arctia villica*, STEPHENS) lies in chrysalis only
three weeks, and therefore does not require a strong web. It
is figured above, along with another, which is still slighter,

though more ingeniously woven, being regularly meshed like net-work.

A very prettily-netted cocoon is constructed by the grub of a very small grey weevil (*Hypera rumicis*), which is not uncommon in July, on the seed spikes of docks (*Rumices*). This cocoon is globular, and not larger than a garden pea, though it appears to be very large in proportion to the pupa of the insect, reminding us not a little of the carved ivory balls from China. The meshes of the net-work are also large, but the materials are strong and of a waxy consistence. Upon remarking that no netting was ever spun over the part of the plant to which the cocoon was attached, we endeavoured to make them spin cocoons perfectly globular by detaching them when nearly finished; but though we tried four or five in this way, we could not make them add a single mesh after removal, all of them making their escape through the opening, and refusing to re-enter in order to complete their structure. (J. R.)

The silk, if it may be so termed, spun by many species of larvæ is of a still stronger texture than the waxy silk of the little weevil just mentioned. We recently met with a remarkable instance of this at Lee, in the cocoons of one of the larger ichneumons (*Ophion Vinulæ?* STEPHENS), inclosed in that of a puss-moth (*Cerura Vinula*)—itself remarkable for being composed of sand as well as wood, the fibres of which had been scooped out of the under-ground cross-bar of an old paling, to which it was attached. But the most singular portion of this was the junction of the outer wall with the edges of the hollow thus scooped out, which was formed of fibres of wood placed across the fibres of the bar nearly at right angles, and strongly cemented together, as if to form a secure foundation for the building.

In this nest were formed, surreptitiously introduced into the original building, five empty cells of a black colour, about an inch long, and a sixth of an inch in diameter; nearly cylindrical in form, but somewhat flattened; vertical and parallel to one another, though slightly curved on the inner side. The cells are composed of strong and somewhat coarse fibres, more like the carbonized rootlets of a tree than silk, and resembling in texture a piece of coarse milled cloth or felt, such as is used for the bases of plated hats. It is worthy of remark, that all these cells opened towards one end, as if the caterpillars which constructed them had been aware that the wall of the puss-moth, in which the flies would have to make a breach, was very hard, and would require their united efforts to effect an escape. The importance of such a precaution will appear more strikingly, when we compare it with the instance formerly mentioned (page 215), in which only one ichneumon had been able to force its way out. (J. R.)

Nest of Puss-moth, inclosing five cocoons of an Ichneumon. Natural size.

It appears indispensable to some grubs to be confined within a certain space in order to construct their cocoons. We saw this well exemplified in the instance of a grub of one of the mason-bees (*Osmia bicornis*), which we took from its nest, and put into a box with the pollen paste which the mother bee had provided for its subsistence. (See pages 45, 46.) When it had completed its growth, it began to spin, but in a very awkward manner—attaching threads, as if at random, to the bits of pollen which remained undevoured, and afterwards tumbling about to another part of the box, as if dissatisfied with what it had done. It sometimes persevered to spin in one place till it had formed a little vaulted

wall; but it abandoned at the least three or four of these in
order to begin others, till at length, as if compelled by the
extreme urgency of the stimulus of its approaching change,
it completed a shell of shining brown silk, woven into a close
texture.   Had the grub remained within the narrow clay cell
built for it by the mother bee, it would, in all probability,
not have thus exhausted itself in vain efforts at building,
which were likely to prevent it from ever arriving at the perfect state—a circumstance which often happens in the artificial breeding of insects.   This bee, however, made its
appearance the following spring.   (J. R.)

Besides silk, the cocoons of many insects are composed of
other animal secretions, intended to strengthen or otherwise
perfect their texture.   We have already seen that some caterpillars pluck off their own hair to interweave amongst their
silk; there are others which produce a peculiar substance for
the same purpose.   The lackey caterpillar (*Clisiocampa
neustria*, CURTIS) in this manner lines its cocoon with pellets
of a downy substance, resembling little tufts of the flowers of
sulphur.   The small egger, again (*Eriogaster lanestris*, GER-
MAR), can scarcely be said to employ silk at all,—the cocoon
being of a uniform texture, looking, at first sight, like dingy
Paris plaster, or the shell of a pheasant's egg; but upon being
broken, and inspected narrowly, a few threads of silk may be
seen interspersed through the whole.   In size it is not larger
than the egg of the gold-crested wren.   It has been considered by Brahm a puzzling circumstance, that this cocoon
is usually perforated with one or two little holes, as if made
by a pin from without; and Kirby and Spence tell us that
their use has not been ascertained.[*]   May they not be left as
air-holes for the included chrysalis, as the close texture of the
cocoon might, without this provision, prove fatal to the animal?   Yet, on comparing one of these with a similar cocoon
of the large egger-moth (*Lasiocampa quercus*), we find no air-
holes in the latter, as we might have been led to expect from
the closeness of its texture.   We found a cocoon of a saw-fly
(*Trichiosoma*), about the same size as that of the egger,

[*] Brahm's Ins. Nat. 289, and Kirby and Spence's Intr. iii. 223.

attached to a hawthorn-twig, in a hedge at New Cross, Deptford, but of a leathery texture, and, externally, exactly the colour of the bark of the tree. During the summer of 1830 we found a considerable number of the same cocoons. These were all without air-holes. The egger, we may remark, unlike the dock-weevil or the bee-grub just mentioned, can work her cocoon without any point of attachment. We had a colony of these caterpillars in the summer of 1825, brought from Epping Forest, and saw several of them work their cocoons, and we could not but admire the dexterity with which they avoided filling up the little pin-holes. The supply of their building material was evidently measured out to them in the exact quantity required; for when we broke down a portion of their wall, by way of experiment, they did not make it above half the thickness of the previous portion, though they plainly preferred having a thin wall to leaving the breach unclosed. (J. R.)

Several species of caterpillars, that spin only silk, are social, like some of those we formerly mentioned, which unite to form a common tent of leaves (see pages 351, 352, &c.). The most common instance of this is in the caterpillars which feed on the nettle—the small tortoise-shell (*Vanessa urticæ*), and the peacock's eye (*V. I.*). Colonies of these may be seen, after midsummer, on almost every clump of nettles, inhabiting a thin web of an irregular oval shape, from which they issue out to feed on the leaves, always returning when their appetite is satisfied, to assist their companions in extending their premises. Other examples, still more conspicuous from being seen on fruit-trees and in hedges, occur in the caterpillars of the small ermine-moth (*Yponomeuta padella*), and of the lackey (*Clisiocampa neustria*), which in some years are but too abundant, though in others they are seldom met with. In the summer of 1826, every hedge and fruit-tree around London swarmed with colonies of the ermine, though it has not since been plentiful; and in the same way, during the summer of 1829, the lackeys were to be seen everywhere. We mention this irregularity of appearance that our readers may not disappoint themselves by looking for what is not always to be found. It is probable that in 1830 the lackeys will be few, for, notwithstanding the myriads of caterpillars last summer, we saw only a single moth of this species, and out of a number of chrysalides which a young friend had in his nurse-boxes, not one moth was bred.

The small ermine does not, besides, feed so indiscriminately as many others, but when the bird-cherry (*Prunus padus*), its peculiar food, is not to be had, it will put up with blackthorn, plum-tree, hawthorn, and almost any sort of orchard fruit-tree. With respect to such caterpillars as feed on different plants, Réaumur and De Geer make the singular remark, that in most cases they would only eat the sort of plant upon which they were originally hatched.[*] We verified this, in the case of the caterpillar in question, upon two different nests which we took, in 1806, from the bird-cherry at Crawfordland, in Ayrshire. Upon bringing these to Kilmarnock, we could not readily supply them with the leaves of this tree; and having then only a slight acquaintance with the habits of insects, and imagining they would eat any sort of leaf, we tried them with almost everything green in the vicinity of the town; but they refused to touch any which we offered them. After they had fasted several days, we at length procured some fresh branches of the bird-cherry, with which they gorged themselves so that most of them died. Last summer (1829) we again tried a colony of these caterpillars, found on a seedling plum-tree at Lee, in Kent, with blackthorn, hawthorn, and many other leaves, and even with those of the bird-cherry; but they would touch nothing except the seedling plum, refusing the grafted varieties. (J. R.)

A circumstance not a little remarkable in so very nice a feeder is, that in some cases the mother moth will deposit her eggs upon trees not of indigenous growth, and not even of the same genus with her usual favourites. Thus, in 1825, the cherry-apple, or Siberian crab (*Pyrus prunifolia*, WILLDENOW), so commonly grown in the suburbs of London, swarmed with them. On a single tree at Islington we

* De Geer, Mém. i. 319.

counted above twenty nests, each of which would contain
from fifty to a hundred caterpillars; and though these do
not grow thicker than a crow-quill, so many of them scarcely
left a leaf undevoured, and, of course, the fruit, which
showed abundantly in spring, never came to maturity. The
summer following they were still more abundant on the
hawthorn hedges, particularly near the Thames, by Battersea
and Richmond. Since then we have only seen them

Encampment of the caterpillar of the small ermine (*Yponomeuta padella*) on the
Siberian crab.

sparingly; and last summer we could only find the single
nest upon which we tried the preceding experiment. (J. R.)

The caterpillars of other moths, which are in some years
very common— such as the brown-tail (*Porthesia auriflua*),
and the golden-tail (*P. chrysorrhœa*), are also social; and,
as the eggs are hatched late in the summer, the brood passes
the winter in a very closely-woven nest of warm silk. This
is usually represented as composed of leaves which have had
their pulpy parts eaten as food by the colonists; but from
minute observation of at least twenty of these nests in the

winter of 1828-9, we are quite satisfied that leaves are only
an accidental, and not a necessary, part of the structure.
When a leaf happens to be in the line of the walls of the
nest, it is included; but there is no apparent design in
pressing it into the service, nor is a branch selected because
it is leafy. On the contrary, by far the greater number of
these nests do not contain a single leaf, but are composed
entirely of grey silk. In external form, no two of these
nests are alike; as it depends entirely upon the form of the
branch. When, therefore, there is only one twig, it is somewhat egg-shaped; but when there are several twigs, it
commonly joins each, assuming an angular shape, as may be
seen in the left-hand figure.

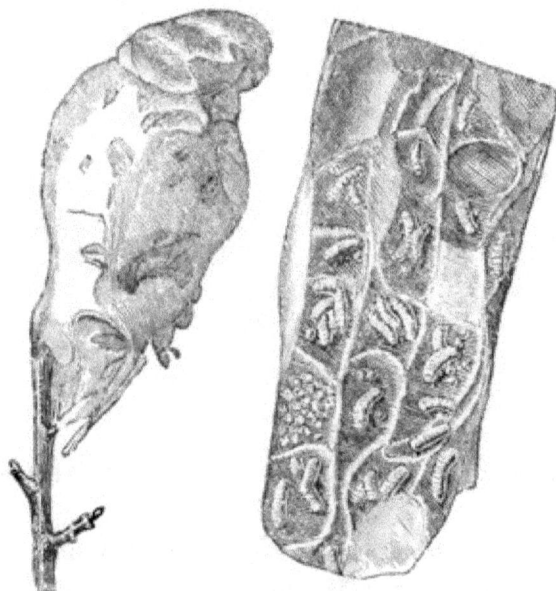

Winter nests of *Porthesia chrysorrhœa*, one being cut open to show the chambers.
The dots represent the egesta of the caterpillars.

This irregularity arises from the circumstance of each
individual acting on its own account, without the direction
or superintendence of the others. The interior of the
structure is, for the same reason, more regular, being divided
into compartments, each of which forms a chamber for one
or more individuals. Previous to the cold weather, these

chambers have but slight partitions; but before the frosts set in the whole is made thick and warm.

None of the preceding details, however, appear so striking as what is recorded of the brown-tail moth (*Porthesia auriflua*) by Mr. W. Curtis,* whose multitudinous colonies spread great alarm over the country in the summer of 1782. This alarm was much increased by the exaggeration and ignorant details which found their way into the newspapers. The actual numbers of these caterpillars must have been

Winter nest of the Social Caterpillars of the Brown-tail Moth (*Porthesia auriflua*), figured from specimen.

immense, since Curtis says, "in many of the parishes near London subscriptions have been opened, and the poor people employed to cut off the webs at one shilling per bushel, which have been burnt under the inspection of the church-wardens, overseers, or beadle of the parish: at the first onset of this business fourscore bushels, as I was most credibly informed, were collected in one day in the parish of Clapham."

It is not, therefore, very much to be wondered at, that the

* Curtis, Hist. of Brown-tail Moth, 4to, London, 1782.

ignorant, who are so prone to become the victims of groundless fears, should have taken serious alarm on having so unusual a phenomenon forced upon their attention. Some alarmists accordingly asserted that the caterpillars "were the usual presage of the plague;" and others that they not only presaged it, but would actually . cause it, for "their numbers were great enough to render the air pestilential;" while, to add to the mischief, "they would destroy every kind of vegetation, and starve the cattle in the fields." "Almost every one," adds Curtis, "ignorant of their history, was under the greatest apprehensions concerning them; so that even prayers were offered up in some churches to deliver the country from the apprehended approaching calamity."

It seems to have been either the same caterpillar, or one very nearly allied to it, probably that of the golden-tail (*Porthesia chrysorrhœa*), which in 1731-2 produced a similar alarm in France. Réaumur, on going from Paris to Tours, in September, 1730, found every oak, great and small, literally swarming with them, and their leaves parched and brown as if some burning wind had passed over them; for when newly hatched, like the young buff-tips, they only eat one of the membranes of the leaf, and of course the other withers away. These infant legions, under the shelter of their warm nests, survived the winter in such numbers, that they threatened the destruction not only of the fruit-trees, but of the forests,—every tree, as Réaumur says, being overrun with them. The Parliament of Paris thought that ravages so widely extended loudly called for their interference, and they accordingly issued an edict, to compel the people to uncaterpillar (*décheniller*) the trees; which Réaumur ridiculed as impracticable, at least in the forests. About the middle of May, however, a succession of cold rains produced so much mortality among the caterpillars, that the people were happily released from the edict; for it soon became difficult to find a single individual of the species.* In the same way the cold rains, during the

* Réaumur, ii. p. 137.

summer of 1829, seem to have nearly annihilated the lackeys, which in the early part of the summer swarmed on every hedge around London. The ignorance displayed in France, at the time in question, was not inferior to that recorded by Curtis; for the French journalists gravely asserted that part of the caterpillars were produced by spiders; and that these spiders, and not the caterpillars, constructed the webs of the slime of snails, which they were said to have been seen collecting for the purpose! "Verily," exclaims Réaumur, "there is more ignorance in our age than one might believe."

It is justly remarked by Curtis, that the caterpillar of the brown-tail moth is not so limited a feeder as some, nor so indiscriminate as others; but that it always confines itself to trees or shrubs, and is never found on herbaceous plants, whose low growth would seldom supply a suitable foundation for its web. Hence the absurdity of supposing it would attack the herbage of the field, and produce a famine among cattle. Curtis says, it is found on the "hawthorn most plentifully, oak the same, elm very plentifully, most fruit-trees the same, blackthorn plentifully, rose-trees the same, bramble the same, on the willow and poplar scarce. None have been noticed on the elder, walnut, ash, fir, or herbaceous plants. With respect to fruit-trees the injuries they sustain are most serious, as, in destroying the blossoms as yet in the bud, they also destroy the fruit in embryo: the owners of orchards, therefore, have great reason to be alarmed."

The sudden appearance of great numbers of these caterpillars in particular years, and their scarcity in others, is in some degree explained by a fact stated by Mr. Salisbury. "A gentleman of Chelsea," he says, "has informed me that he once took a nest of moths and bred them; that some of the eggs came the first year, some the second, and others of the same nest did not hatch till the third season."* We reared, during 1829, several nests both of the brown-tails and of the golden-tails, and a number of the females

* Salisbury, Hints on Orchards, p. 53

deposited their eggs in our nurse-cages; but, contrary to the experiment just quoted, all of these were hatched during the same autumn. (J. R.) The difference of temperature and moisture in particular seasons may produce this diversity.

A no less remarkable winter nest, of a small species of social caterpillar, is described by M. Bonnet, which we omitted to introduce when treating of the Glanville fritillary. The nest in question is literally pendulous, being hung from the branch of a fruit tree by a strong silken thread. It consists of one or two leaves neatly folded, and held together with silk, in which the caterpillars live harmoniously together.

Pendulous leaf-nests, from Bonnet.

In a recently-published volume of 'Travels in Mexico,' we find a very remarkable account of some pendulous nests of caterpillars, which appear to be almost as curious as the nests of the pasteboard-making wasps, described at p. 177. The author of these Travels does not define the species of caterpillar whose constructions attracted his observation. He says, "After having ascended for about an hour, we came to the region of oaks and other majestically tall trees, the names of which I could not learn. Suspended from their stately branches were innumerable nests, enclosed, apparently, in white paper bags, in the manner of bunches of grapes in England, to preserve them from birds and flies.

I had the curiosity to examine one of them, which I found to contain numberless caterpillars. The texture is so strong that it is not easily torn; and the interior contained a quantity of green leaves, to support the numerous progeny within."*

[We will now give a brief account of several foreign insects that are remarkable for the pendulous nests which they make.

[The first of these is built by a hymenopterous insect belonging to the genus Larrada. It is fastened to the under side of a leaf, and is made of vegetable fibres, cut up very short, and masticated by the insect, much like the materials used by

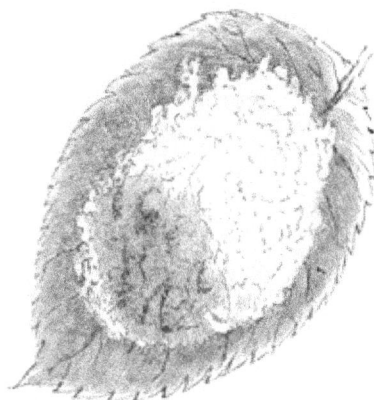

Nest of Larrada.

the Fungus Ant, described on page 311. The insect which forms this nest is black in colour and has very thick legs. The wings are clouded. It is but a small insect, being only three-eighths of an inch in length. Both nest and insect are in the collection of Mr. F. Smith, of the British Museum.

[In the next illustration are seen two nests built by hymenopterous insects belonging to the hymenopterous genus Polybia. The left-hand figure represents a nest made by Polybia sedula, a Brazilian insect. It is fixed to a large leaf, and, as may be seen, has the entrance at the end of a long neck. The exterior of the nest is a very thin sheet

* Hardy's Travels in the Interior of Mexico, p. 32.

of pale-brown substance, almost identical with the paper
with which our British wasps make their nests.   It is,
however, very much stronger and very much thinner, and
is very close in texture, so that it effectually excludes rain.

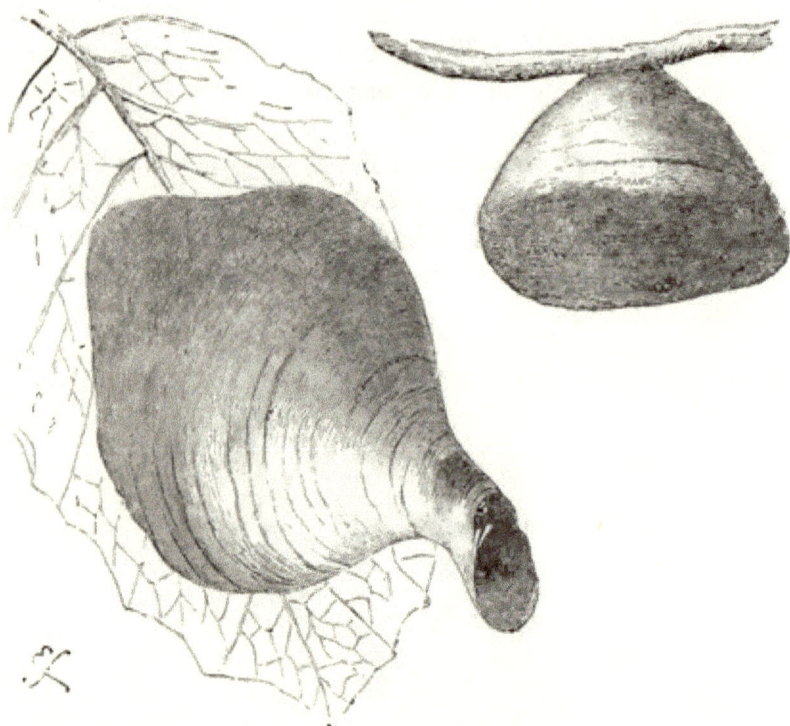

Nests of Polybia.

[The right-hand figure shows a nest also brought from
Brazil.   It has no neck, the opening being a mere hole
beneath.   The name Polybia is derived from two Greek
words signifying that many insects live together, and has
been given to the genus on account of the social habits of
its members.

[Our next illustration contains some very remarkable
nests.

[The large central nest is the cocoon of a moth belonging to the genus Oiketicus, or Housebuilder.   There
are many species of Housebuilder moths, and all are re-

markable for the fact that the larva never exhibits itself,
but builds a dwelling in which it conceals itself, just as does
the well-known caddis-worm. Indeed the nests of several
Oiketici look exactly as if the dwelling of a caddis-worm
had been greatly enlarged and hung up in a tree.

[The nest of this species, however, differs from that of the
common Oiketicus by being covered with a coating of greyish

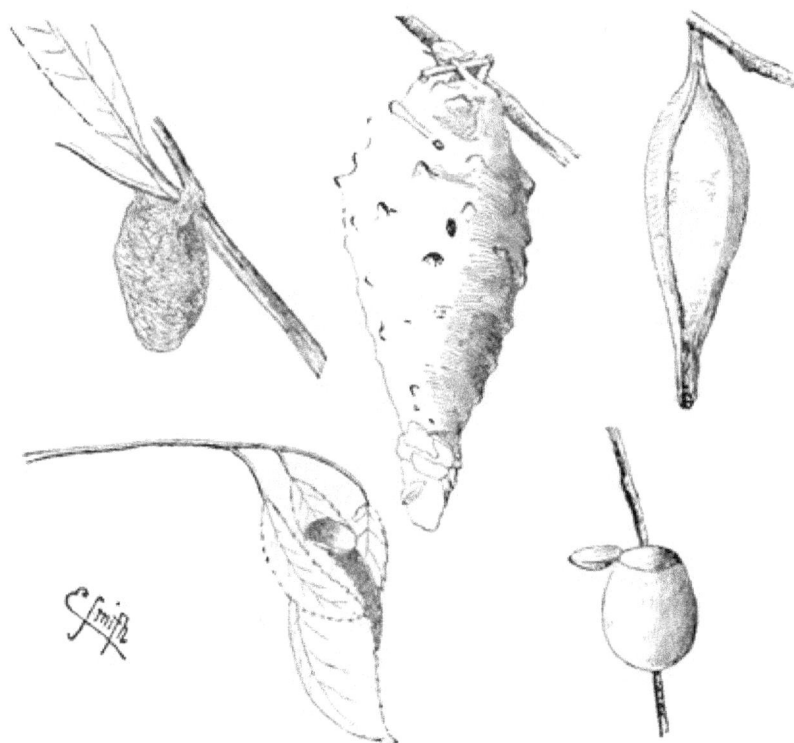

Nests of Oiketicus, &c.

silk. If we cut open the silk, we find a great number of
little sticks and leaf-stems crossed on each other, and
showing their ends through the silken cover. Within these
defences there is a layer of leaves cut into small pieces, and
lastly comes the cell inhabited by the caterpillar. It is
lined with a silken web similar in character to that on the
outside, but finer, stronger, and whiter. The caterpillar is
therefore defended by four distinct barriers. First comes

the strong silken web which lines the cell, and next is the layer of leaf fragments. Outside them comes the *chevaux de frise* of crossed sticks, and lastly we have the grey silken web. This outer wrapper has no connection with the interior of the cell, and is only lightly attached to the ends of the cross sticks.

[Within this curious dwelling the caterpillar conceals the whole of its body, clinging to the branch or leaf by its feet, and if alarmed drawing itself up so that the mouth of the cocoon is pressed tightly against the branch, and effectually conceals even the feet which hold it.

[The other figure on the right hand represents the dwelling of another Housebuilder caterpillar. It looks very much as if it were made of drab cloth. The most remarkable point about it is the lower end. When the insect is within the dwelling the extremity has a spiral twist, but when the moth has escaped the spiral form is destroyed, and it appears as represented in the illustration. The female Oikcticus never attains the winged state nor leaves her house, but lives and dies in it, almost unchanged in shape. In fact, the adult female is even more undeveloped in appearance than the caterpillar, and looks like a large, fat, unwieldy grub, covered with down. The male, on the contrary, is a tolerably active moth, with sharply-pointed wings and beautiful feathered antennæ.

[Another kind of Housebuilder's residence is shown in the lower left-hand figure, enveloped and almost concealed by leaves.

[The remaining figures represent the dwellings of two unknown insects, both from Australia. The upper left-hand nest is made wonderfully like that of the weaver-bird, being composed of fibres like cow-hairs woven loosely together. It is brown outside and white in the interior.

[The last nest is made of some substance which is smooth, and hard as horn, brown within, and dark grey on the outside. The circular lid by which the enclosed insect escapes is shown open.

[In the accompanying illustration, we have five remarkable pensile nests of insects, some British, and others exotic.

[Fig. 1 represents the nest of a Pelopæus from Natal. It is made of dried cow-dung, and is fixed to straws. The length is from three to five inches, and there are sometimes found three or more in a row upon a single straw. The insect is about an inch in length, black-blue in colour, and with clouded wings. The abdomen is small, sharply pointed, and placed on a long footstalk.

[At Fig. 2 is seen the nest of *Pelopæus Flavipes*, a North

Nests of Pelopæus (1, 2); Anthidium (3); Trypoxylon (4); and Eumenes (5).

American insect, which is also fixed along its whole length to the supporting object, which is sometimes a wall, and sometimes, as in the illustration, a branch. It is made of mud, and the insects seem to have a sort of gregarious instinct, loving to fix their nests in rows, one above the other. There is only one larva in each cell. The Pelopæi are, by the way, allied to the English genus *Ammophila*.

[Fig. 3 shows the nest of *Anthidium cordatum*, one of the solitary bees of Natal. It is made of vegetable fibres. The insect as well as the nest is represented of the natural size. It is black and shining, with the under part and sides and legs yellowish.

[At Fig. 4 are seen three of the nests of *Trypoxylon aurifrons*, a Brazilian insect. They are built of mud, and are remarkable for their elegant shape, which looks as if it had been formed by the hand of the potter, and for the manner in which the mouth is turned over so as to form a distinct neck. The larvæ is fed with a store of spiders. The insect is represented of the natural size; its colour is black, and the face is covered with short golden hairs, a fact which has gained for it the name of aurifrons, or golden-fronted.

[Our last example, Fig. 5, is the nest of an English insect, *Eumenes coarctata*. The insect is represented of its natural size. It is very pretty in colour as well as elegant in shape, being black, diversified with yellow bands and spots. The nest is made of clay, and is found upon the heath twigs. The larvæ of the Eumenes are fed with those of a species of *Crambus*. The insect is tolerably common in Surrey and Hampshire, and appears in July and August.

[The three figures in the next illustration represent the cocoons of three species of the Bombycidæ, and are given in order to show the different modes by which they are fastened. The upper nest is hung by a slight cord, which spreads into a broad silken band wrapped round the branch for some distance. The right-hand figure shows a very remarkable cocoon suspended by a long footstalk affixed to a ring. The remarkable point in the construction of this ring is that it is very hard and horny, and is not fastened to the branch, but passes loosely round it, so that the cocoon swings backwards and forwards in the breeze. The cocoon is about two inches in length, and is covered with thick black veinings. The lowermost cocoon is most curiously fixed to the branch by bending the leaves round the exterior of the dwelling, and fixing them to it with silk. All these specimens were brought from Northern India.]

In all the nests of social caterpillars, care is taken to leave apertures for passing out and in. It is remarkable, also, that however far they may ramble from their nest, they never fail to find their way back when a shower of rain or nightfall renders shelter necessary. It requires no great shrewdness to discover how they effect this : for by looking closely at their track it will be found that it is carpeted with silk—no individual moving an inch without constructing such a pathway, both for the use of his com-

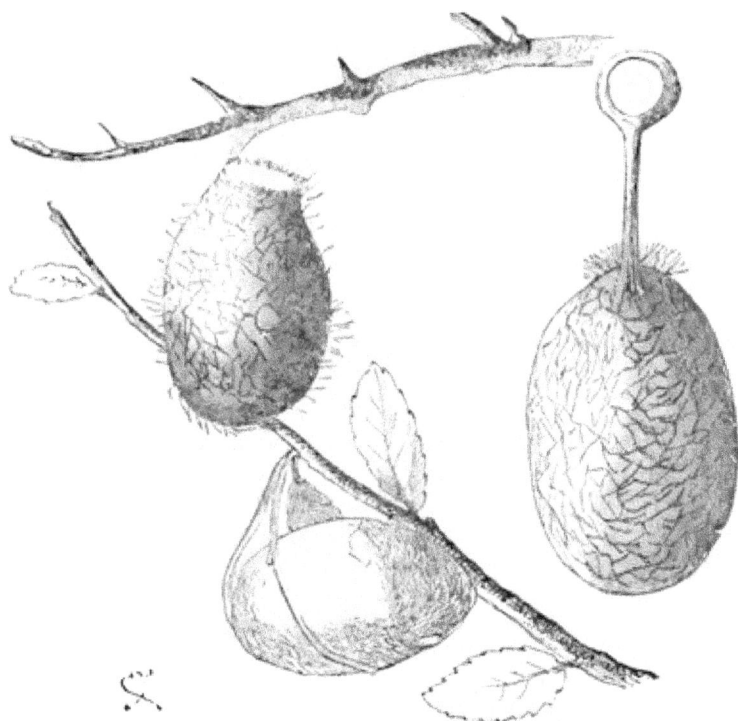

Bombycidæ.

panions and to facilitate his own return. All these social caterpillars, therefore, move more or less in processional order, each following the road which the first chance traveller has marked out with his strip of silk carpeting.

There are some species, however, which are more remarkable than others in the regularity of their processional marchings, particularly two which are found in the south of Europe, but are not indigenous in Britain. The one named by Réaumur the Processionary (*Cnethocampa processionea*, STEPHENS) feeds upon the oak; a brood dividing, when newly hatched, into one or more parties of several hundred individuals, which afterwards unite in constructing a common nest nearly two feet long, and from four to six inches in diameter. As it is not divided like that of the brown-tails into chambers, but consists of one large ball, it is not necessary that there should be more openings than one;

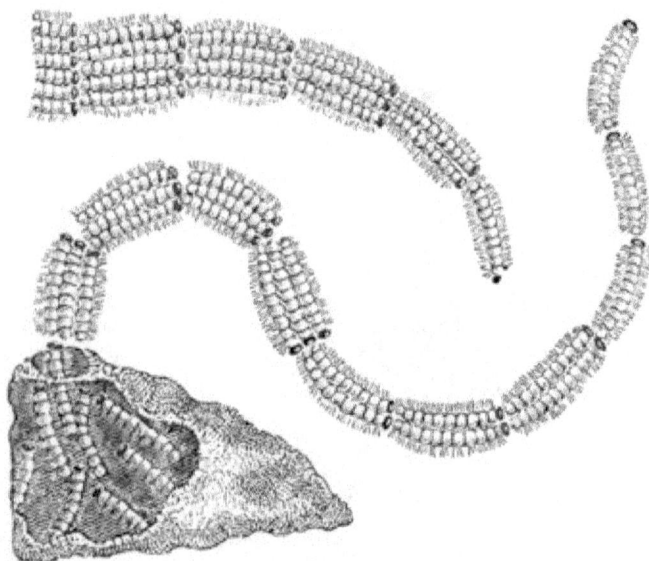

Nest and order of marching of the Processionary Caterpillars of the oak (*Cnethocampa processionea*).

and accordingly, when an individual goes out and carpets a path, the whole colony instinctively follow in the same track, though from the immense population they are often compelled to march in parallel files from two to six deep. The procession is always headed by a single caterpillar; sometimes the leader is immediately followed by one or two in single file, and sometimes by two abreast, as represented in the cut. A similar procedure is followed by a species of social caterpillars which feed on the pine in Savoy and Languedoc; and though their nests are not half the size of

the preceding, they are more worthy of notice, from the
strong and excellent quality of their silk, which Réaumur was
of opinion might be advantageously manufactured. Their
nest consists of more chambers than one, but is furnished
with a main entrance, through which the colonists conduct
their foraging processions.

[In the accompanying illustration is shown a nest of the

Nest of Processionary Caterpillars *C e'.iocampa* and *Calosoma*).

Processionary caterpillar, part of which has been torn away
to show the interior. Inside may be seen the larva of a certain
beetle (*Calosoma sycophanta*), which feeds on these caterpillars,
and one of the beetles is seen below in the act of ascending
the tree. This beetle, although exceedingly scarce in England,
is very common on the Continent, and trees have been cleared
of Processionary caterpillars by the simple process of putting
a few female beetles upon the branches.]

CHAPTER XVIII.

STRUCTURES OF SPIDERS.

MODERN naturalists do not rank spiders among insects, because they have no antennæ, and no division between the head and the shoulders. They breathe by leaf-shaped gills, situated under the belly, instead of spiracles in the sides; have a heart connected with these; have eight legs instead of six; and eight fixed eyes. But as spiders are popularly considered insects, it will sufficiently suit our purpose to introduce them here as such.

The apparatus by which spiders construct their ingenious fabrics is much more complicated than that which we have described as common to the various species of caterpillars. Caterpillars have only two reservoirs for the materials of their silk; but spiders, according to the dissections of M. Treviranus, have four principal vessels, two larger and two smaller, with a number of minute ones at their base. Several small tubes branch towards the reservoirs, for carrying to them, no doubt, a supply of the secreted material. Swammerdam describes them as twisted into many coils of an agate colour.* We do not find them coiled, but nearly straight, and of a deep-yellow colour. From these, when broken, threads can be drawn out like those spun by the spider, though we cannot draw them so fine by many degrees.

From these little flasks or bags of gum, situated near the apex of the abdomen, and not at the mouth, as in caterpillars, a tube originates, and terminates in the external spinnerets, which may be seen by the naked eye in the larger spiders, in the form of five little teats surrounded by a circle, as represented in the following figure.

We have seen that the silken thread of a caterpillar is com-

* Hill's Swammerdam, part i. p. 23.

posed of two united within the tube of the spinneret, but the spider's thread would appear, from the first view of its five spinnerets, to be quintuple, and in some species which have six teats, so many times more. It is not safe, however, in our interpretations of nature to proceed upon conjecture, however plausible, nor to take anything for granted which we have not actually seen; since our inferences in such cases are almost certain to be erroneous. If Aristotle, for example, had ever looked narrowly at a spider when spinning, he could

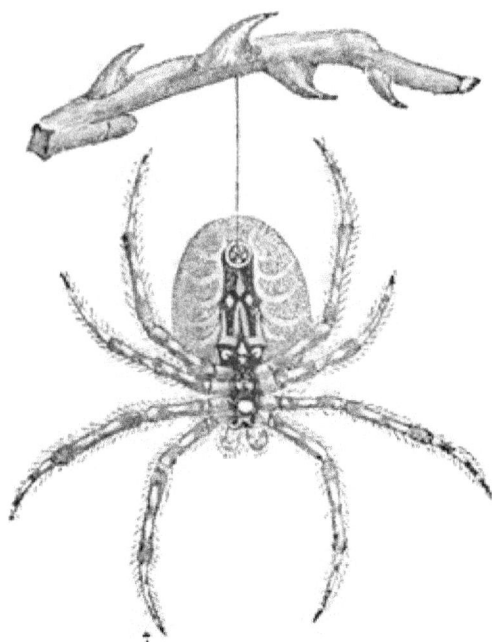

Garden Spider (*Epeira diadema*), suspended by a thread proceeding from its spinneret.

not have fancied, as he does, that the materials which it uses are nothing but wool stripped from its body. On looking, then, with a strong magnifying glass, at the teat-shaped spinnerets of a spider, we perceive them studded with regular rows of minute bristle-like points, about a thousand to each teat, making in all from five to six thousand. These are minute tubes which we may appropriately term *spinnerules*, as each is connected with the internal reservoirs, and emits a thread of inconceivable fineness. In the following figure, this wonderful apparatus is represented as it appears in the microscope.

We do not recollect that naturalists have ventured to assign any cause for this very remarkable multiplicity of the spinnerules of spiders, so different from the simple spinneret of caterpillars. To us it appears to be an admirable provision for their mode of life. Caterpillars neither require such strong materials, nor that their thread should dry as quickly. It is well known in our manufactures, particularly in rope-spinning, that in cords of equal thickness, those which are composed of many smaller ones united are greatly stronger than those which are spun at once. In the instance of the

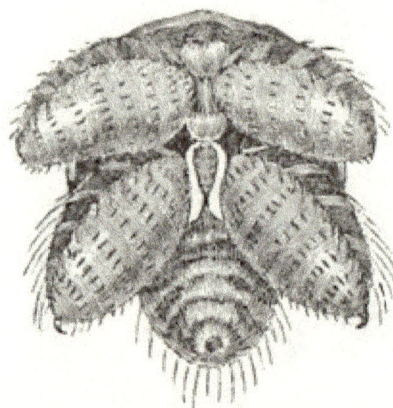

Spinnerets of a Spider magnified to show the *Spinnerules.*

spider's thread, this principle must hold still more strikingly, inasmuch as it is composed of fluid materials that require to be dried rapidly, and this drying must be greatly facilitated by exposing so many to the air separately before their union, which is effected at the distance of about a tenth of an inch from the spinnerets. In the following figure each of the threads represented is reckoned to contain one hundred minute threads, the whole forming only one of the spider's common threads.

Leeuwenhoeck, in one of his extraordinary microscopical observations on a young spider not bigger than a grain of sand, upon enumerating the threadlets in one of its threads, calculated that it would require four millions of them to be as thick as a hair of his beard.

Another important advantage derived by the spider from the multiplicity of its threadlets is, that the thread affords a

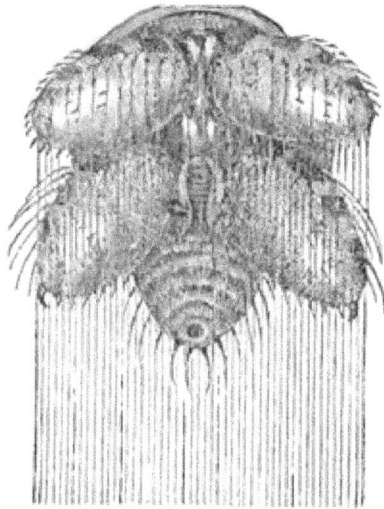

A single thread of a Spider, greatly magnified, so that, for the small space represented, the lines are shown as parallel.

much more secure attachment to a wall, a branch of a tree, or any other object, than if it were simple ; for, upon pressing the spinneret against the object, as spiders always do when they fix a thread, the spinnerules are extended over an area

Attached end of a Spider's thread magnified.

of some diameter, from every hair's-breadth of which a *strand*, as rope-makers term it, is extended to compound the main cord. The preceding figure exhibits this ingenious contrivance.

Those who may be curious to examine this contrivance will see it best when the line is attached to any black object, for the threads, being whitish, are, in other cases, not so easily perceived.

SHOOTING OF THE LINES.

It has long been considered a curious though a difficult investigation, to determine in what manner spiders, seeing that they are destitute of wings, transport themselves from tree to tree, across brooks, and frequently through the air itself, without any apparent starting point. On looking into the authors who have treated upon this subject, it is surprising how little there is to be met with that is new, even in the most recent. Their conclusions, or rather their conjectural opinions, are, however, worthy of notice; for by unlearning error, we the more firmly establish truth.

1. One of the earliest notions upon this subject is that of Blancanus, the commentator on Aristotle, which is partly adopted by Redi, by Henricus Regius of Utrecht, by Swammerdam,* by Lehmann, and by Kirby and Spence.† "The spider's thread," says Swammerdam, "is generally made up of two or more parts, and after descending by such a thread, it ascends by one only, and is thus enabled to waft itself from one height or tree to another, even across running waters; the thread it leaves loose behind it being driven about by the wind, and so fixed to some other body." "I placed," says Kirby, "the large garden spider (*Epeira diadema*) upon a stick about a foot long, set upright in a vessel containing water. . . . . It let itself drop, not by a single thread, but by *two*, each distant from the other about the twelfth of an inch, guided, as usual, by one of its hind feet, and one apparently smaller than the other. When it had suffered itself to descend nearly to the surface of the water, it stopped short, and by some means, which I could not distinctly see, broke off, close to the spinners, the smallest thread, which

* Swammerdam, part i. p. 24.   † Intr., vol. i. p. 415.

still adhering by the other end to the top of the stick, floated in the air, and was so light as to be carried about by the slightest breath. On approaching a pencil to the loose end of this line, it did not adhere from mere contact. I therefore twisted it once or twice round the pencil, and then drew it tight. The spider, which had previously climbed to the top of the stick, immediately pulled at it with one of its feet, and finding it sufficiently tense, crept along it, strengthening it as it proceeded by another thread, and thus reached the pencil."

We have repeatedly witnessed this occurrence, both in the fields and when spiders were placed for experiment, as Kirby has described; but we very much doubt that the thread broken is ever intended as a bridge cable, or that it would have been so used in that instance, had it not been artificially fixed and accidentally found again by the spider. According to our observations, a spider never abandons, for an instant, the thread which she despatches in quest of an attachment, but uniformly keeps trying it with her feet, in order to ascertain its success. We are, therefore, persuaded that when a thread is broken in the manner above described, it is because it has been spun too weak, and spiders may often be seen breaking such threads in the process of netting their webs. (J. R.)

The plan, besides, as explained by these distinguished writers, would more frequently prove abortive than successful, from the cut thread not being sufficiently long. They admit, indeed, that spiders' lines are often found "a yard or two long, fastened to twigs of grass not a foot in height. . . . . Here, therefore, some other process must have been used."*

2. Our celebrated English naturalist, Dr. Lister, whose treatise upon our native spiders has been the basis of every subsequent work on the subject, maintains that "some spiders shoot out their threads in the same manner that porcupines do their quills;† that whereas the quills of the

---

* Kirby and Spence, vol. i. Intr. p. 416.

† Porcupines do not shoot out their quills, as was once generally believed.

latter are entirely separated from their bodies, when thus shot out, the threads of the former remain fixed to their anus, as the sun's rays to its body."* A French periodical writer goes a little farther, and says, that spiders have the power of shooting out threads, and directing them at pleasure towards a determined point, judging of the distance and position of the object by some sense of which we are ignorant.† Kirby also says, that he once observed a small garden spider (*Aranea reticulata*) "standing midway on a long perpendicular fixed thread, and an appearance caught" his "eye, of what seemed to be the emission of threads." "I, therefore," he adds, "moved my arm in the direction in which they apparently proceeded, and, as I had suspected, a floating thread attached itself to my coat, along which the spider crept. As this was connected with the spinners of the spider, it could not have been formed" by breaking a "secondary thread."‡ Again, in speaking of the gossamer-spider, he says, " it first extends its thigh, shank, and foot, into a right line, and then, elevating its abdomen till it becomes vertical, *shoots its thread* into the air, and flies off from its station."§

Another distinguished naturalist, Mr. White of Selborne, in speaking of the gossamer-spider, says, "Every day in fine weather in autumn do I see these spiders shooting out their webs, and mounting aloft : they will go off from the finger, if you will take them into your hand. Last summer, one alighted on my book as I was reading in the parlour; and running to the top of the page, and *shooting out a web*, took its departure from thence. But what I most wondered at was, that it went off with considerable velocity in a place where no air was stirring; and I am sure I did not assist it with my breath."‖

Having so often witnessed the thread set afloat in the air by spiders, we can readily conceive the way in which those eminent naturalists were led to suppose it to be ejected by

* Lister, Hist. Animalia Angliæ, 4:o. p. 7.
† Phil. Mag., ii. p. 275.
‡ Vol. i. Intr., p. 417.   § Ibid., ii. p. 339.
‖ Nat. Hist. of Selborne, vol. i. p. 327.

some animal force acting like a syringe; but as the statement can be completely disproved by experiment, we shall only at present ask, in the words of Swammerdam—" how can it be possible that a thread so fine and slender should be shot out with force enough to divide and pass through the air?—is it not rather probable that the air would stop its progress, and so entangle it and fit it to perplex the spider's operations?"* The opinion, indeed, is equally improbable with another, suggested by Dr. Lister, that the spider can retract her thread within the abdomen, after it has been omitted.† De Geer‡ very justly joins Swammerdam in rejecting both of these fancies, which, in our own earlier observations upon spiders, certainly struck us as plausible and true. There can be no doubt, indeed, that the animal has a voluntary power of permitting the material to escape, or stopping it at pleasure, but this power is not projectile.

3. "There are many people," says the Abbé de la Pluche, "who believe that the spider flies when they see her pass from branch to branch, and even from one high tree to another; but she transports herself in this manner: she places herself upon the end of a branch, or some projecting body, and there fastens her thread; after which, with her two hind feet, she squeezes her dugs (*spinnerets*), and presses out one or more threads of two or three ells in length, which she leaves to float in the air till it be fixed to some particular place."§ Without pretending to have observed this, Swammerdam says, " I can easily comprehend how spiders, without giving themselves any motion, may, by only compressing their spinnerets, force out a thread, which being driven by the wind, may serve to waft them from one place to another."‖ Others, proceeding upon a similar notion, give a rather different account of the matter. "The spider," says Bingley, "fixes one end of a thread to the place where she stands, and then with her hind paws *draws out* several other

* Book of Nature, part i. p. 25.   † Hist. Anim. Angliæ, 4to.
‡ Mémoires, vol. vii. p. 189.   § Spectacle de la Nature, vol. i.
‖ Book of Nature, part i. p. 25.

threads from the nipples, which, being lengthened out and driven by the wind to some neighbouring tree or other object, are by their natural clamminess fixed to it."*

Observation gives some plausibility to the latter opinion, as the spider always actively uses her legs, though not to draw out the thread, but to ascertain whether it has caught upon any object. The notion of her pressing the spinneret with her feet must be a mere fancy; at least it is not countenanced by anything which we have observed.

4. An opinion much more recondite is mentioned, if it was not started, by M. D'Isjonval, that the floating of the spider's thread is electrical. "Frogs, cats, and other animals," he says, "are affected by natural electricity, and feel the change of weather; but no other animal more than myself and my spiders." During wet and windy weather he accordingly found that they spun very short lines, "but when a spider spins a long thread, there is a certainty of fine weather for at least ten or twelve days afterwards."† A periodical writer, who signs himself Carolan,‡ fancies that in darting out her thread the spider emits a stream of air, or some subtle electric fluid, by which she guides it as if by magic.

A living writer (Mr. John Murray), whose learning and skill in conducting experiments give no little weight to his opinions, has carried these views considerably farther. "The aëronautic spider," he says, "can propel its thread both horizontally and vertically, and at all relative angles, in motionless air, and in an atmosphere agitated by winds; nay more, the aërial traveller can even dart its thread, to use a nautical phrase, in the 'wind's eye.' My opinion and observations are based on many hundred experiments. . . . . The entire phenomena are electrical. When a thread is propelled in a vertical plane, it remains perpendicular to the horizontal plane, always upright, and when others are projected at angles more or less inclined, their

* Animal Biography, vol. iii. p. 475, 3rd edition.
† Brez, Flore des Insectophiles. Notes, Supp. p. 134.
‡ Thomson's Ann. of Philosophy, vol. iii. p. 306.

direction is invariably preserved; the threads never intermingle, and when a pencil of threads is propelled, it ever presents the appearance of a divergent brush. These are electrical phenomena, and cannot be explained but on electrical principles."

"In clear, fine weather, the air is invariably positive; and it is precisely in such weather that the aëronautic spider makes its ascent most easily and rapidly, whether it be in summer or in winter." "When the air is weakly positive, the ascent of the spider will be difficult, and its altitude extremely limited, and the threads propelled will be but little elevated above the horizontal plane. When negative electricity prevails, as in cloudy weather, or on the approach of rain, and the index of De Saussure's hygrometer rapidly advancing towards humidity, the spider is unable to ascend."[*]

Mr. Murray had previously told us, that "when a stick of excited sealing-wax is brought near the thread of suspension, it is evidently repelled; consequently, the electricity of the thread is of a negative character," while "an excited glass tube brought near, seemed to attract the thread, and with it the aëronautic spider."[†] His friend, Mr. Bowman, further describes the aërial spider as "shooting out four or five, often six or eight, extremely fine webs several yards long, which waved in the breeze, diverging from each other like a pencil of rays." One of them "had two distinct and widely-diverging fasciculi of webs," and "a line uniting them would have been at right angles to the direction of the breeze."[‡]

Such is the chief evidence in support of the electrical theory; but though we have tried these experiments, we have not succeeded in verifying any one of them. The following statements of Mr. Blackwall come nearer our own observations.

5. "Having procured a small branched twig," says

* Loudon's Mag. of Nat. Hist., vol. i. p. 322.
† Experim. Researches in Nat. Hist., p. 136.
‡ Mag. Nat. Hist., vol. i. p. 324.

Mr. Blackwall, "I fixed it upright in an earthen vessel containing water, its base being immersed in the liquid, and upon it I placed several of the spiders which produce gossamer. Whenever the insects thus circumstanced were exposed to a current of air, either naturally or artificially produced, they directly turned the thorax towards the quarter whence it came, even when it was so slight as scarcely to be perceptible, and elevating the abdomen, they emitted from their spinners a small portion of glutinous matter, which was instantly carried out in a line, consisting of four finer ones, with a velocity equal, or nearly so, to that with which the air moved, as was apparent from observations made on the motion of detached lines similarly exposed. The spiders, in the next place, carefully ascertained whether their lines had become firmly attached to any object or not, by pulling at them with the first pair of legs; and if the result was satisfactory, after tightening them sufficiently, they made them pass to the twig; then discharging from their spinners, which they applied to the spot where they stood, a little more of their liquid gum, and committing themselves to these bridges of their own constructing, they passed over them in safety, drawing a second line after them, as a security in case the first gave way, and so effected their escape.

"Such was invariably the result when spiders were placed where the air was liable to be sensibly agitated: I resolved, therefore, to put a bell-glass over them; and in this situation they remained seventeen days, evidently unable to produce a single line by which they could quit the branch they occupied, without encountering the water at its base; though, on the removal of the glass, they regained their liberty with as much celerity as in the instances already recorded.

"This experiment, which, from want of due precaution, has misled so many distinguished naturalists, I have tried with several geometric spiders, and always with the same success."*

* Linn. Trans., vol. xv. p. 456.

Mr. Blackwall, from subsequent experiments, says he is
"confident in affirming, that in motionless air spiders have
not the power of darting their threads even through the
space of half an inch."* The following details are given in
confirmation of this opinion. Mr. Blackwall observed, the
1st Oct., 1826, a little before noon, with the sun shining
brightly, no wind stirring, and the thermometer in the shade
ranging from 55°·5 to 64°, a profusion of shining lines
crossing each other at every angle, forming a confused net-
work, covering the fields and hedges, and thickly coating
his feet and ankles, as he walked across a pasture. He was
more struck with the phenomenon, because on the previous
day a strong gale of wind had blown from the south, and as
gossamer is only seen in calm weather, it must have been all
produced within a very short time.

"What more particularly arrested my attention," says
Mr. Blackwall, "was the ascent of an amazing quantity of
webs, of an irregular, complicated structure, resembling
ravelled silk of the finest quality and clearest white; they
were of various shapes and dimensions, some of the largest
measuring upwards of a yard in length, and several inches
in breadth in the widest part; while others were almost as
broad as long, presenting an area of a few square inches
only.

"These webs, it was quickly perceived, were not formed
in the air, as is generally believed, but at the earth's surface.
The lines of which they were composed, being brought into
contact by the mechanical action of gentle airs, adhered
together, till, by continual additions, they were accumulated
into flakes or masses of considerable magnitude, on which
the ascending current, occasioned by the rarefaction of the
air contiguous to the heated ground, acted with so much
force as to separate them from the objects to which they
were attached, raising them in the atmosphere to a perpendicular height of at least several hundred feet. I collected a
number of these webs about mid-day, as they rose; and
again in the afternoon, when the upward current had ceased,

* Mag. Nat. Hist., vol. ii. p. 397.

and they were falling; but scarcely one in twenty contained a spider: though, on minute inspection, I found small winged insects, chiefly aphides, entangled in most of them.

"From contemplating this unusual display of gossamer, my thoughts were naturally directed to the animals which produced it, and the countless myriads in which they swarmed almost created as much surprise as the singular occupation that engrossed them. Apparently actuated by the same impulse, all were intent upon traversing the regions of air: accordingly, after gaining the summits of various objects, as blades of grass, stubble, rails, gates, &c., by the slow and laborious process of climbing, they raised themselves still higher by straightening their limbs; and elevating the abdomen, by bringing it from the usual horizontal position into one almost perpendicular, they emitted from their spinning apparatus a small quantity of the glutinous secretion with which they construct their webs. This viscous substance being drawn out by the ascending current of rarefied air into fine lines several feet in length, was carried upward, until the spiders, feeling themselves acted upon with sufficient force in that direction, quitted their hold of the objects on which they stood, and commenced their journey by mounting aloft.

"Whenever the lines became inadequate to the purpose for which they were intended, by adhering to any fixed body, they were immediately detached from the spinners, and so converted into terrestrial gossamer, by means of the last pair of legs, and the proceedings just described were repeated; which plainly proves that these operations result from a strong desire felt by the insects to effect an ascent."* Mr. Blackwall has recently read a paper (still unpublished) in the Linnean Society, confirmatory of his opinions.

6. Without going into the particulars of what agrees or disagrees in the above experiments with our own observations, we shall give a brief account of what we have actually seen in our researches. (J. R.) So far as we have determined, then, all the various species of spiders, how different

* Linn. Trans., vol. xv. p. 453.

soever the form of their webs may be, proceed in the circumstance of shooting their lines precisely alike; but those
which we have found the most manageable in experimenting,
are the small gossamer spider (*Aranea obtextrix*, BECHSTEIN),
known by its shining blackish-brown body and reddish-brown
semi-transparent legs; but particularly the long-bodied
spider (*Tetragnatha extensa*, LATR.), which varies in colour
from green to brownish or grey—but has always a black line
along the belly, with a silvery white or yellowish one on
each side. The latter is chiefly recommended by being a
very industrious and persevering spinner, while its movements
are easily seen, from the long cylindrical form of its body
and the length of its legs.

We placed the above two species with five or six others,
including the garden, the domestic and the labyrinthic
spiders, in empty wine-glasses, set in tea-saucers filled with
water to prevent their escape. When they discovered, by
repeated descents from the brims of the glasses, that they
were thus surrounded by a wet ditch, they all set themselves to the task of throwing their silken bridges across.
For this purpose they first endeavoured to ascertain in what
direction the wind blew, or rather (as the experiment was
made in our study) which way any current of air set,—by
elevating their arms as we have seen sailors do in a dead
calm. But, as it may prove more interesting to keep to one
individual, we shall first watch the proceedings of the
gossamer spider.

Finding no current of air on any quarter of the brim of
the glass, it seemed to give up all hopes of constructing its
bridge of escape, and placed itself in the attitude of repose;
but no sooner did we produce a stream of air, by blowing
gently towards its position, than, fixing a thread to the
glass, and laying hold of it with one of its feet, by way of
security, it placed its body in a vertical position, with its
spinnerets extended outwards; and immediately we had
the pleasure of seeing a thread streaming out from them
several feet in length, on which the little aëronaut sprung
up into the air. We were convinced, from what we thus

observed, that it was the double or bend of the thread which was blown into the air; and we assigned as a reason for her previously attaching and drawing out a thread from the glass, the wish to give the wind a *point d'appui*—something upon which it might have a *purchase*, as a mechanic would say of a lever. The bend of the thread, then, on this view of the matter, would be carried out by the wind,—would form the point of impulsion,—and, of course, the escape bridge would be an ordinary line doubled.

Such was our conclusion, which was strongly corroborated by what we subsequently found said by M. Latreille—than whom no higher authority could be given. "When the animal," says he, "desires to cross a brook, she fixes to a tree or some other object one of the ends of her first threads, in order that the wind or a current of air may carry the other end beyond the obstacle;"[*] and as one end is always attached to the spinnerets, he must mean that the double of the thread flies off. In his previous publications, however, Latreille had contented himself with copying the statement of Dr. Lister.

In order to ascertain the fact, and put an end to all doubts, we watched, with great care and minuteness, the proceedings of the long-bodied spider above mentioned, by producing a stream of air in the same manner, as it perambulated the brim of the glass. It immediately, as the other had done, attached a thread, and raised its body perpendicularly, like a tumbler, standing on his hands with his head downwards; but we looked in vain for this thread bending, as we had at first supposed, and going off double. Instead of this it remained tight, while another thread, or what appeared to be so, streamed off from the spinners, similar to smoke issuing through a pin-hole, sometimes in a line, and sometimes at a considerable angle, with the first, according to the current of the air,—the first thread, extended from the glass to the spinnerets, remaining all the

* "———L'un des bouts de ces premiers fils, afin que le vent ou un courant d'air pousse l'autre extrémité de l'un d'eux au de là de l'obstacle,"— Dict. Classique d'Hist. Nat., vol. i. p. 510.

while tight drawn in a right line. It further appeared to us, that the first thread proceeded from the pair of spinnerets nearest the head, while the floating thread came from the outer pair,—though it is possible in such minute objects we may have been deceived. That the first was continuous with the second, without any perceptible joining, we ascertained in numerous instances, by catching the floating line and pulling it tight, in which case the spider glides along without attaching another line to the glass; but if she have to coil up the floating line to tighten it, as usually happens, she gathers it into a packet and glues the two ends tight together. Her body, while the floating line streamed out, remained quite motionless, but we distinctly saw the spinnerets not only projected, as is always done when a spider spins, but moved in the same way as an infant moves its lips when sucking. We cannot doubt, therefore, that this motion is intended to emit (if *eject* or *project* be deemed too strong words) the liquid material of the thread; at the same time, we are quite certain that it cannot throw out a single inch of thread without the aid of a current of air. A long-bodied spider will thus throw out in succession as many threads as we please, by simply blowing towards it; but not one where there is no current, as under a bell-glass, where it may be kept till it die, without being able to construct a bridge over water of an inch long. We never observed more than one floating thread produced at the same time; though other observers mention several.

The probable commencement, we think, of the floating line, is by the emission of little globules of the glutinous material to the points of the spinnerules—perhaps it may be dropped from them, if not ejected, and the globules being carried off by the current of air, drawn out into a thread. But we give this as only a conjecture, for we could not bring a glass of sufficient power to bear upon the spinnerules at the commencement of the floating line.

In subsequent experiments we found that it was not indispensable for the spider to rest upon a solid body when producing a line, as she can do so while she is suspended in the air by another line. When the current of air also is strong, she will sometimes commit herself to it by swinging from the end of the line. We have even remarked this when there was scarcely a breath of air.

We tried another experiment. We pressed pretty firmly upon the base of the spinnerets, so as not to injure the spider, blowing obliquely over them; but no floating line appeared. We then touched them with a pencil and drew out several lines an inch or two in length, upon which we blew in order to extend them; but in this also we were unsuccessful, as they did not lengthen more than a quarter of an inch. We next traced out the reservoirs of a garden-spider (*Epeira diadema*), and immediately taking a drop of the matter from one of them on the point of a fine needle, we directed upon it a strong current of air, and succeeded in blowing out a thick yellow line, as we might have done with gum-water, of about an inch and a half long.

When we observed our long-bodied spider eager to throw a line by raising up its body, we brought within three inches of its spinnerets an excited stick of sealing-wax, of which it took no notice, nor did any thread extend to it, not even when brought almost to touch the spinnerets. We had the same want of success with an excited glass rod; and indeed we had not anticipated any other result, as we have never observed that these either attract or repel the floating threads, as Mr. Murray has seen them do; nor have we ever seen the end of a float-ing thread separated into its component threadlets and diverging like a brush, as he and Mr. Bowman describe. It may be proper to mention that Mr. Murray, in con-formity with his theory, explains the shooting of lines in a current of air by the electric state produced by motion in consequence of the mutual friction of the gaseous par-ticles. But this view of the matter does not seem to affect our statements.

Nests, Webs, and Nets of Spiders.

The neatest, though the smallest spider's nest which we have seen, was constructed in the chink of a garden post, which we had cut out in the previous summer in getting at the cells of a carpenter-bee. The architect was one of the large hunting-spiders, erroneously said by some naturalists to be incapable of spinning. The nest in question was about two inches high, composed of a very close satin-like texture. There were two parallel chambers placed perpendicularly, in which position also the inhabitant reposed there during the day, going, as we presume, only abroad to prey during the night. But the most remarkable circumstance was, that the openings (two above and two below) were so elastic, that they shut almost as closely as the boat cocoon of the *Tortrix Chlorana.* We observed this spider for several months, but at last it disappeared, and we took the nest out, under the notion that it might contain eggs; but we found none, and therefore conclude that it was only used as a day retreat. (J. R.) The account which Evelyn has given of these hunting-spiders is so interesting, that we must transcribe it.

"Of all sorts of insects," says he, "there is none has afforded me more divertisement than the *renatores* (hunters), which are a sort of *lupi* (wolves) that have their dens in rugged walls and crevices of our houses; a small brown and delicately-spotted kind of spiders, whose hinder legs are longer than the rest. Such I did frequently observe at Rome, which, espying a fly at three or four yards' distance, upon the balcony where I stood, would not make directly to her, but crawl under the rail till, being arrived at the antipodes, it would steal up, seldom missing its aim; but if it chanced to want anything of being perfectly opposite, would, at first peep, immediately slide down again,—till, taking better notice, it would come the next time exactly upon the fly's back : but if this happened not to be within a competent leap, then would this insect move so softly, as the very shadow of the gnomon seemed not to be more imperceptible, unless the fly moved; and then would the spider move also in the same proportion, keeping that just time with her motion, as if the same soul had animated both these little bodies; and whether it were forwards, backwards, or to either side, without at all turning her body like a well-managed horse: but if the capricious fly took wing and pitched upon another place behind our huntress, then would the spider whirl its body so nimbly about, as nothing could be imagined more swift: by which means she always kept the head towards her prey, though, to appearance, as immovable as if it had been a nail driven into the wood, till by that indiscernible progress (being arrived within the sphere of her reach) she made a fatal leap, swift as lightning, upon the fly, catching him in the pole, where she never quitted hold till her belly was full, and then carried the remainder home."

One feels a little sceptical, however, when he adds, "I have beheld them instructing their young ones how to hunt, which they would sometimes discipline for not well observing; but when any of the old ones did (as sometimes) miss a leap, they would run out of the field and hide themselves in their crannies, as ashamed, and haply not to be seen abroad for four or five hours after; for so long have I watched the nature of this strange insect, the contemplation of whose so wonderful sagacity and address has amazed me; nor do I find in any chase whatsoever more cunning and stratagem observed. I have found some of these spiders in my garden, when the weather, towards spring, is very hot, but they are nothing so eager in hunting as in Italy."[*]

We have only to add to this lively narrative, that the hunting-spider, when he leaps, takes good care to provide against accidental falls by always swinging himself from a good strong cable of silk, as Swammerdam correctly states,[†] and which anybody may verify, as one of the

---

[*] Evelyn's Travels in Italy.　　[†] Book of Nature, part i. p. 24.

small hunters (*Salticus scenicus*), known by having its back striped with black and white like a zebra, is very common in Britain.

Mr. Weston, the editor of 'Bloomfield's Remains,' falls into a very singular mistake about hunting-spiders, imagining them to be web-weaving ones which have exhausted their materials, and which are therefore compelled to hunt. In proof of this he gives an instance which fell under his own observation !*

As a contrast to the little elastic satin nest of the hunter, we may mention the largest with which we are acquainted,— that of the labyrinthic spider (*Agelena labyrinthica*, WALCKE-NAER). Our readers must often have seen this nest spread out like a broad sheet in hedges, furze, and other low bushes, and sometimes on the ground. The middle of this sheet, which is of a close texture, is swung like a sailor's hammock, by silken ropes extended all around to the higher branches; but the whole curves upwards and backwards, sloping down to a long funnel-shaped gallery which is nearly horizontal at the entrance, but soon winds obliquely till it becomes quite perpendicular. This curved gallery is about a quarter of an inch in diameter, is much more closely woven than the sheet part of the web, and sometimes descends into a hole in the ground, though oftener into a group of crowded twigs, or a tuft of grass. Here the spider dwells secure, frequently resting with her legs extended from the entrance of the gallery, ready to spring out upon whatever insect may fall into her sheet net. She herself can only be caught by getting behind her and forcing her out into the web; but though we have often endeavoured to make her construct a nest under our eye, we have been as unsuccessful as in similar experiments with the common house spider (*Aranea domestica*). (J. R.)

The house spider's proceedings were long ago described by Homberg, and the account has been copied, as usual, by almost every subsequent writer. Goldsmith has, indeed, given some strange misstatements from his own observations,

* Bloomfield's Remains, vol. ii. p. 64, *note*.

2 c

and Bingley has added the original remark, that, after fixing its first thread, creeping along the wall, and joining it as it proceeds, it "*darts itself to the opposite side*, where the other end is to be fastened!"[*] Homberg's spider took the more circuitous route of travelling to the opposite wall, carrying in one of the claws the end of the thread previously fixed, lest it should stick in the wrong place. This we believe to be the correct statement, for as the web is always horizontal, it would seldom answer to commit a floating thread to the wind, as is done by other species. Homberg's spider, after stretching as many lines by way of *warp* as it deemed sufficient between the two walls of the corner which it had chosen, proceeded to cross this in the way our weavers do in adding the *woof*, with this difference, that the spider's threads were only laid on, and not interlaced.[†] The domestic spiders, however, in these modern days, must have forgot this mode of weaving, for none of their webs will be found to be thus regularly constructed!

The geometric, or net-working spiders (*Tendeuses*, Latr.), are as well known in most districts as any of the preceding; almost every bush and tree in the gardens and hedge-rows having one or more of their nets stretched out in a vertical position between adjacent branches. The common garden spider (*Epeira diadema*), and the long-bodied spider (*Tetragnatha extensa*), are the best known of this order.

The chief care of a spider of this sort is, to form a cable of sufficient strength to bear the net she means to hang upon it; and, after throwing out a floating line as above described, when it catches properly she doubles and redoubles it with additional threads. On trying its strength she is not contented with the test of pulling it with her legs, but drops herself down several feet from various points of it, as we have often seen, swinging and bobbing with the whole weight of her body. She proceeds in a similar manner with the rest of the framework of her wheel-shaped net; and it may be remarked that some of the ends of these lines are not

---

[*] Animal Biography, iii. 470–1.

[†] Mém. de l'Acad. des Sciences pour 1707, p. 339.

simple, but in form of a Y, giving her the additional security of two attachments instead of one.

In constructing the body of the net, the most remarkable circumstance is her using her limbs as a measure, to regulate the distances of her *radii* or wheel-spokes, and the circular meshes interwoven into them. These are consequently always proportional to the size of the spider. She often

Geometric Net of *Epeira diadema.*

takes up her station in the centre, but not always, though it is so said by inaccurate writers; for she as frequently lurks in a little chamber constructed under a leaf or other shelter at the corner of her web, ready to dart down upon whatever prey may be entangled in her net. The centre of the net is said also to be composed of more viscid materials than its suspensory lines,—a circumstance alleged to be proved by the former appearing under the microscope studded with globules of gum.* We have not been able to verify this distinction, having seen the suspensory lines as often studded in this manner as those in the centre. (J. R.)

MASON-SPIDERS.

A no less wonderful structure is composed by a sort of spiders, natives of the tropics and the south of Europe, which have been justly called mason-spiders by M. Latreille. One of these (*Mygale nidulans*, WALCKN.), found in the West Indies, " digs a hole in the earth obliquely downwards about three inches in length, and one in diameter. This cavity she lines with a tough thick web, which, when taken out, resembles a leathern purse ; but, what is most curious, this house has a door with hinges, like the operculum of some sea-shells, and herself and family, who tenant this nest, open and shut the door whenever they pass and repass. This history was told me," says Darwin, "and the nest, with its door, shown me by the late Dr. Butt, of Bath, who was some years physician in Jamaica."†

The nest of a mason-spider, similar to this, has been obligingly put into our hands by Mr. Riddle, of Blackheath. It came from the West Indies, and is probably that of Latreille's clay-kneader (*Mygale cratiens*), and one of the smallest of the genus. We have since seen a pair of these spiders in possession of Mr. William Mello, of Blackheath. The nest is composed of very hard argillaceous clay, deeply tinged with brown oxide of iron. It is in form of a tube, about one inch in diameter, between six and seven inches long, and slightly bent towards the lower extremity— appearing to have been mined into the clay rather than built. The interior of the tube is lined with a uniform tapestry of silken web, of an orange-white colour, with a texture inter- mediate between India paper and very fine glove leather. But the most wonderful part of this nest is its entrance, which we look upon as the perfection of insect architecture. A circular door, about the size of a crown piece, slightly

---

* Kirby and Spence, Intr. i. 419.

† Darwin's Zoonomia, i. 253, 8vo. ed.

concave on the outside and convex within, is formed of more
than a dozen layers of the same web which lines the interior,
closely laid upon one another, and shaped so that the inner
layers are the broadest, the outer being gradually less in
diameter, except towards the hinge, which is about an inch
long ; and in consequence of all the layers being united there,
and prolonged into the tube, it becomes the thickest and
strongest part of the structure. The elasticity of the

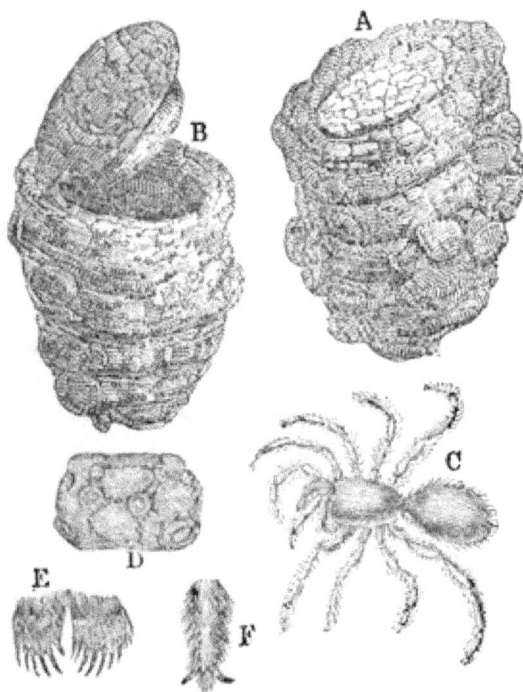

Nest of the Mason-Spider.
A. The nest shut. B. The nest open. C. The spider, *Mygale cæmentaria*. D. The
eyes magnified. E, F. Parts of the foot and claw magnified.

materials, also, gives to this hinge the remarkable peculiarity
of acting like a spring, and shutting the door of the nest
spontaneously. It is, besides, made to fit so accurately to
the aperture, which is composed of similar concentric layers of
web, that it is almost impossible to distinguish the joining
by the most careful inspection. To gratify curiosity, the
door has been opened and shut hundreds of times, without in

the least destroying the power of the spring. When the door is shut, it resembles some of the lichens (*Lecidea*), or the leathery fungi, such as *Polyporus versicolor* (MICHELI), or, nearer still, the upper valve of a young oyster shell. The door of the nest, the only part seen above ground, being of a blackish-brown colour, it must be very difficult to discover. (J. R.)

Another mason-spider (*Mygale cœmentaria*, LATR.), found in the south of France, usually selects for her nest a place bare of grass, sloping in such a manner as to carry off the water, and of a firm soil, without rocks or small stones. She digs a gallery a foot or two in depth, and of a diameter (equal throughout) sufficient to admit of her easily passing. She lines this with a tapestry of silk glued to the walls. The door, which is circular, is constructed of many layers of earth kneaded, and bound together with silk. Externally, it is flat and rough, corresponding to the earth around the entrance, for the purpose, no doubt, of concealment : on the inside it is convex, and tapestried thickly with a web of fine silk. The threads of this door-tapestry are prolonged, and strongly attached to the upper side of the entrance, forming an excellent hinge, which, when pushed open by the spider, shuts again by its own weight, without the aid of spring hinges. When the spider is at home, and her door forcibly opened by an intruder, she pulls it strongly inwards, and even when half-opened often snatches it out of the hand ; but when she is foiled in this, she retreats to the bottom of her den, as her last resource.[*]

Rossi ascertained that the female of an allied species (*Mygale sauvagesii*, LATR.), found in Corsica, lived in one of these nests, with a numerous posterity. He destroyed one of these doors to observe whether a new one would be made, which it was ; but it was fixed immovably, without a hinge ; the spider, no doubt, fortifying herself in this manner till she thought she might reopen it without danger.[†]

[The accompanying illustration shows one of these nests,

---

[*] Mém. Soc. d'Hist. Nat. de Paris. An. vii.
[†] Ibid., p. 125, and Latreille, Hist. Nat. Génér. viii. p. 163.

which is in my own collection. It was brought from Jamaica, together with the spider that made it

[The nest is nearly six inches in length, and is made of a double layer of silken web. The inner layer is yellowish, with a tinge of red, and although fine, very tough and strong. The outer layer is thick, coarse, dark brown, and rather flaky, the dark colour being probably caused by the earth which is mixed with it. The lid is made of eight or ten layers of coarse web, overlapping each other like the tiles of a house-roof, and the entrance of the nest is formed after the same fashion. If the lid be opened, the inside of the nest is seen to be of a different make from the exterior, being greyish-white, smooth, close-textured, and looking much like the finest kid leather.

[The smaller illustration shows the spider in the act of emerging from its home.]

" The Rev. Revett Shepherd has often noticed, in the fen ditches of Norfolk, a very large spider (the species not yet determined) which actually forms a *raft* for the purpose of obtaining its prey with more facility. Keeping its station upon a ball of weeds about three inches in diameter, probably held together by slight silken cords, it is wafted along the surface of the water upon this floating island, which it quits the moment it sees a drowning insect. The booty thus seized it devours at leisure upon its raft, under which it retires when alarmed by any danger."* In the spring of 1830, we found a spider on some reeds in the Croydon Canal, which agreed in appearance with Mr. Shepherd's.

* Kirby and Spence, Intr. i, 425.

Among our native spiders there are several besides this one, which, not contented with a web like the rest of their congeners, take advantage of other materials to construct cells where, "hushed in grim repose," they "expect their insect prey." The most simple of those spider-cells is constructed by a longish-bodied spider (*Aranea holosericea,* LINN.), which is a little larger than the common hunting-spider. It rolls up a leaf of the lilac or poplar, precisely in the same manner as is done by the leaf-rolling cater-pillars, upon whose cells it sometimes seizes to save itself trouble, having first expelled, or perhaps devoured, the rightful owner. The spider, however, is not satisfied with the tapestry of the caterpillar, but always weaves a fresh set of her own, much more close and substantial.

Another spider, common in woods and copses (*Epeira quadrata ?*), weaves together a great number of leaves to form a dwelling for herself, and in front of it she spreads her toils for entrapping the unwary insects which stray thither. These, as soon as caught, are dragged into her den, and stored up for a time of scarcity. Here also her eggs are deposited and hatched in safety. When the cold weather approaches, and the leaves of her edifice wither, she abandons it for the more secure shelter of a hollow tree, where she soon dies; but the continuation of the species depends upon eggs, deposited in the nest before winter, and remaining to be hatched with the warmth of the ensuing summer.

The spider's den of united leaves, however, which has just been described, is not always useless when withered and deserted, for the dormouse usually selects it as a ready-made roof for its nest of dried grass. That those old spiders' dens are not accidentally chosen by the mouse, appears from the fact, that out of about a dozen mouse-nests of this sort found during winter in a copse between Lewisham and Bromley, Kent, every second or third one was furnished with such a roof. (J. R.)

DIVING WATER-SPIDER.

Though spiders require atmospheric air for respiration, yet one species well known to naturalists is aquatic in its habits, and lives not only upon the surface but below the surface of the water, contriving to carry down with it a sufficiency of air for the support of life during a considerable period of time. Its subaqueous nest is in fact a sort of diving-bell, and constitutes a secure and most ingenious habitation. This spider does not like stagnant water, but prefers low running streams, canals, and ditches, where she may often be seen in the vicinity of London and elsewhere, living in her diving-bell, which shines through the water like a little globe of silver: her singular economy was first, we believe, described by Clerck,* L. M. de Lignac,† and De Geer.

" The shining appearance," says Clerck, " proceeds either from an inflated globule surrounding the abdomen, or from the space between the body and the water. The spider, when wishing to inhale the air, rises to the surface, with its body still submersed, and only the part containing the spinneret rising just to the surface, when it briskly opens and moves its four teats. A thick coat of hair keeps the water from approaching or wetting the abdomen. It comes up for air about four times an hour or oftener, though I have good reason to suppose it can continue without it for several days together.

" I found in the middle of May one male and ten females, which I put into a glass filled with water, where they lived together very quietly for eight days. I put some duckweed (*Lemna*) into the glass to afford them shelter, and the females began to stretch diagonal threads in a confused manner from it to the sides of the glass about half-way down. Each of the females afterwards fixed a close bag to the edge of the glass, from which the water was expelled by the air from the spinneret, and thus a cell was formed capable of containing

* Aranei Suecici, Stockholm, 1757.
† Mém. des Araign. Aquat., 12mo. Paris, 1799.

the whole animal. Here they remained quietly, with their abdomens in their cells, and their bodies still plunged in the water ; and in a short time brimstone-coloured bags of eggs appeared in each cell, filling it about a fourth part. On the 7th of July several young ones swam out from one of the bags. All this time the old ones had nothing to eat, and yet they never attacked one another as other spiders would have been apt to do."*

" These spiders," says De Geer, "spin in the water a cell of strong, closely-woven white silk, in the form of half the shell of a pigeon's egg, or like a diving-bell. This is some-times left partly above water, but at others is entirely sub-mersed, and is always attached to the objects near it by a great number of irregular threads. It is closed all round. but has a large opening below, which, however, I found closed on the 15th of December, and the spider living quietly within, with her head downwards. I made a rent in this cell, and expelled the air, upon which the spider came out; yet, though she appeared to have been laid up for three months in her winter quarters, she greedily seized upon an insect and sucked it. I also found that the male as well as the female constructs a similar subaqueous cell, and during summer no less than in winter."† We have recently kept one of these spiders for several months in a glass of water, where it built a cell half under water, in which it laid its eggs.

Cleanliness of Spiders.

When we look at the viscid material with which spiders construct their lines and webs, and at the rough, hairy cover-ing (with a few exceptions) of their bodies, we might con-clude that they would be always stuck over with fragments of the minute fibres which they produce. This, indeed, must often happen, did they not take careful precautions to avoid it ; for we have observed that they seldom, if ever, leave a thread to float at random, except when they wish to form a

* Clerck, Aranei Suecici, cap. viii.
† De Geer, Mém. des Insectes, vii. 312.

bridge. When a spider drops along a line, for instance, in order to ascertain the strength of her web, or the nature of the place below her, she invariably, when she reascends, coils it up into a little ball, and throws it away. Her claws are admirably adapted for this purpose, as well as for walking along the lines, as may be readily seen by a magnifying glass.

There are three claws, one of which acts as a thumb, the others being toothed like a comb, for gliding along the lines. This structure, however, unfits it to walk, as flies can do, upon any upright polished surface like glass; although the contrary* is erroneously asserted by the Abbé de la Pluche. Before she can do so, she is obliged to con-

Triple-clawed foot of a Spider, magnified.

struct a ladder of ropes, as Mr. Blackwell remarks,† by elevating her spinneret as high as she can, and laying down a step upon which she stands to form a second, and so on; as any one may try by placing a spider at the bottom of a very clean wine-glass.

The hairs of the legs, however, are always catching bits of web and particles of dust; but these are not suffered to remain long. Most people may have remarked that the house-fly is ever and anon brushing its feet upon one another to rub off the dust, though we have not seen it remarked in authors that spiders are equally assiduous in keeping themselves clean. They have, besides, a very efficient instrument in their mandibles or jaws, which, like their claws, are furnished with teeth; and a spider which appears to a

* Spectacle de la Nature, i. 58.　　　　　† Linn. Trans. vol. xv.

careless observer as resting idly, in nine cases out of ten will be found slowly combing her legs with her mandibles, beginning as high as possible on the thigh, and passing down to the claws. The flue which she thus combs off is regularly tossed away.

With respect to the house-spider (*A. domestica*), we are told in books, that " she from time to time clears away the dust from her web, and sweeps the whole by giving it a shake with her paw, so nicely proportioning the force of her blow, that she never breaks anything."* That spiders may be seen shaking their webs in this manner, we readily admit; though it is not, we imagine, to clear them of dust, but to ascertain whether they are sufficiently sound and strong.

We recently witnessed a more laborious process of cleaning a web than merely shaking it. On coming down the Maine by the steamboat from Frankfort, in August, 1829, we observed the geometric-net of a conic-spider (*Epeira conica*, WALCK.) on the framework of the deck, and as it was covered with flakes of soot from the smoke of the engine, we were surprised to see a spider at work on it; for, in order to be useful, this sort of net must be clean. Upon observing it a little closely, however, we perceived that she was not constructing a net, but dressing up an old one; though not, we must think, to save trouble, so much as an expenditure of material. Some of the lines she dexterously stripped of the flakes of soot adhering to them; but in the greater number, finding that she could not get them sufficiently clean, she broke them quite off, bundled them up, and tossed them over. We counted five of these packets of rubbish which she thus threw away, though there must have been many more, as it was some time before we discovered the manoeuvre, the packets being so small as not to be readily perceived, except when placed between the eye and the light. When she had cleared off all the sooty lines, she began to replace them in the usual way ; but the arrival of the boat at Mentz put an end to our observations. (J. R.) Bloomfield,

* Spectacle de la Nature, i. p. 61.

the poet, having observed the disappearance of these bits of ravelled web, imagined that the spider swallowed them ; and even says that he observed a garden spider moisten the pellets before swallowing them!* Dr. Lister, as we have already seen, thought the spider retracted the threads within the abdomen.

---

* Remains, ii. 62-5. It is a remarkable fact, as recorded from personal observation by Mr. Bell (British Reptiles), that the toad swallows the cuticle detached from its body during the moult which it undergoes.

CHAPTER XIX.

STRUCTURES OF GALL-FLIES AND APHIDES.

MANY of the processes which we have detailed bear some resemblance to our own operations of building with materials cemented together; but we shall now turn our attention to a class of insect-architects, who cannot, so far as we know, be matched in prospective skill by any of the higher orders of animals. We refer to the numerous family which have received the name of gall-flies,—a family which, as yet, is very imperfectly understood, their economy being no less difficult to trace than their species is to arrange in the established systems of classification; though the latter has been recently much improved by Mr. Westwood.

Small berry-shaped galls of the oak leaf, produced by *Cynips quercus folii?*

One of the most simple and very common instances of the nests constructed by gall-insects, may be found in abundance during the summer, on the leaves of the rose-tree, the oak, the poplar, the willow (*Salix viminalis*), and many other trees, in the globular form of a berry, about the size of a currant, and usually of a green colour, tinged with red, like a ripe Alban or Baltimore apple.

When this psuedo-apple in miniature is cut into, it is found to be fresh, firm, juicy, and hollow in the centre, where there is either an egg or a grub safely lodged, and protected from all ordinary accidents. Within this hollow ball the egg is hatched, and the grub feeds securely on its substance, till it prepares for its winter sleep, before changing into a gall-fly (*Cynips*) in the ensuing summer. There is a mystery as to the manner in which this gall-fly contrives to produce the hollow miniature apples, each enclosing one of her eggs; and the doubts attendant upon the subject cannot, so far as our present knowledge extends, be solved, except by plausible conjecture. Our earlier naturalists were of opinion that it was the grub which produced the galls, by eating, when newly hatched, through the cuticle of the leaf, and remaining till the juices flowing from the wound enveloped it, and acquired consistence by exposure to the air. This opinion, however, plausible as it appeared to be, was at once disproved by finding unhatched eggs on opening the galls.

There can be no doubt, indeed, that the mother gall-fly makes a hole in the plant for the purpose of depositing her

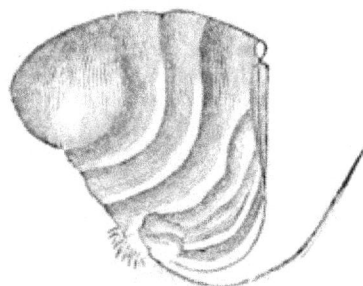

Ovipositor of Gall-fly, greatly magnified.

eggs. She is furnished with an admirable ovipositor for that express purpose, and Swammerdam actually saw a gall-fly thus depositing her eggs, and we have recently witnessed

the same in several instances. In some of these insects the
ovipositor is conspicuously long, even when the insect is at
rest; but in others, not above a line or two of it is visible,
till the belly of the insect be gently pressed. When this is
done to the fly that produces the currant-gall of the oak,
the ovipositor may be seen issuing from a sheath in form of
a small curved needle, of a chestnut-brown colour, and of
a horny substance, and three times as long as it at first
appeared.

What is most remarkable in this ovipositor is, that it is
much longer than the whole body of the insect, in whose
belly it is lodged in a sheath, and, from its horny nature, it

Gall-fly, and mechanism of ovipositor, greatly magnified.

cannot be either shortened or lengthened. It is on this
account that it is bent into the same curve as the body of
the insect. The mechanism by which this is effected is
similar to that of the tongue of the woodpeckers (*Picidæ*),
which, though rather short, can be darted out far beyond
the beak, by means of a forked bone at the root of the
tongue, which is thin and rolled up like the spring of a
watch. The base of the ovipositor of the gall-fly is, in a
similar way, placed near the anus, runs along the curvature
of the back, makes a turn at the breast, and then, following

the curve of the belly, appears again near where it originates. We copy from Réaumur his accurate sketch of this remarkable structure.

With this instrument the mother gall-fly pierces the part of a plant which she selects, and, according to our older naturalists, "ejects into the cavity a drop of her corroding liquor, and immediately lays an egg or more there; the circulation of the sap being thus interrupted, and thrown, by the poison, into a fermentation that burns the contiguous parts and changes the natural colour. The sap, turned from its proper channel, extravasates and flows round the eggs, while its surface is dried by the external air, and hardens into a vaulted form."* Kirby and Spence tell us, that the parent fly introduces her egg "into a puncture made by her curious spiral sting, and in a few hours it becomes surrounded with a fleshy chamber."† M. Virey says, the gall tubercle is produced by irritation, in the same way as an inflamed tumor in an animal body, by the swelling of the cellular tissue and the flow of liquid matter, which changes the organization, and alters the natural external form.‡ This seems to be the received doctrine at present in France.§

Sprengel, speaking of the rose-willow, says, the insect in spring deposits its eggs in the leaf buds. "The new stimulus attracts the sap,—the type of the part becomes changed, and from the prevailing acidity of the animal juice, it happens, that in the rose and stock-shaped leaves which are pushed out, a red instead of a green colour is evolved."‖

Without pretending positively to state facts which are, perhaps, beyond human penetration, we may view the process in a rather different light. (J. R.) Following the analogy of what is *known* to occur in the case of the saw-flies,

* Spectacle de la Nature, i. 119.
† Introduction, ii. 449.
‡ Hist. des Mœurs et de l'Instinct, vol. ii.
§ Entomologie, par B. A. E., p. 242. Paris, 1826.
‖ Elements of the Philosophy of Plants, Eng. Trans., p. 285.

after the gall-fly has made a puncture and pushed her
egg into the hole, we may suppose that she covers it
over with some adhesive gluten or gum, or the egg itself,
as is usual among moths, &c., may be coated over with
such a gluten. In either of these two cases, the gluten
will prevent the sap that flows through the puncture from
being scattered over the leaf and wasted; and the sap, being
thus confined to the space occupied by the eggs, will expand and force outwards the pellicle of gluten that confines

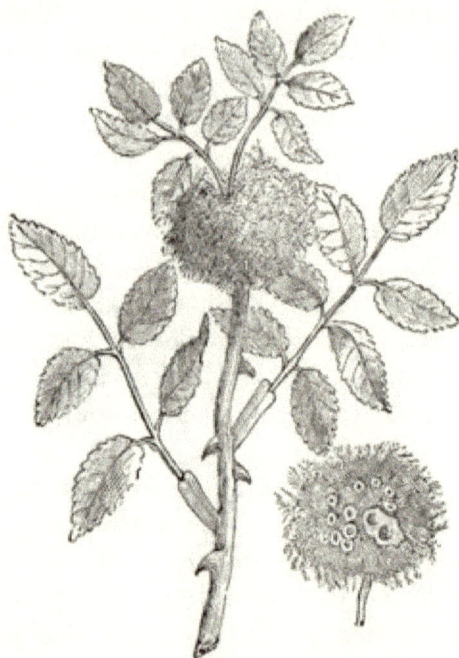

Bedeguar Gall of the Rose, produced by *Cynips rosœ*.

it, till becoming thickened by evaporation and exposure to
the air, it at length shuts up the puncture, stops the further
escape of the sap, and the process is completed. This
explanation will completely account for the globular form of
the galls alluded to; that is, supposing the egg of the gall-
fly to be globular, and covered or coated with a pellicle of
gluten of uniform thickness, and consequently opposing uniform resistance, or rather uniform expansibility, to the sap
pressing from within. It will also account for the remark-

able uniformity in the size of the gall apples; for the punctures and the eggs being uniform in size, and the gluten, by supposition, uniform in quantity, no more than the same quantity of sap can escape in such circumstances.

But though this explanation appears to be plausible, it is confessedly conjectural; for though Swammerdam detected a gall-fly in the act of depositing her eggs, he did not attend to this circumstance; and in the instances which we have observed, some unlucky accident always prevented us from following up our observations. The indefatigable Réaumur, on one occasion, thought he would make sure of tracing the steps of the process in the case of the gall-fly which produces the substance called *bedeguar* on the wild rose-tree.

One of the bristles of the Bedeguar of the Rose magnified.

and to which we shall presently advert. His plan was to enclose in a box, in which a brood of flies had just been produced from a bedeguar, a living branch from a wild rose-tree; but, to his great disappointment, no eggs were laid, and no bedeguar formed. Upon further investigation, he discovered that the brood of flies produced from the bedeguar were not the genuine bedeguar insects at all, but one of the parasite ichneumons (*Callimone bedeguaris*, STEPHENS), which had surreptitiously deposited their eggs there, in order to supply their young with the bedeguar grubs, all of which they appeared to have devoured. It may prove interesting to look into the remarkable structure of the bedeguar itself, which is very different from the globular galls above described.

The gall-fly of the willow (*Cynips viminalis*) deposits, as we have just seen, only a single egg on one spot; but the bedeguar insect lays a large cluster of eggs on the extremity of a growing branch of the wild rose-tree, making, probably, a proportionate number of punctures to procure materials for the future habitation of her young progeny. As in the former case, also, each of these eggs becomes (as we may suppose) surrounded with the sap of the rose, enclosed in a pellicle of gluten. The gluten, however, of the bedeguar insect is not, it would appear, sufficiently tenacious to confine the flowing sap within the dimensions of any of the little clustered globes containing the eggs, for it oozes out from numerous cracks or pores in the pellicle; which cracks or pores, however, are not large enough to admit a human hair. But this, so far from being a defect in the glutinous pellicle of the bedeguar fly, is, as we shall presently see, of great utility. The sap which issues from each of these pores, instead of being evaporated and lost, shoots out into a reddish-coloured, fibrous bristle.

It is about half an inch long, and, from the natural tendency of the sap of the rose-tree to form prickles, these are all over studded with weak pricklets. The bedeguar, accordingly, when fully formed, has some resemblance, at a little distance, to a tuft of reddish-brown hair or moss stuck upon the branch. Sometimes this tuft is as large as a small apple, and of a rounded but irregular shape; at other times it is smaller, and in one instance mentioned by Réaumur, only a single egg had been laid on a rose-leaf, and, consequently, only one tuft was produced. Each member of the congeries is furnished with its own tuft of bristles, arising from the little hollow globe in which the egg or the grub is lodged.

The prospective wisdom of this curious structure is admirable. The bedeguar grubs live in their cells through the winter, and as their domicile is usually on one of the highest branches, it must be exposed to every severity of the weather. But the close, non-conducting, warm, mossy collection of bristles, with which it is surrounded, forms for the soft, tender grubs a snug protection against the winter's cold, till, through the influence of the warmth of the succeeding summer, they undergo their final change into the winged state; preparatory to which they eat their way with their sharp mandibles through the walls of their little cells, which are now so hard as to be cut with difficulty by a knife. (J. R.)

Another structure, similar in principle, though different in appearance, is very common upon oak-trees, the termination of a branch being selected as best suited for the purpose. This structure is rather larger than a filbert, and is composed of concentric leaves diverging from the base, and expanding upwards, somewhat like an artichoke. Whether this leafy

Artichoke Gall of the Oak-bud, with Gall-fly (*Cynips quercus gemmæ*), natural size, and its ovipositor (*a*) magnified.

structure is caused by a superinduced disease, as the French think, or by the form of the pores in the pellicle of gluten surrounding the eggs, or rather by the tendency of the exuding sap of the oak to form leaves, has not been ascertained ; but that it is intended, as in the case of the bedeguar, to afford an efficient protection against the weather to the included eggs or grubs, there can be no doubt.

From the very nature of the process of forming willow-galls, bedeguar, and the artichoke of the oak, whatever

theory be adopted, it will be obvious that their growth must
be rapid; for the thickening of the exuded sap, which is
quickly effected by evaporation, will soon obstruct and
finally close the orifice of the puncture made by the parent
insect. It is accordingly asserted by Réaumur and other
observers, that all the species of galls soon reach their full
. growth.

A very minute reddish-coloured grub feeds upon dyer's
broom (*Genista*), producing a sort of gall, frequently globular,
but always studded with bristles, arising from the amor-

Leafy Gall of Dyer's Broom, produced by *Cynips genistæ?* A, gall, natural size;
B, a leaflet magnified.

phous leaves. The stem of the shrub passes through this
ball, which is composed of a great number of leaves, shorter
and broader than natural, and each rolled into the form of a
horn, the point of which ends in a bristle. In the interior
we find a thick fleshy substance, serving to sustain the leaves,
and also for the nourishment of the grubs, some of which are
within and some between the leaves. They are in prodigious numbers,—hundreds being assembled in the small
gall, and so minute as scarcely to be perceived without the
aid of a magnifying glass. The bud of the plant attacked

by those grubs, instead of forming a shoot, pushes out nothing but leaves, and these are all rolled and turned round the stem. Some shrubs have several of these galls, which are of various sizes, from that of a filbert to that of a walnut.

A similar but still more beautiful production is found upon one of the commonest of our indigenous willows (*Salix purpurea*), which takes the name of *rose-willow*, more probably from this circumstance than from the red colour of its twigs. The older botanists, not being aware of the cause of such excrescences, considered the plants so affected as distinct species; and old Gerard accordingly figures and describes the rose-willow as "not only making a gallant show, but also yielding a most cooling air in the heat of summer, being set up in houses for decking the same." The production in question, however, is nothing more than the effect produced by a species of gall-fly (*Cynips salicis*) depositing its eggs in the terminal shoot of a twig, and, like the bedeguar and the oak artichoke, causing leaves to spring out, of a shape totally different from the other leaves of the tree, and arranged very much like the petals of a rose. Decandolle says it is found chiefly on the *Salix helix*, *S. alba*, and *S. riparia.**

A production very like that of the rose-willow may be commonly met with on the young shoots of the hawthorn, the growth of the shoot affected being stopped, and a crowded bunch of leaves formed at the termination. These leaves, besides being smaller than natural, are studded with short bristly prickles, from the sap (we may suppose) of the hawthorn being prevented from rising into a fresh shoot, and thrown out of its usual course in the formation of the arms. These bristles appear indiscriminately on both sides of the leaves, some of which are bent inwards, while others diverge in their natural manner.

This is not caused by the egg or grub of a true gall-fly, but by the small white tapering grub of some dipterous insect, of which we have not ascertained the species, but

* Flore Franç. Disc. Préliminaire.

which is, probably, a *cecidomyia*.   Each terminal shoot is inhabited by a number of these—not lodged in cells, however,
but burrowing indiscriminately among the half-withered
brown leaves which occupy the centre of the production.
(J. R.)

A more remarkable species of gall than any of the above
we discovered, in June, 1829, on the twig of an oak in the
grounds of Mr. Perkins, at Lee, in Kent.   When we first
saw it, we imagined that the twig was beset with some
species of the lanigerous aphides, similar to what is vulgarly
called the American or white blight (*Aphis lanata*); but on

Semi-Gall of the Hawthorn, produced by *Cecidomyia?* drawn from a specimen.

closer examination we discarded this notion.   The twig was
indeed thickly beset with a white downy, or rather woolly,
substance around the stem at the origin of the leaves, which
did not appear to be affected in their growth, being well
formed, healthy, and luxuriant.   We could not doubt that
the woolly substance was caused by some insect; but though
we cut out a portion of it, we could not detect any egg or
grub, and we therefore threw the branch into a drawer,
intending to keep it as a specimen, whose history we might
complete at some subsequent period.

A few weeks afterwards, on opening this drawer, we were
surprised to see a brood of several dozens of a species of

gall-fly (*Cynips*), similar in form and size to that whose eggs cause the bedeguar of the rose, and differing only in being of a lighter colour, tending to a yellowish brown. We have since met with a figure and description of this gall in Swammerdam. We may remark that the above is not the first instance which has occurred in our researches, of gall insects outliving the withering of the branch or leaf from which they obtain their nourishment.

The woolly substance on the branch of the oak which we

Woolly Gall of the Oak, less than the natural size, caused by a *Cynips*, and drawn from a specimen.

have described was similarly constituted with the bedeguar of the rose, with this difference, that instead of the individual cells being diffused irregularly through the mass, they were all arranged at the off-goings of the leaf-stalks, each cell being surrounded with a covering of the vegetable wool, which the stimulus of the parent egg, or its gluten, had caused to grow, and from each cell a perfect fly had issued. We also remarked that there were several small groups of individual cells, each of which groups was contained in a species of calyx or cup of leaf-scales, as occurs also in the well-known gall called the oak-apple.

We were anxious to watch the proceedings of these flies in the deposition of their eggs, and the subsequent developments of the gall-growths; and endeavoured for that purpose to procure a small oak plant in a garden-pot; but we did not succeed in this: and though they alighted on rose and sweet-briar trees, which we placed in their way, we never observed that they deposited any eggs upon them. In a week or two the whole brood died, or disappeared. (J. R.)

There are some galls, formed on low-growing plants, which are covered with down, hair, or wool, though by no means so copiously as the one which we have just described. Among the plants so affected are the germander speedwell, wild thyme, ground-ivy, and others to which we shall afterwards advert.

Oak-apple Galls, one being cut open to show the vessels running to granules.

The well-known oak-apple is a very pretty example of the galls formed by insects; and this, when compared with other galls which form on the oak, shows the remarkable difference produced on the same plant by the punctures of insects of different species. The oak-apple is commonly as large as a walnut or small apple, rounded, but not quite spherical, the surface being irregularly depressed in various places. The skin is smooth, and tinged with red and yellow, like a ripe apple; and at the base there is, in the earlier part of the summer, a calyx or cup of five or six small brown scaly leaves; but these fall off as the season advances. If an oak-apple be cut transversely, there is brought into view a number of oval granules, each containing a grub, and embedded in a fruit-looking fleshy substance, having fibres running through it. As these fibres, however, run in the direction of the stem, they are best exhibited by a vertical section of the gall; and this also shows the remarkable peculiarity of each fibre terminating in one of the granules, like a foot-stalk, or rather like a vessel for carrying nourishment. Réaumur, indeed, is of opinion that these fibres are the diverted nervures of the leaves, which would have sprung from the bud in which the gall-fly had inserted her eggs, and actually do carry sap-vessels throughout the substance of the gall.

Root Galls of the Oak, produced by *Cynips quercus inferus?* drawn from a specimen.

Réaumur says the perfect insects (*Cynips quercus*) issued from his galls in June and the beginning of July, and were of a reddish-amber colour. We have procured insects, agreeing with Réaumur's description, from galls formed on the bark or wood of the oak, at the line of junction between the root and the stem. These galls are precisely similar in structure to the oak-apple, and are probably formed at a season when the fly perceives, instinctively, that the buds of the young branches are unfit for the purpose of nidification.

There is another oak-gall, differing little in size and appearance from the oak-apple, but which is very different in structure, as, instead of giving protection and nourishment to a number of grubs, it is only inhabited by one. This sort of gall, besides, is hard and woody on the outside, resembling a little wooden ball of a yellowish colour, but internally of a soft, spongy texture. The latter substance, however, encloses a small hard gall, which is the immediate residence of the included insect. Galls of this description are often found in clusters of from two to seven, near the extremity of a branch, not incorporated, however, but distinctly separate.

We have obtained a fly very similar to this from a very common gall, which is formed on the branches of the willow. Like the one-celled galls just described, this is of a hard, ligneous structure, and forms an irregular protuber-

Woody Gall on a Willow branch, drawn from a specimen.

ance, sometimes at the extremity, and sometimes on the body, of a branch. But instead of one, this has a considerable number of cells, irregularly distributed through its substance. The structure is somewhat spongy, but fibrous; and externally the bark is smoother than that of the branch upon which it grows. (J. R.)

The currant-galls (as the French call them) of the oak are exactly similar, when formed on the leaves, to those which we have first described as produced on the leaves of the willow and other trees. But the name of currant-gall seems still more appropriate to an excrescence which grows on the catkins of the oak, giving them very much the appearance of a straggling branch of currants or bird-cherries. The galls resemble currants which have fallen from the tree before being ripe. These galls do not seem to differ from those formed on the leaves of the oak; and are probably the production of the same insect, which selects the catkin in preference, by the same instinct that the oak-apple gall-fly, as we have seen, sometimes deposits its eggs in the bark of the oak near the root.

Currant Gall of the catkins of the Oak, produced by *Cynips quercus pedunculi?*

The gall of the oak, which forms an important dye-stuff, and is used in making writing-ink, is also produced by a *Cynips*, and has been described in the 'Library of Entertaining Knowledge' (Vegetable Substances, p. 16). The employment of the *Cynips psenes* for ripening figs is described in the same volume, p. 214.

GALL OF A HAWTHORN WEEVIL.

In May, 1829, we found on a hawthorn at Lee, in Kent, the leaves at the extremity of a branch neatly folded up in a bundle, but not quite so closely as is usual in the case of leaf-rolling caterpillars. On opening them, there was no caterpillar to be seen, the centre being occupied with a roundish, brown-coloured, woody substance, similar to some excrescences made by gall-insects (*Cynips*). Had we been aware of its real nature, we should have put it immediately under a glass or in a box, till the contained insect had developed itself; but instead of this, we opened the ball, where we found a small yellowish grub coiled up, and feeding on the exuding juices of the tree. As we could not replace the grub in its cell, part of the walls of which we had unfortunately broken, we put it in a small pasteboard box with a fresh shoot of hawthorn, expecting that it might construct a fresh cell. This, however, it was probably incompetent to perform: it did not at least make the attempt, and neither did it seem to feed on the fresh branch, keeping in preference to the ruins of its former cell. To our great surprise, although it was thus exposed to the air, and deprived of a

Gall of the Hawthorn Weevil, drawn from specimen. *a*, Opened to show the grub.

considerable portion of its nourishment, both from the part of the cell having been broken off, and from the juices of the branch having been dried up, the insect went through its regular changes, and appeared in the form of a small greyish-brown beetle of the weevil family. The most remarkable circumstance in the case in question, was the apparent inability of the grub to construct a fresh cell after the first was injured,—proving, we think, beyond a doubt, that it is the puncture made by the parent insect when the egg is deposited that causes the exudation and subsequent concretion of the juices forming the gall. These galls were very abundant during the summer of 1830. (J. R.)

A few other instances of beetles producing galls are recorded by naturalists. Kirby and Spence have ascertained, for example, that the bumps formed on the roots of kedlock or charlock (*Sinapis arvensis*) are inhabited by the larvæ of a weevil (*Curculio contractus*, MARSHAM; and *Rhynchænus assimilis*, FABR.); and it may be reasonably supposed that either the same or similar insects cause the clubbing of the roots of cabbages, and the knob-like galls on turnips, called in some places the *anbury*. We have found them also infesting the roots of the hollyhock (*Alcea rosea*). They are evidently beetles of an allied genus which form the woody galls sometimes met with on the leaves of the guelder-rose (*Viburnum*), the lime-tree (*Tilia Europæa*), and the beech (*Fagus sylvatica*).

There are also some two-winged flies which produce woody galls on various plants, such as the thistle-fly (*Tephritis cardui*, LATR.). The grubs of this pretty fly produce on the leaf-stalks of thistles an oblong woody knob. On the common white briony (*Bryonia dioica*) of our hedges may be found a very pretty fly of this genus, of a yellowish-brown colour, with pellucid wings, waved much like those of the thistle-fly with yellowish brown. This fly lays its eggs near a joint of the stem, and the grubs live upon its substance. The joint swells out into an oval form, furrowed in several places, and the fly is subsequently disclosed. In its perfect state, it feeds on the blossom of the briony. (J. R.)

Flies of another minute family, the gall-gnats (*Cecidomyiæ*, LATR.), pass the first stage of their existence in the small globular cottony galls which abound on germander speedwell (*Veronica chamædrys*), wild thyme (*Thymus serpyllum*), and ground-ivy (*Glechoma hederacea*). The latter is by no means uncommon, and may be readily recognised.

Certain species of plant-lice (*Aphides*), whose complete history would require a volume, produce excrescences upon plants which may with some propriety be termed galls, or semi-galls. Some of these are without any aperture, whilst others are in form of an inflated vesicle, with a narrow opening on the under side of a leaf, and expanding (for the most part irregularly) into a rounded knob on its upper surface. The mountain-ash (*Pyrus aucuparia*) has its leaves and young shoots frequently affected in this way, and sometimes exhibits galls larger than a walnut or even than a man's fist; at other times they do not grow larger than a filbert. Upon opening one of these, they are found to be filled with the *aphides sorbi*. If taken at an early stage of their growth, they are found open on the under side of the leaf, and inhabited only by a single female aphis, pregnant with a numerous family of young. In a short time the aperture becomes closed, in consequence of the insect making repeated punctures round its edge, from which sap is exuded and forms an additional portion of the walls of the cell.

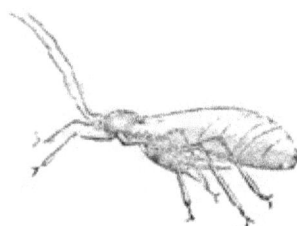

A Plant-Louse (*Aphis*), magnified.

In this early stage of its growth, however, the gall does not, like the galls of the cynips, increase very much in dimensions. It is after the increase of the inhabitants by the young brood that it grows with considerable rapidity; for each additional insect, in order to procure food, has to puncture the wall of the chamber and suck the juices, and from the punctures thus made the sap exudes, and enlarges the walls. As those galls are closed all round in the more advanced state, it does not appear how the insects can ever effect an exit from their imprisonment.

A much more common production, allied to the one just described, may be found on the poplar in June and July. Most of our readers may have observed, about midsummer, a small snow-white tuft of downy-looking substance floating about on the wind, as if animated. Those tufts of snow-white down are never seen in numbers at the same time, but generally single, though some dozens of them may be observed in the course of one day This singular object is a four-winged fly (*Eriosoma populi*, LEACH), whose body is thickly covered with long down—a covering which seems to impede its flight, and make it appear more like an inani-

Galls produced on the leaves and leaf-stalks of the Poplar by *Eriosoma populi*, with the various forms of the insects, winged, not winged, and covered with wool, both of the natural size and magnified.

mate substance floating about on the wind, than impelled by the volition of a living animal. This pretty fly feeds upon the fresh juices of the black poplar, preferring that of the leaves and leaf-stalks, which it punctures for this purpose with its beak. It fixes itself with this design to a suitable place upon the principal nervure of the leaf, or

upon the leaf-stalk, and remains in the same spot till the
sap, exuding through the punctures, and thickening by
contact with the air, surrounds it with a thick fleshy wall of
living vegetable substance, intermediate in texture between
the wood and the leaf, being softer than the former and
harder than the latter. In this snug little chamber, secure
from the intrusion of lady-birds and the grubs of aphidivorous flies (*Syrphi*), she brings forth her numerous brood
of young ones, who immediately assist in enlarging the
extent of their dwelling, by puncturing the walls. In one
respect, however, the galls thus formed differ from those of
the mountain-ash just described,—those of the poplar having
always an opening left into some part of the cell, and
usually in that portion of it which is elongated into an
obtuse beak. From this opening the young, when arrived
at the winged state, make their exit, to form new colonies;
and, during their migrations, attract the attention of the
most incurious by the singularity of their appearance.
(J. R.)

On the black poplar there may be found, later in the
season than the preceding, a gall of a very different form,
though, like the other, it is for the most part on the leaf-
stalk. The latter sort of galls are of a spiral form; and
though they are closed, they open upon slight pressure, and
appear to be formed of two laminæ, twisted so as to unite.
It is at this opening that an aperture is formed spontaneously
for the exit of the insects, when arrived at a perfect state.
In galls of this kind we find aphides, but of a different species
from the lanigerous ones, which form the horn-shaped galls
above described.

LEAF-ROLLING APHIDES.

It may not be improper to introduce here a brief sketch
of some other effects, of a somewhat similar kind, produced
on leaves by other species of the same family (*Aphidæ*). In
all the instances of this kind which we have examined, the
form which the leaf takes serves as a protection to the

insects, both from the weather and from depredators. That there is design in it appears from the circumstance of the aphides crowding into the embowering vault which they have formed; and we are not quite certain whether they do not puncture certain parts of the leaf for the very purpose of making it arch over them; at least, in many cases, such as that of the hop-fly (*Aphis humuli*), though the insects are in countless numbers, no arching of the leaves follows. The rose-plant louse, again (*Aphis rosæ*), sometimes arches the leaves, but more frequently gets under the protecting folds of the half-expanded leaf-buds. (J. R.)

Leaf of the Currant-bush, bulged out by the *Aphis ribis.*

One of the most common instances of what we mean occurs on the leaves of the currant-bush, which may often be observed raised up into irregular bulgings, of a reddish-brown colour. On examining the under side of such a leaf there will be seen a crowd of small insects, some with and some without wings, which are the *Aphides ribis* in their different stages, feeding securely and socially on the juices of the leaf.

The most remarkable instance of this, however, which we have seen, occurs on the leaves of the elm, and is caused by the *Aphis ulmi.* The edge of an elm-leaf inhabited by those aphides is rolled up in an elegant convoluted form, very much like a spiral shell; and in the embowered chamber thus formed the insects are secure from rain, wind, and partially from the depredations of carnivorous insects. One of their greatest enemies, the lady-bird (*Coccinella*), seldom ventures, as we have remarked, into concealed corners except in cold weather, and contrives to find food enough among the aphides which feed openly and unprotected, such as the zebra aphides of the alder (*Aphides sambuci*). The grubs, however, of the lady-bird, and also those of the aphidivorous flies (*Syrphi*), may be found prying into the most secret recesses of a leaf to prey upon the inhabitants, whose slow movements disqualify them from effecting an escape. (J. R.)

The effect of the puncture of aphides on growing plants is strikingly illustrated in the shoots of the lime-tree and several other plants, which become bent and contorted on the side attacked by the insects, in the same way that a shoot might warp by the loss of its juices on the side exposed to a brisk fire. The curvings thus effected become very advantageous to the insects, for the leaves sprouting from the twig, which naturally grow at a distance from each other, are brought close together in a bunch, forming a kind of nosegay, that conceals all the colour of the sprig, as well as the insects which are embowered under it, protecting them against the rain and the sun, and at the same time hiding them from observation. It is only requisite, however, where they have formed bowers of this description, to raise the leaves, in order to see the little colony of the aphides,—or the remains of those habitations which they have abandoned. We have sometimes observed sprigs of the lime-tree, of a thumb's thickness, portions of which resembled spiral screws; but we could not certainly have assigned the true cause for this twisting, had we not been acquainted with the manner in which aphides contort the young shoots of this tree.* The shoots of the gooseberry and the willow are sometimes contorted in the same way, but not so strikingly as the shoots of the lime.

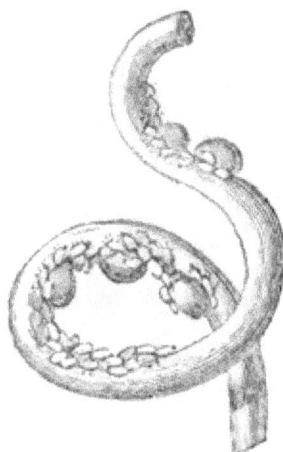

Shoot of the Lime-tree contorted by the punctures of the *Aphis tiliæ.*

PSEUDO-GALLS.

It may not be out of place to mention here certain anomalous excrescences upon trees and other plants, which, though they much resemble galls, are not so distinctly traceable to the operations of any insect. In our researches after galls, we have not unfrequently met with excrescences which so very much resemble them, that before dissection we should not hesitate to consider them as such, and predict that they formed the nidus of some species of insects. In more instances than one we have felt so strongly assured of this, that we have kept several specimens for some months, in nurse-boxes, expecting that in due time the perfect insect would be disclosed.

One of these pseudo-galls occurs on the common bramble (*Rubus fruticosus*), and bears some resemblance to the bedeguar of the rose when old and changed by weather. It clusters round the branches in the form of irregular granules, about the size of a pea, very much crowded, the whole

* Réaumur, vol. iii.

excrescence being rather larger than a walnut. We expected to find this excrescence full of grubs, and were much surprised to discover, upon dissection, that it was only a diseased

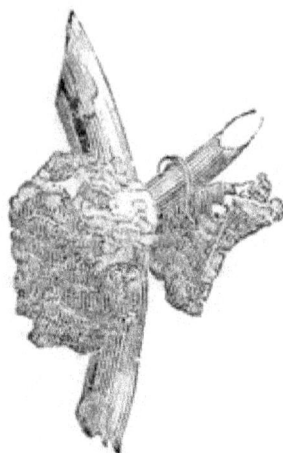

Pseudo-gall of the Bramble, drawn from a specimen.

growth of the plant, caused (it might be) by the puncture of an insect, but not for the purpose of a nidus or habitation. (J. R.)

Another sort of excrescence is not uncommon on the terminal shoots of the hawthorn. This is in general irregularly oblong, and the bark which covers it is of an iron colour, similar to the scoriæ of a blacksmith's forge. When dissected, we find no traces of insects, but a hard, ligneous, and rather porous texture. It is not improbable that this excrescence may originate in the natural growth of a shoot being checked by the punctures of aphides, or of those grubs which we have described.

Many of these excrescences, however, are probably altogether unconnected with insects, and are simply hypertrophic diseases, produced by too much nourishment, like the wens produced on animals. Instances of this may be seen at the roots of the hollyhock (*Althea rosea*) of three or four years' standing; on the stems of the elm and other trees, immediately above the root; and on the upper branches of the birch, where a crowded cluster of twigs sometimes grows, bearing no distant resemblance to a rook's nest in miniature, and provincially called witch-knots.

Pseudo-galls of the Hawthorn, drawn from specimens.

One of the prettiest of these pseudo-galls with which we are acquainted, is produced on the Scotch fir (*Pinus sylvestris*), by the *Aphis pini*, which is one of the largest species of our indigenous aphides. The production we allude to may be found, during the summer months, on the terminal shoots of this tree, in the form of a small cone, much like the fruit of the tree in miniature, but with this difference, that the fruit terminates in a point, whereas the pseudo-gall is nearly globular. Its colour also, instead of being green, is reddish; but it exhibits the tiled scales of the fruit cone.

Pseudo-gall produced by *Aphis pini* on the Scotch fir, drawn from a specimen.

We have mentioned this the more willingly that it seems to confirm the theory which we have hazarded respecting the formation of the bedeguar of the rose and other true galls —by which we ascribed to the sap, diverted from its natural course by insects, a tendency to form leaves, &c., like those of the plant from which it is made to exude.

CHAPTER XX.

ANIMAL GALLS,* PRODUCED BY BREEZE-FLIES AND SNAIL-
BEETLES.

THE structures which we have hitherto noticed have all
been formed of inanimate materials, or at the most of
growing vegetables; but those to which we shall now advert
are actually composed of the flesh of living animals, and
seem to be somewhat akin to the galls already described as
formed upon the shoots and leaves of plants. These were
first investigated by the accurate Vallisnieri, and subsequently by Réaumur, De Geer, and Linnæus; but the best
account which has hitherto been given of them is by our
countryman Mr. Braccy Clark, who differs essentially from
his predecessors as to the mode in which the eggs are deposited. As, in consequence of the extreme difficulty, if not the
impossibility, of personal observation, it is no easy matter to
decide between the conflicting opinions, we shall give such
of the statements as appear most plausible.

The mother breeze-fly (*Oestrus bovis*, CLARK;—*Hypoderma
bovis*, LATR.), which produces the tumours in cattle called
wurbles or *wormuls* (*quasi, worm-holes*), is a two-winged insect,
smaller, but similar in appearance and colour to the carder-
bee (p. 75), with two black bands, one crossing the shoulders
and the other the abdomen, the rest being covered with
yellow hair. This fly appears to have been first discovered
by Vallisnieri, who has given a curious and interesting
history of his observations upon its economy. "After having
read this account," says Réaumur, "with sincere pleasure, I
became exceedingly desirous of seeing with my own eyes

---

* In order to prevent ambiguity, it is necessary to remark that the excrescences thus called must not be confounded with the true galls, which are occasionally found in the gall-bladder.

what the Italian naturalist had reported in so erudite and pleasing a manner. I did not then imagine that it would ever be my lot to speak upon a subject which had been treated with so much care and elegance; but since I have enjoyed more favourable opportunities than M. Vallisnieri, it was not difficult for me to investigate some of the circumstances better, and to consider them under a different point of view. It is not, indeed, very wonderful to discover something new in an object, though it has been already carefully inspected with very good eyes, when we sit down to examine it more narrowly, and in a more favourable position; while it sometimes happens, also, that most indifferent observers have detected what had been previously unnoticed by the most skilful interpreters of nature."[*]

From the observations made by Réaumur, he concluded that the mother-fly, above described, deposits her eggs in the flesh of the larger animals, for which purpose she is furnished with an ovipositor of singular mechanism. We have seen that the ovipositors in the gall-flies (*Cynips*) are rolled up within the body of the insect somewhat like the spring of a watch, so that they can be thrust out to more than double their apparent length. To effect the same purpose, the ovipositor of the ox-fly lengthens, by a series of sliding tubes, precisely like an opera-glass. There are four of these tubes, as may be seen by pressing the belly of the fly till they come into view. Like other ovipositors of this sort, they are composed of a horny substance; but the terminal piece is very different indeed from the same part in the gall-flies, the tree-hoppers (*Cicadæ*), and the ichneumons, being composed of five points, three of which are longer than the other two, and at first sight not unlike a *fleur-de-lis*, though, upon narrower inspection, they may be discovered to terminate in curved points, somewhat like the claw of a cat. The two shorter pieces are also pointed, but not curved; and by the union of the five, a tube is composed for the passage of the eggs.

It would be necessary, Réaumur confesses, to see the fly employ this instrument to understand in what manner it acts, though he is disposed to consider it fit for boring through the hides of cattle. "Whenever I have succeeded," he adds, "in seeing these insects at work, they have usually shown that they proceeded quite differently from what I had

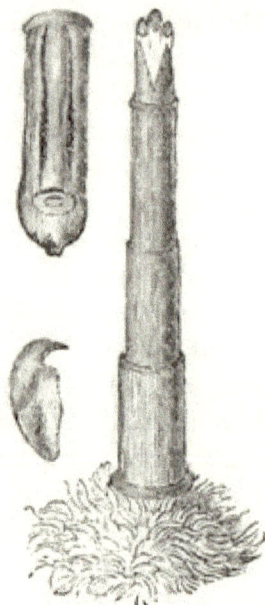

Ovipositor of the Breeze-fly, greatly magnified, with a claw and part of the tube, distinct.

imagined; but unfortunately I have never been able to see one of them pierce the hide of a cow under my eyes."*

Mr. Bracey Clark, taking another view of the matter, is decidedly of opinion that the fly does not pierce the skin of cattle with its ovipositor at all, but merely glues its eggs to the hairs, while the grubs, when hatched, eat their way under the skin. If this be the fact, as is not improbable, the three curved pieces of the ovipositor, instead of acting, as Réaumur imagined, like a centre-bit, will only serve to prevent the eggs from falling till they are firmly glued to the hair, the opening formed by the two shorter points permitting this to be effected. This account of the matter is rendered more plausible, from Réaumur's statement that the deposition of

* Mém. iv. 538.

the egg is not attended by much pain, unless, as he adds, some very sensible nervous fibres have been wounded. According to this view, we must not estimate the pain produced by the thickness of the instrument; for the sting of a wasp, or a bee, although very considerably smaller than the ovipositor of the ox-fly, causes a very pungent pain. It is, in the latter case, the poison infused by the sting, rather than the wound, which occasions the pain; and Vallisnieri is of opinion that the ox-fly emits some acrid matter along with her eggs, but there is no proof of this beyond conjecture.

It ought to be remarked, however, that cattle have very thick hides, which are so far from being acutely sensitive of pain, that in countries where they are put to draw ploughs and waggons, they find a whip ineffectual to drive them, and have to use a goad, in form of an iron needle, at the end of a stick. Were the pain inflicted by the fly very acute, it would find it next to impossible to lay thirty or forty eggs without being killed by the strokes of the ox's tail; for though Vallisnieri supposes that the fly is shrewd enough to choose such places as the tail cannot reach, Réaumur saw a cow repeatedly flap its tail upon a part full of the gall-bumps; and in another instance he saw a heifer beat away a party of common flies from a part where there were seven or eight gall-bumps. He concludes, therefore, with much plausibility, that these two beasts would have treated the ox-flies in the same way, if they had given them pain when depositing their eggs.

The extraordinary effects produced upon cattle, on the appearance of one of these flies, would certainly lead us to conclude that the pain inflicted is excruciating. Most of our readers may recollect to have seen, in the summer months, a whole herd of cattle start off across a field in full gallop, as if they were racing,—their movements indescribably awkward —their tails being poked out behind them as straight and stiff as a post, and their necks stretched to their utmost length. All this consternation has been known, from the earliest times, to be produced by the fly we are describing.

Virgil gives a correct and lively picture of it in his Georgics,[*] of which the following is a translation, a little varied from Trapp :

> Round Mount Alburnus, green with shady oaks,
> And in the groves of Silarus, there flies
> An insect pest (named *Œstrus* by the Greeks,
> By us *Asilus*) : fierce with jarring hum
> It drives, pursuing, the affrighted herd
> From glade to glade ; the air, the woods, the banks
> Of the dried river echo their loud bellowing.

Had we not other instances to adduce, of similar terror caused among sheep, deer, and horses, by insects of the same genus, which are ascertained not to penetrate the skin, we should not have hesitated to conclude that Vallisnieri and Réaumur are right, and Mr. Bracey Clark wrong. In the strictly similar instance of Reindeer-fly (*Œstrus tarandi*, LINN.), we have the high authority of Linnæus for the fact, that it lays its eggs *upon* the skin.

"I remarked," he says, "with astonishment how greatly the reindeer are incommoded in hot weather, insomuch that they cannot stand still a minute, no not a moment, without changing their posture, starting, puffing and blowing continually, and all on account of a little fly. Even though amongst a herd of perhaps five hundred reindeer, there were not above ten of those flies, every one of the herd trembled and kept pushing its neighbour about. The fly, meanwhile, was trying every means to get at them ; but it no sooner touched any part of their bodies, than they made an immediate effort to shake it off. I caught one of these insects as it was flying along with its tail protruded, which had at its extremity a small linear orifice perfectly white. The tail itself consisted of four or five tubular joints, slipping into

---

[*] Est lucos Silari circa ilicibusque virentem
Plurimus Alburnum volitans, cui nomen asilo
Romanum est, Œstrum Graii vertere vocantes,
Asper, acerba sonans ; quo tota exterrita silvis
Diffugiunt armenta ; furit mugitibus æther
Concussus, sylvæque et sicci ripa Tanagri.

Georg. lib. iii. 146.

each other like a pocket spying-glass, which this fly, like others, has a power of contracting at pleasure."*

In another work he is still more explicit. "This well-known fly," he says, "hovers the whole day over the back of the reindeer, with its tail protruded and a little bent, upon the point of which it holds a small white egg, scarcely so large as a mustard-seed, and when it has placed itself in a perpendicular position, it drops its egg, which rolls down amongst the hair to the skin, where it is hatched by the natural heat and perspiration of the reindeer, and the grub eats its way slowly under the skin, causing a bump as large as an acorn."† The male and female of the reindeer breeze-fly are figured in the 'Library of Entertaining Knowledge, Menageries,' vol. i. p 405.

There is one circumstance which, though it appears to us to be of some importance in the question, has been either overlooked or misrepresented in books. "While the female fly," say Kirby and Spence, "is performing the operation of oviposition, the animal attempts to lash her off as it does other flies, with its tail;"‡ though this is not only at variance with their own words in the page but one preceding. where they most accurately describe "the herd with their tails in the air, or turned upon their backs, or stiffly stretched out in the direction of the spine,"§ but with the two facts mentioned above from Réaumur, as well as with common observation. If the ox then do not attempt to lash off the breeze-fly, but runs with its tail stiffly extended, it affords a strong presumption that the fly terrifies him by her buzzing (*asper, accrba sonans*), rather than pains him by piercing his hide: her buzz, like the rattle of the rattle-snake, being instinctively understood, and intended, it may be, to prevent an over-population, by rendering it difficult to deposit the eggs.

The horse breeze-fly (*Gasterophilus equi*, Leach), which

* Linnæus, Lachesis Lapponica, July 19th.
† Linnæus, Flora Lapponica, p. 378, ed. Lond. 1792.
‡ Kirby and Spence, Introd. i. 151.
§ Ibid. p. 149.

produces the maggots well known by the name of *botts* in horses, is ascertained beyond a doubt to deposit her eggs upon the hair ; and as insects of the same genus almost invariably proceed upon similar principles, however much they may vary in minute particulars, it may be inferred with justice, that the breeze-flies which produce galls do the same. The description given by Mr. Bracey Clark, of the proceedings of the horse breeze-fly, is exceedingly interesting.

" When the female has been impregnated, and her eggs sufficiently matured, she seeks among the horses a subject for her purpose, and approaching him on the wing, she carries her body nearly upright in the air, and her tail, which is *lengthened for the purpose,** curved inwards and upwards ; in this way she approaches the part where she designs to deposit the egg ; and suspending herself for a few seconds before it, suddenly darts upon it and leaves the egg adhering to the hair ; she hardly appears to settle, but merely touches the hair with the egg *held out on the projected point of the abdomen.** The egg is made to adhere by means of a glutinous liquor secreted with it. She then leaves the horse at a small distance, and prepares a second egg, and poising herself before the part, deposits it in the same way. The liquor dries, and the egg becomes firmly glued to the hair : this is repeated by these flies till four or five hundred eggs are sometimes placed on one horse."

Mr. Clark farther tells us, that the fly is careful to select a part of the skin which the horse can easily reach with his tongue, such as the inside of the knee, or the side and back part of the shoulder. It was at first conjectured, that the horse licks off the eggs thus deposited, and that they are by this means conveyed into its stomach ; but Mr. Clark says, " I do not find this to be the case, or at least only by accident ; for when they have remained on the hair four or five days, they become ripe, after which time the slightest

---

* These circumstances afford, we think, a complete answer to the query of Kirby and Spence—"There can be little doubt (or else what is the use of such an apparatus?) that it bores a hole in the skin."—Introd. i. 162, 2nd edit.

application of warmth and moisture is sufficient to bring
forth, in an instant, the latent larva. At this time, if the
tongue of the horse touches the egg, its operculum is thrown
open, and a small, active worm is produced, which readily
adheres to the moist surface of the tongue, and is thence
conveyed with the food to the stomach." He adds, that "a
horse which has no ova deposited on him may yet have botts,
by performing the friendly office of licking another horse
that has."* The irritations produced by common flies
(*Anthomyiæ meteoricæ*, MEIGEN) are alleged as the incitement
to licking.

The circumstance, however, of most importance to our
purpose, is the agitation and terror produced both by this
fly and by another horse breeze-fly (*Gasterophilus hæmor-
rhoidalis*, LEACH), which deposits its eggs upon the lips of
the horse as the sheep breeze-fly (*Œstrus ovis*) does on that
of the sheep. The first of these is described by Mr. Clark
as "very distressing to the animal, from the excessive
titillation it occasions; for he immediately after rubs his
mouth against the ground, his fore-feet, or sometimes against
a tree, with great emotion; till, finding this mode of defence
insufficient, he quits the spot in a rage, and endeavours to
avoid it by galloping away to a distant part of the field, and
if the fly still continues to follow and teaze him, his last
resource is in the water, where the insect is never observed
to pursue him. These flies appear sometimes to hide themselves in the grass, and as the horse stoops to graze they
dart upon the mouth or lips, and are always observed to poise
themselves during a few seconds in the air, while the egg is
prepared *on the extended point of the abdomen.*"†

The moment the second fly just mentioned touches the
nose of a sheep, the animal shakes its head and strikes the
ground violently with its fore-feet, and at the same time
holding its nose to the earth, it runs away, looking about on
every side to see if the flies pursue. A sheep will also smell
the grass as it goes, lest a fly should be lying in wait, and if
one be detected, it runs off in terror. As it will not, like a

* Linn. Trans. iii. 305.   † Ibid.

horse or an ox, take refuge in the water, it has recourse
to a rut or dry dusty road, holding its nose close to the
ground, thus rendering it difficult for the fly to get at the
nostril.

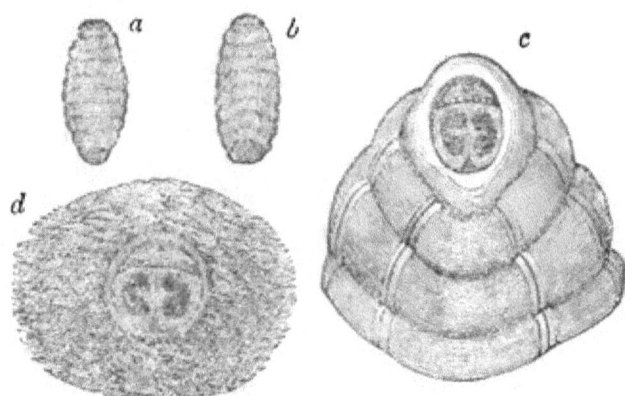

a, The belly of the grub. b, Its back. c, The tail of the grub, greatly magnified.
d, The bump, or gall, having its external aperture filled with the tail of the grub.

When the egg of the ox breeze-fly (*Hypoderma bovis*, Latr.)
is hatched, it immediately (if Mr. Bracey Clark be correct)
burrows into the skin; while, according to Réaumur, it is
hatched there. At all events, the grub is found in a bump
on the animal's back, resembling a gall on a tree,—"a
place," says Réaumur, "where food is found in abundance,
where it is protected from the weather, where it enjoys at
all times an equal degree of warmth, and where it finally
attains maturity."* When in an advanced stage, the bumps
appear much like the swellings produced upon the forehead by
a smart blow. These, with the grubs, are represented in the
foregoing figure, and also at page 434.

Every bump, according to Réaumur, has in its inside a
cavity, which is a lodging proportionate to the size of the
insect. The bump and cavity also increase in proportion to
the growth of the grub. It is not until about the middle
of May that these bumps can be seen full grown. Owing
to particular circumstances, they do not all attain an equal

* Mém. iv. 540.

size. The largest of them are sixteen or seventeen lines in diameter at their base, and about an inch high ; but they are scarcely perceptible before the beginning or during the course of the winter.

It is commonly upon young cattle, such, namely, as are two or three years old, that the greatest number of bumps is found ; it being rare to observe them upon very old animals. The fly seems to be well aware that such skins will not

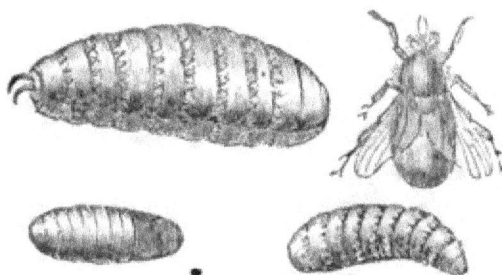

Fly, maggot, and grub of the Ox breeze-fly, with a microscopic view of the maggot.

oppose too much resistance, and seems to know, also, that tender flesh is the most proper for supplying good nourishment to its progeny. "And why," asks Réaumur, "should not the instinct which conducts it to confide its eggs to the flesh of certain species only, lead it to prefer the flesh of animals of the same species which is most preferable ?" The number of bumps which are found upon a beast is equal to the number of eggs which have been deposited in its flesh ; or, to speak more correctly, to the number of eggs which have succeeded, for apparently all are not fertile ; but this number is very different upon different cattle. Upon one cow only three or four bumps may be observed, while upon another there will appear from thirty to forty. They are not always placed on the same parts, nor arranged in the same manner : commonly, they are near the spine, but sometimes upon or near the thighs and shoulders. Sometimes they are at remote distances from each other ; at other times they are so near that their circumferences meet. In certain places, three or four tumors may be seen touching each other ; and more

2 F

than a dozen sometimes occur arranged as closely together as possible.

It is very essential to the grub that the hole of the tumor should remain constantly open; for by this aperture a communication with the air necessary for respiration is preserved; and the grub is thence placed in the most favourable position for receiving air. Its spiracles for respiration, like those of many other grubs, are situated immediately upon the posterior extremity of the body. Now, being almost always placed in such a situation as to have this part above,

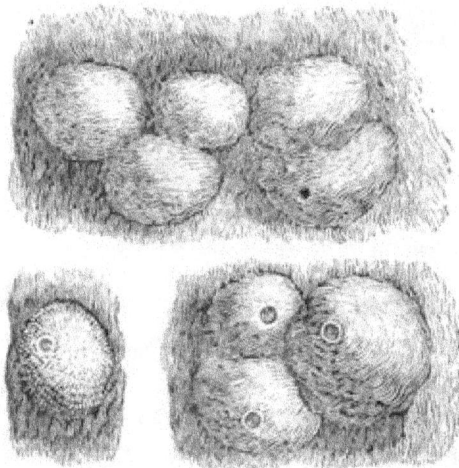

Bumps or warbles produced on cattle by the Ox breeze-fly.

or upon a level with the external aperture, it is enabled to respire freely.*

We have not so many examples of galls of this kind as we have of vegetable galls; and when we described the surprising varieties of the latter, we did not perceive that it was essential to the insects inhabiting them to preserve a communication with the external air: in the galls of trees, openings expressly designed or kept free for the admission of air are never observed. Must the grub, then, which inhabits the latter have less need of respiring air than the

* Réaumur, iv. 549.

grub of the breeze-flies in a flesh-gall? Without doubt, not ; but the apertures by which the air is admitted to the inhabitants of the woody gall, although they may escape our notice, in consequence of their minuteness, are not, in fact, less real. We know that, however careful we may be in inserting a cork into a glass, the mercury with which it is filled is not sheltered from the action of the air, which weighs upon the cork; we know that the air passes through, and acts upon the mercury in the tube. The air can also, in the same way, penetrate through the obstruction of a gall of wood, though it have no perceptible opening or crack ; but the air cannot pass in this manner so readily through the skins and membranes of animals.

In order to see the interior of the cavity of an animal gall, Réaumur opened several, either with a razor or a pair of scissors ; the operation, however, cannot fail to be painful to the cow, and consequently renders it impatient under the process. The grub being confined in a tolerably large fistulous ulcer, a part of the cavity must necessarily be filled with pus or matter. The bump is a sort of cautery, which has been opened by the insect, as issues are made by caustic : the grub occupies this issue, and prevents it from closing. If the pus or matter which is in the cavity, and that which is daily added to it, had no means of escaping, each tumor would become a considerable abscess, in which the grub would perish ; but the hole of the bump, which admits the entrance of the air, permits the pus or matter to escape ; that pus frequently mats the hairs together which are above the small holes, and this drying around the holes acquires a consistency, and forms in the interior of the opening a kind of ring. This matter appears to be the only aliment allowed for the grub, for there is no appearance that it lives, like the grubs of flesh-flies, upon putrescent meat. Mandibles, indeed, similar to those with which other grubs break their food, are altogether wanting. A beast which has thirty, forty, or more of these bumps upon its back, would be in a condition of great pain and suffering, terrible indeed in the extreme, if its flesh were torn and devoured by as many large grubs; but there is every appearance that they do not at all afflict, or only afflict it with little pain. For this reason cattle most covered with bumps are not considered by the farmer as injured by the presence of the fly, which generally selects those in the best condition.

A fly, evidently of the same family with the preceding, is described in Bruce's 'Travels,' under the name of zimb, as burrowing during its grub state in the hides of the elephant, the rhinoceros, the camel, and cattle. " It resembles," he says, " the gad-fly in England, its motion being more sudden and rapid than that of a bee. There is something peculiar in the sound or buzzing of this insect; it is a jarring noise together with a humming, which as soon as it is heard all the cattle forsake their food and run wildly about the plain till they die, worn out with fatigue, fright, and hunger. I have found," he adds, " some of these tubercles upon almost every elephant and rhinoceros that I have seen, and attribute them to this cause. When the camel is attacked by this fly, his body, head, and legs break out into large bosses, which swell, break, and putrefy, to the certain destruction of the creature."* That camels die under such symptoms, we do not doubt; but we should not, without more minutely-accurate observation, trace all this to the breeze-fly.

MM. Humboldt and Bonpland discovered, in South America, a species, probably of the same genus, which attacks man himself. The perfect insect is about the size of our common house-fly (*Musca domestica*), and the bump formed by the grub, which is usually on the belly, is similar to that caused by the ox breeze-fly. It requires six months to come to maturity; and if it is irritated it eats deeper into the flesh, sometimes causing fatal inflammations.

GRUB PARASITE IN THE SNAIL.

During the summer of 1829, we discovered in the hole of a garden-post, at Blackheath, one of the larger grey snail shells (*Helix aspersa*, MULLER), with three white soft-bodied grubs

* Bruce's Travels, i. 5, and v. 191.

burrowing in the body of the snail. They evidently, from their appearance, belonged to some species of beetle, and we carefully preserved them in order to watch their economy. It appeared to us that they had attacked the snail in its stronghold while it was laid up torpid for the winter; for more than half of the body was already devoured. They constructed for themselves little cells attached to the inside of the shell, and composed of a sort of fibrous matter, having no distant resemblance to shag tobacco, both in form and smell, and which could be nothing else than the remains of the snail's body. Soon after we took them, appearing to have devoured all that remained of the poor snail, we furnished them with another, which they devoured in the same manner. They formed a cocoon of the same fibrous materials during the autumn, and in the end of October appeared in their perfect form, turning out to be *Drilus flavescens*, the grub of which was first discovered in France in 1824. The time of their appearance, it may be remarked, coincides with the period when snails become torpid. (J. R.)

In the following autumn, we found a shell of the same species with a small pupa-shaped egg deposited on the lid. From this a caterpillar was hatched, which subsequently devoured the snail, spun a cocoon within the shell, and was transformed into a small moth (of which we have not ascertained the species) in the spring of 1830.

[Before concluding the account of the parasite insects, it will be necessary to mention two of our British Ichneumonidæ, which not only deposit their eggs in the larvæ of other insects, but make for themselves cells of very beautiful structure. In the accompanying illustration are shown the cells of one of our commonest and most useful ichneumonidæ (*Microgaster glomeratus*), together with the insect itself. At Fig. 1a (p. 438) is shown the little insect of the natural size, and the same is given at 1 much magnified.

[This creature lays its eggs in the body of the cabbage caterpillar, forty or fifty eggs being deposited in the same larva. They soon hatch into little transparent grubs, which lie under the skin, and live on the fatty parts of the caterpillar, which continues to grow, and seems to thrive, whereas its bulk is largely made up of the ichneumon larvæ.

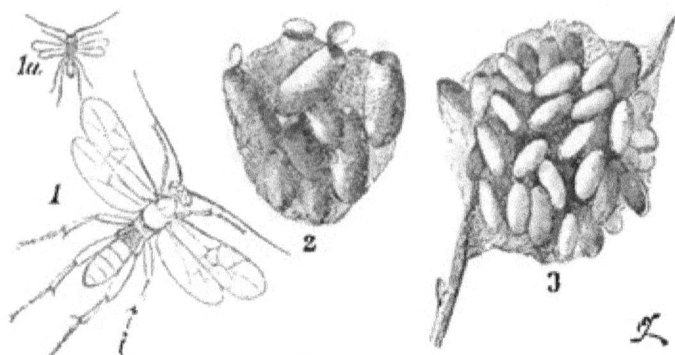

Microgaster glomeratus.

[After the caterpillar ceases from feeding, it crawls aside for the purpose of assuming the pupal state. But, before it can do so, the ichneumon larvæ, which have also ceased from feeding, burst their way through the sides of the caterpillar, and immediately begin to spin their cocoon. These are oval, very small, and covered with yellow silk. A group of these cocoons is shown at Fig. 3. The innumerable fibres of these cocoons hamper the caterpillars so much that, in most cases, it seldom is able to stir from the spot, but dies in the midst of its enemies. Groups of these yellow cocoons can be found in every wall or paling near cabbage gardens. In a few days, the larvæ have passed through their pupal stage, assuming the winged state, and emerge from the cocoons through little circular doors, as seen in Fig. 2.

[Our second illustration represents another species, *Microgaster alveolarius*, together with its cocoons. As before, the insect is shown of its natural size at 1a, and magnified at 1. The preliminary life of this insect is exactly the same as that of the preceding; but, instead of making a number of independent and separate cocoons, the insects spin so closely together that they form an edifice very much resembling a beecomb. Fig. 5 represents one of these cell-groups of the natural size, and the edge of another group is shown at Fig. 4.

A longitudinal section, slightly enlarged, is given at 3, in order to show the hexagonal shape assumed by the aggre-

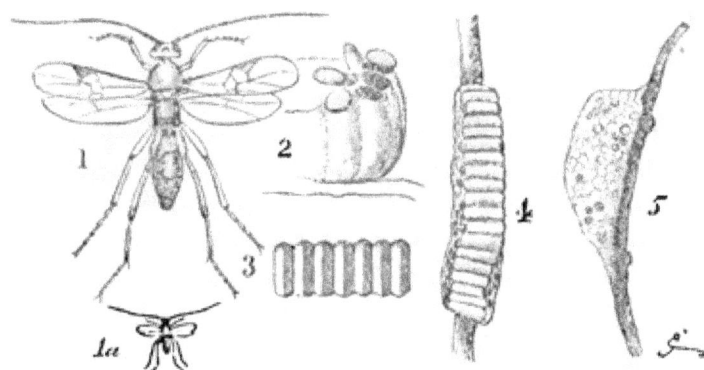

Microgaster alveolarius.

gated cells; and Fig. 2 shows the little lids which open to give egress to the insect. All these figures are drawn from specimens in my collection.]

THE END.

LONDON: PRINTED BY W. CLOWES AND SONS, STAMFORD STREET AND CHARING CROSS.